Alchemy Tried in the Fire

Alchemy Tried in The Fire

Starkey, Boyle, and the Fate of Helmontian Chymistry

William R. Newman and Lawrence M. Principe

THE UNIVERSITY OF CHICAGO PRESS

CHICAGO AND LONDON

William R. Newman is professor in the Department of History and Philosophy of Science at Indiana University. He is author of *Gehennical Fire: The Lives of George Starkey, An American Alchemist in the Scientific Revolution* (1994) and *The Summa Perfectionis of Pseudo-Geber: A Critical Edition, Translation, and Study* (1991). Lawrence M. Principe is professor in the Department of the History of Science, Medicine, and Technology and the Department of Chemistry at Johns Hopkins University. He is author of *The Aspiring Adept: Robert Boyle and His Alchemical Quest* (1998) and coeditor of the six-volume *Correspondence of Robert Boyle* (2001, with Michael Hunter and Antonio Clericuzio).

The University of Chicago Press, Chicago 60637
The University of Chicago Press, Ltd., London
© 2002 by The University of Chicago
All rights reserved. Published 2002
Printed in the United States of America

11 10 09 08 07 06 05 04 03 02   1 2 3 4 5

ISBN: 0-226-57711-2 (cloth)

Library of Congress Cataloging-in-Publication Data

Newman, William R.
    Alchemy tried in the fire : Starkey, Boyle, and the fate of Helmontian chymistry / William R. Newman and Lawrence M. Principe.
        p.   cm.
    Includes bibliographical references and index.
    ISBN 0-226-57711-2 (cloth : alk. paper)
        1. Chemistry—England—History—17th century.   2. Boyle, Robert, 1627–1691.   3. Starkey, George, 1627–1665.   4. Alchemy—History.   I. Principe, Lawrence.   II. Title.
    QD18 .G7 N48   2002
    540'.942'09032—dc21   2002012922

We are astonished, in reading this Treatise [Joan Baptista Van Helmont's *Ortus medicinae*], to find an infinite number of facts, which we are accustomed to consider as more modern, and we cannot forebear to acknowledge, that Van Helmont has related, at that period, almost every thing, which we are now acquainted with, on this subject [i.e., "airs"] . . . It is easy to see that almost all the discoveries of this kind, which we have usually attributed to Mr. Boyle, really belong to Van Helmont, and that the latter has even carried his theory much farther. Antoine Laurent Lavoisier, *Essays Physical and Chemical*

CONTENTS

LIST OF FIGURES

ACKNOWLEDGMENTS

The collaborative research presented in this book was supported by research grants from the National Science Foundation (SBR-9510135) and the National Endowment for the Humanities (RH-21301-95). Research on the Daniel Sennert material in chapter 1 was supported by other grants (to WRN) also made by the National Science Foundation (SES-9906126), the John Simon Guggenheim Foundation, and the Institute for Advanced Study. Research for part of the material in chapter 6 dealing with Wilhelm Homberg was supported by a separate grant (to LMP) from the National Science Foundation (SBR-9984106). The actual writing of this book was enormously facilitated by Senior Fellowships granted to each of us by the Dibner Institute for the History of Science and Technology in 1999–2000. These fellowships allowed us to be in the same place at the same time for an extended period and in an environment extremely conducive to writing and research.

Our text has benefited from the comments of several scholars who shared their expertise either in conversations or by critiquing early drafts of various chapters; we wish therefore to thank Mordechai Feingold, Guido Giglioni, Anthony Grafton, Robert Halleux, Frederic L. Holmes, Michael Hunter, Michael Koplow, K. D. Kuntz, Thomas L. Leng, John E. Murdoch, Margaret J. Osler, Rose-Mary Sargent, and Walt Woodward. We also wish to thank Bruce Janacek, who drew our attention to contemporaneous references to Starkey recorded in John Ward's diaries at the Folger Shakespeare Library.

This study depends on a considerable amount of archival material, for access to which we thank the British Library, the Bodleian Library, the Harvard University Archives, the Royal Society of London, the University of Glasgow, the Folger Shakespeare Library, and the Massachusetts Historical Society. We also thank Christine Woollett at the Royal Society and the staff of the British Library for arranging the photography of illustrative pages from the Starkey manuscripts in their collections that are reproduced by permission here.

The collaboration that resulted in this volume has also produced several re-lated items, and this book should be read in their context. In previous and separate publications, both of us tried to tackle the problems of the terms "alchemy" and "chemistry" and—less explicitly—the problems posed by incorrect perceptions of alchemy that are extremely widespread among the educated public and often still in evidence even among some scholars. Ac-cording to our first intent, the present book was to have begun with this material, but it soon took on a life of its own, first splitting off into a sepa-rate paper, and then, by virtue of an overgrown footnote, into two rather lengthy papers. The first of these papers—"Alchemy vs. Chemistry: The Etymological Origins of a Historiographic Mistake" (in *Early Science and Medicine* 3 [1998]: 32–65)—shows that throughout the seventeenth cen-tury the words "alchemy" and "chemistry" were not consistently distin-guished and that the widespread restriction of the definition of "alchemy" to "gold making" was a development of the period around 1700, which took advantage of an etymological error propagated through the seven-teenth-century textbook tradition. The second paper—"Some Problems with the Historiography of Alchemy" (in *Secrets of Nature: Astrology and Alchemy in Early Modern Europe*, ed. William R. Newman and Anthony Grafton [Cambridge: MIT Press, 2001])—recounts the historiographic fortunes of alchemy from the eighteenth century to the present and dem-onstrates that the most widespread and popular interpretations of alchemy (as a subject radically distinct from chemistry) devolve from anachronistic constructions, particularly those of Victorian occultism, and consequently have little resemblance to the topic as known and practiced in the early modern period. Accordingly, a great deal of further work is necessary in or-der to correct the popular view. The results contained in these publications are used widely throughout this volume, and readers are encouraged to consider these two papers as studies linked to this one, and propaedeutic to much of the further material detailed here.

Beyond the background provided by these two papers, there is also a companion volume to the present study. Originally, we had intended to provide the reader with an appendix containing extended selections from Starkey's notebooks. The original sources, however, proved to be so rich and so compelling when seen in their entirety that we thought it better to preserve them whole and to make them more widely available in this form for further work by other scholars. Starkey was, after all, the most widely read American author on the natural sciences before Benjamin Franklin. Thus we have prepared a companion volume consisting of Starkey's laboratory notebooks in a scholarly edition with transcriptions and annotated translations, and have also included his surviving correspondence (including five letters to Robert Boyle), daybook entries and other hitherto unpublished Starkey/Philalethes materials (*The Laboratory Notebooks and Correspondence of George Starkey* [forthcoming]).

ABBREVIATIONS

BCC	*Bibliotheca chemica curiosa,* ed. J. J. Manget, 2 vols. (Geneva, 1702; reprint, Sala Bolognese: Arnoldo Forni, 1976)
BL	Royal Society Boyle Letters
Boyle, *Correspondence*	*The Correspondence of Robert Boyle,* ed. Michael Hunter, Antonio Clericuzio, and Lawrence M. Principe, 6 vols. (London: Pickering and Chatto, 2001)
Boyle, *Works*	*The Works of Robert Boyle,* ed. Michael Hunter and Edward B. Davis, 14 vols. (London: Pickering and Chatto, 1999–2000)
BP	Royal Society Boyle Papers
Ferguson	University of Glasgow, Ferguson Manuscripts
HP	Hartlib Papers
Newman, *Gehennical Fire*	William R. Newman, *Gehennical Fire: The Lives of George Starkey, an American Alchemist in the Scientific Revolution* (Cambridge: Harvard University Press, 1994)
Notebooks and Correspondence	William R. Newman and Lawrence M. Principe, *The Laboratory Notebooks and Correspondence of George Starkey* (forthcoming)
Principe, *Aspiring Adept*	Lawrence M. Principe, *The Aspiring Adept: Robert Boyle and His Alchemical Quest* (Princeton: Princeton University Press, 1998)
RSMS	Royal Society Manuscript
Sloane	British Library Sloane Manuscripts
Van Helmont, *Opuscula*	Joan Baptista Van Helmont, *Opuscula medica inaudita* (Amsterdam, 1648; reprint, Brussels: Culture et Civilisation, 1966)
Van Helmont, *Ortus*	Joan Baptista Van Helmont, *Ortus medicinae* (Amsterdam, 1648; reprint, Brussels: Culture et Civilisation, 1966)

Introduction

In 1698, a curious treatise on the customs of the masters of alchemy was published. The anonymous author of this essay, who styles himself simply "Philadept," tells his readers of the secret ways of the *adepti* and warns them that if they want to contact these possessors of the grand arcana of Nature, they must beware of being too open in their ways. To underline his warning, Philadept tells a cautionary anecdote about the seventeenth century's most celebrated adept, the great Eirenaeus Philalethes.

> *Irenaeus Philalethes* came to *London* a purpose to impart the Art to the Honorable Mr. *B.* who undoubtedly might have done a great deal of good with it, doing much good with what he had, and, besides his good disposition, being in a capacity to render great services to the Publick by reason of his quality and fortune, his Credit and Learning. But that Honourable Person, who had such Natural and acquired accomplishments, and was so well qualified for Philosophy, and extremely desired to see an *Adept,* yet took no notice of a very great failing; he was too communicative, and his house was so disposed, that nothing was done in it could be Secret. He ordinarily spent Two Hundred pounds a Year in making Experiments, and kept several men to that purpose; But his men knew as well as himself all that was done, and they were perpetually about him. *Philalethes* presently perceived this dangerous disposition.[1]

As a result of this "dangerous disposition" Philalethes decided not to contact "Mr. *B.*" at all, but instead met a far more cautious man, and gave him "as much of the *White Elixir,* as sufficed to make 60000 l. Sterling worth of *Silver.*" There can be no doubt that the unfortunate Mr. B. of this story is none other than Robert Boyle, who had died less than seven years before this *Essay Concerning Adepts* was published. Indeed, Boyle was extremely keen to contact alchemical masters in the hope of sharing in their

---

1. [Philadept], *An Essay Concerning Adepts* (London, 1698), 45–46.

special knowledge, a fact that was known during Boyle's life and that has recently been brought to light again.

Imagine for a moment that the shadowy Eirenaeus Philalethes was actually a real person; imagine further that he did come "a purpose" to London to visit Robert Boyle and to teach him his secret knowledge. What would have happened at that meeting between the two?

Would Philalethes have required Boyle to purify himself mentally or spiritually before the operation, or to recite prayers and oaths to ensure the success of a process? For that matter, would Philalethes have even emphasized chemical operations at all, or would he have viewed alchemy as a mostly spiritual exercise, elevating or healing the practitioner rather than (or along with) materials in a flask? Or if his secret knowledge really did focus on laboratory operations, could Philalethes have simply taught Boyle a bare receipt for the Great Work to be carried out by tossing handfuls of substances into a flask or crucible to be stewed up like some witches' brew? Or might the adept have had a theoretical system to explain the reasons behind the process? What sort of concern would Philalethes have shown for the proper identity of the substances he used, the way he combined them, or the kinds of vessels he used? Would he have asked Boyle for a balance to weigh out ingredients and products, or would he have been content with a "more or less" approach? Would there have been a series of tests—quantitative or qualitative—that this master of transmutation would have provided to Boyle in order to monitor the progress of the Great Work and to determine its final success?

These questions about an imagined encounter between Philalethes and Boyle all involve issues relating to *the content and practice of laboratory chymistry in the early modern period*. Using Philalethes as an exemplar of the doctrines, goals, and practices of "alchemy," considering how he would have taught and demonstrated his chymistry evokes a multitude of questions regarding the actualities of the chymical laboratory in Boyle's day. While Boyle's chymical experimentalism has long been a subject of study—indeed, it has traditionally been seen as a crucial development in the history of chemistry—we still know relatively little about the practice of laboratory chymistry during or before Boyle's lifetime. Hence our knowledge of the sources upon which Boyle (and his chymical successors) were able to draw is likewise curtailed. As a result, questions about the content of seventeenth-century chymical laboratory practice form the basis of this book.

Our imagined scene of the adeptus Philalethes teaching chymical arcana to Boyle not only conjures up questions about the nature of early laboratory practice, but also—most remarkably—contains more than a kernel of truth. For we now know that Eirenaeus Philalethes and his influential writings were the creation of the American chymist George Starkey, and that Starkey himself communicated and collaborated with Boyle on chymical projects in the 1650s. Thus "Philalethes" did in fact interact with Boyle in a very real sense, even if Starkey hid his identity as Philalethes so well that even Boyle never unraveled the disguise. It is not only the fact that the "real Philalethes" actually interacted with Robert Boyle, however, that inspires excitement. More exciting still is the array of intricate and detailed laboratory notebooks that Starkey compiled, several of which survive today. Thanks to these records of Starkey's laboratory activities we can gain clear and otherwise unattainable answers to our questions about what "Philalethes" was doing from day to day in his laboratory, how he thought about his laboratory activities, and how he set about designing, executing, and evaluating them. While Starkey's notebooks may not (alas!) tell us how to prepare the Philosophers' Stone and how to employ it in the transmutation of base metals into gold, they do give us an unprecedented glimpse into the mind and labors of a prominent seventeenth-century chymist renowned not only for his chrysopoetic (gold-making) endeavors but also for his Helmontian iatrochemistry and his potentially lucrative "industrial" processes.[2]

The first chapter of this book begins by introducing two of the main characters of our story, George Starkey and Robert Boyle. These two men have been set up on occasion as exemplars of two hemispheres in the history of chemistry; the former as the last of the alchemists and the latter as the first of the chemists. Here we examine the relative interests of Starkey and Boyle during their youth in the 1640s, and then at the time of their first meeting in early 1651. We also consider how Boyle came to be viewed as standing in the vanguard of the "New Chemistry." This image did not arise spontaneously, but resulted in part from the way in which Boyle chose to present himself and his relationship to his chymical forebears. In the second chapter we explore the background to Starkey's chymical practice by briefly sketching issues of quantitative and qualitative chymical practice from the High Middle Ages down to the seventeenth century. The latter half of this chapter is devoted to the study of Starkey's most important preceptor, the

---

2. Other laboratory notebooks of seventeenth-century chymists also exist, of course, such as the "Aqua vitae: non vitis" of Thomas Vaughan. Vaughan's notebook retains much of the character of a traditional recipe book, however, being written mostly in the imperative mood and lacking the pervasive interaction of theory and practice found in Starkey's notebooks. See Donald R. Dickson, *Thomas and Rebecca Vaughan's Aqua Vitae: Non Vitis* (Tempe: Arizona Center for Medieval and Renaissance Studies, 2001).

Flemish natural philosopher Joan Baptista Van Helmont. Although Van Helmont's importance to seventeenth-century chymical and medical thought has been generally recognized, the details of his influence are not yet fully understood, and the magnitude of his impact remains significantly underappreciated. Here we look especially at Van Helmont's view of the role of mathematics in natural philosophy and his deployment of quantitative techniques—particularly the notion of "mass balance"—in investigating nature and probing the outcome of practical chymical processes. As we show in chapter 6, the Helmontian emphasis on mass balance provided a new focus for chymistry that culminated eventually in the famous balance-sheet method of Antoine Laurent Lavoisier and his predecessors at the Académie Royale des Sciences.

Having reviewed the background to Starkey's chymical practice, we analyze the contents of Starkey's laboratory notebooks in chapters 3 and 4. In the first of these chapters we reconstruct Starkey's methodology of experiment—how he moved from concept through theory to practice and finally to the evaluation of his results. As we demonstrate, Starkey's working laboratory methodology proves to be coherent and sophisticated in several ways. Starkey also developed the quantitative techniques of Helmontian chymistry into labor-saving and cost-monitoring tools intended to improve the industrial efficiency of his laboratory. In chapter 4, we explore further aspects of Starkey's laboratory practice, particularly his evaluation of sources of authority (including the role of divine illumination in chymical practice), and his deployment of the logical investigative techniques that he learned at Harvard College and the laboratory techniques that he acquired elsewhere in the young Massachusetts Bay Colony. One of his notebooks— devoted to discovering the preparation of the transmutatory Philosophers' Stone—allows us also to witness the way a practicing chrysopoeian decoded the wildly allegorical writings of traditional alchemy into laboratory practice. This topic also allows us to showcase how and why Starkey turned his own laboratory experiences into the allegorical, secretive treatises that eventually appeared under the name of Philalethes.

In chapter 5 we return to the topic of Robert Boyle's 1650s collaboration with Starkey and explore the scope and signs of Starkey's impact on Boyle's developing chymistry. In order to gauge the young Boyle's debt to Starkey, as opposed to the wider circle of natural philosophers, reformers, utopians, projectors, and others connected through the intelligencer Samuel Hartlib, we consider as well the chymical endeavors of several "Hartlibians," including Benjamin Worsley, Frederick Clodius, and Sir Kenelm Digby. Despite the fact that this early phase of Boyle's career is the best-charted area of Helmontian influence, we show that even here the sources of Boyle's chymical training have been incompletely understood. Finally, in chapter 6 we look at the continued influence of Helmontian notions and practices after the 1650s, first in Boyle's mature chymistry, and then down to that late eighteenth-century admirer of Van Helmont and reformer of chemistry, Antoine Laurent Lavoisier.

ONE

Worlds Apart

"How strangely unseasonable is this Melancholy weather! and how tedious a Winter have we endur'd this Summer?"[1] Thus complained the twenty-one-year-old Robert Boyle during the dismal summer of 1648, a season made doubly miserable by the cold, rainy weather and the confusions of the Second Civil War. That summer, while the city of Colcester was besieged just a few miles to the north, Boyle was staying with his sister Mary and her female relations at Leez in Essex, at the house of her father-in-law, Richard Rich, second earl of Warwick. Although Boyle complained both of the poor weather and of being obliged to spend his time reading romances to the ladies of the house, he still found some hours to devote to his grand project of amending the moral character of his fellow gentry.

Since his return from a Continental grand tour in 1644, the young Boyle had busied himself with the writing of moral and devotional literature. Troubled by everything in his elite society from rouged cheeks and half-bared bosoms to swearing and idleness, he had completed a comprehensive system of "Ethickal Elements," whose bald and rigorous format had "almost frighted most of those I had design'd them to work the quite contrary effects on."[2] To promote his message more effectively, Boyle had begun to improve his style, borrowing elements from the French romances and popular literature he loved so well and producing moralistic epistolary conceits. During the summer of 1648, Boyle worked on the most ambitious of such projects, a set of fictional letters entitled *Amorous Controversies*. Its intent was to excite devotion to God or, as Boyle termed it, "Seraphicke Love," as a worthy and more valuable substitute for the fragile, fleeting, and ultimately unsatisfying earthly love between the sexes. On the afternoon of Sunday, 6 August 1648, the aristocratic young Robert finished the last strokes of the dedicatory epistle that he wrote to accompany the completed

1. Robert Boyle, *Occasional Reflections,* in *Works,* 1:86.
2. Ibid., 1:167.

version of the final and culminating letter of the work. As he set down his pen on that Sunday afternoon to join "the whole Constelation of faire Ladies" awaiting him "in the Parke," he had no idea that within a very few years his devotion to devotion would be transmuted into a devotion to experimental natural philosophy or that he would be remembered primarily not as a writer of moral and exhortatory tracts but as an experimental natural philosopher.[3]

Another thing that Boyle did not realize was that three thousand miles to the west, in a distant outpost of English civilization on the edge of the vast wilderness of the New World, another young man (only about sixteen months his junior) had himself the previous Wednesday sent off a letter of his own, but one of a very different nature. The writer was the twenty-year-old George Starkey, the Bermuda-born son of a Scottish minister and a recent graduate of the fledgling Harvard College. The recipient was John Winthrop Jr., who would later become the first governor of Connecticut.[4] The content of Starkey's letter was far from the celestial exhortations of Boyle's *Seraphic Love,* for in it Starkey asked Winthrop to send or lend him some mercury and some antimony as well as chymical glassware and books—including Joan Baptista Van Helmont's *De lithiasi* and *De febribus,* the chrysopoeian works of Jean d'Espagnet, and the four-volume collection *Theatrum chymicum.*[5] Indeed, young Starkey was already involved in practical experimentation involving both chymical medicine and the search for the secret of metallic transmutation. We can learn from Starkey's later publications that even while a teenager in Bermuda—before going to Harvard in 1643—the young man had a keen interest in nature, and spent time observing the life cycle of metamorphosing insects.

In that first week of August 1648, those two young men were, in more than one way, an ocean apart. Boyle, the privileged child of Richard, the Great Earl of Cork, had decided to devote his life to amending the morals and piety of his fellow gentry; Starkey, the common Colonial, had devoted

3. On Boyle's early phase as a moralist and devotional writer see, John T. Harwood, *The Early Essays and Ethics of Robert Boyle* (Carbondale: Southern Illinois University Press, 1991); Principe, "Virtuous Romance and Romantic Virtuoso: The Shaping of Robert Boyle's Literary Style," *Journal of the History of Ideas* 56 (1995): 377–97; "Style and Thought of the Early Boyle: Discovery of the 1648 Manuscript of *Seraphic Love,*" *Isis* 85 (1994): 247–60; Michael Hunter, "How Boyle Became a Scientist," *History of Science* 33 (1995): 59–103.

4. For a fuller account of the following details of Starkey's early life, see Newman, *Gehennical Fire,* 14–53.

5. George Starkey to John Winthrop Jr., 2 August 1648, Massachusetts Historical Society, Winthrop Papers; printed in *Notebooks and Correspondence,* document 1, and in *Winthrop Papers* (Boston: Massachusetts Historical Society, 1943–92), 5:241–42.

himself for four years already to book and laboratory in the search after medicinal and chymical *arcana maiora.*

Two years later Starkey was preparing to immigrate to England. For six years he had been laboriously attempting to make do in his laboratory with the implements and materials available to him in New England. Dissatisfied with the results, he set sail from Boston in autumn 1650 to London, where glassware, implements, chemicals, and books were more readily available. At the same time, far to the east, Boyle had only recently discovered what he called the "Elysium" of the chymical laboratory, and set about writing a treatise on the "Morall speculations, with which my Chymicall Practises have entertained mee" and a discourse on the theological uses of natural philosophy.[6] At this time, Boyle's primary interest in devotion and morality remained unchanged, but he had discovered a new set of phenomena that could suggest literary and rhetorical images to him. Since the mid-1640s he had been writing "Occasional Reflections"—devotional meditations provoked by scriptural readings or casual observations of everyday events. The sights (and presumably the smells) of the chymical laboratory seem to have become a new source of such reflections for Boyle, as expressed, for example, in his eventually published meditation on charity provoked by "distilling Spirit of Roses in a Limbeck."[7]

The paths of these two very different young men did not, however, remain separate. Soon after Starkey's arrival in England he was introduced to the Hartlib circle—the group of reformers, utopians, natural philosophers, and others gathered around the German émigré and intelligencer Samuel Hartlib—with which Boyle had been affiliated since the mid-1640s. On 29 November 1650 Benjamin Worsley, one of the Hartlib circle's important chymical enthusiasts, reported to Hartlib that he had recently met Starkey for the first time.[8] By early December, Hartlib himself had met Starkey and was very impressed by him; indeed, he had heard of Starkey's prowess as a physician and chymist even before the young man had left New England, and entered a report on him in his *Ephemerides* in early 1650.[9] It was finally

---

6. Boyle to Katherine, Lady Ranelagh, 31 August 1649, in Boyle, *Correspondence,* 1:82; see also Principe, "Romantic Virtuoso," 392–93. Boyle had purchased a chymical furnace in 1647, but it broke in transit, and nothing came of this earlier, stillborn attempt to carry out chymical operations; see Boyle to Katherine, Lady Ranelagh, in Boyle, *Correspondence,* 1:50, and Hunter, "How Boyle Became a Scientist," 63.

7. Boyle, *Occasional Reflections,* 1:54–55.

8. Samuel Hartlib, *Ephemerides* 1650, HP 28/1/78B. For more on Worsley, see Charles Webster, *The Great Instauration: Science, Medicine, and Reform, 1626–1660* (London: Duckworth; New York: Holmes and Meier, 1975), and below, chapter 5.

9. Hartlib, *Ephemerides* 1650 (March), HP 28/1/57A.

through the mediation of Robert Child—who had conveyed reports on Starkey's expertise from New England to Hartlib—that Starkey and Boyle met for the first time around the beginning of 1651.[10] Perhaps this meeting was the result of Boyle's concerns about his health at this time and his fears of having kidney or bladder stones; thus their first meeting may well have been as patient and physician. Indeed, the earliest of Starkey's surviving letters to Boyle—written in April or May 1651—enclosed a medicine for the stone, and Hartlib recorded that Starkey had prescribed a medicine for Boyle in January 1651, around the time Boyle and Starkey first met.[11] Immediately after their meeting, as historians have known for some time, Boyle and Starkey began corresponding and collaborating on chymical experiments and preparations. We have Boyle's own published testimony in his *Usefulnesse of Experimental Naturall Philosophy* (published 1663) as well as Starkey's congruent testimony in his own *George Starkey's Pill Vindicated* that the two collaborated on the preparation of the *ens Veneris,* a Helmontian pharmaceutical.

At this point, Starkey had been carrying out experimentation since 1644 and so had seven years of experience behind him, while Boyle's work in affairs of the laboratory had been under way for just a little over a year; thus, merely in terms of duration, Starkey was the more experienced laboratory worker at the time of their meeting in 1651. This impression is well corroborated by the vivid depictions of their relative interests and levels of experimental proficiency, experience, and activities recorded in their contemporaneous private writings.

Since the time in his youth when Boyle withdrew to Geneva with his tutor (after the Irish Rebellion brought his father's funding, and consequently his Continental grand tour, to an abrupt end in 1642), Boyle had compiled commonplace books, generally begun on the first day of the year. The earliest of these collections dates from Boyle's 1643 residence in Geneva and survives in a recently rediscovered notebook that the teenaged Boyle used there.[12] Later, more mature examples dating from 1647 to the mid-1650s are preserved among the Boyle Papers. The items from the 1640s reflect Boyle's preoccupations with his moral program; they functioned as

10. Starkey recalls that he and Boyle first met "by the occasion of our mutual Friend, Dr. *Robert Child*"; see George Starkey, *Pyrotechny Asserted* (London, 1658), Epistle Dedicatory, [xiii].

11. Starkey to Boyle, April/May 1651, in *Notebooks and Correspondence,* document 3, and in Boyle, *Correspondence,* 1:90–103; Hartlib, *Ephemerides* 1651 (January), HP 28/2/3A.

12. Principe, "Newly Discovered Boyle Documents in the Royal Society Archive: Alchemical Tracts and His Student Notebook," *Notes and Records of the Royal Society* 49 (1995): 57–70.

repositories of clever rhetorical devices or quotations from his reading that might be either merely memorable or potentially useful in his rhetorically self-conscious moral and devotional writings. For example, Boyle's "Diurnall Collections" of 1647 contains pages of quotations copied from La Calprenède's lengthy romance *Cassandre* (1642–45) along with snippets of text from other sources as well as phrases of Boyle's own composition. A similar collection from 1649 preserves ten pages of quotations copied from the manuscript of his brother Roger's romance *Parthenissa* (the first part of which was published in 1651).[13]

As Michael Hunter has observed in his study of these documents, however, the middle of the year 1649 marks a time of "conversion" in Boyle's interests; the corresponding collection for 1650, entitled "Memorialls Philosophicall," contains no rhetorical material, but is instead entirely devoted to medical receipts.[14] We know from Samuel Hartlib's *Ephemerides* that Boyle also read at least one chymical book in 1649 and collected one chrysopoetic recipe.[15] Later examples of Boyle's collections, dating from 1652, 1654, and 1655, continue to display medical and chymical materials exclusively. We will examine these collections more thoroughly later, particularly in terms of the receipts in them obtained from Starkey and other Hartlibians. At present, however, the key point is that these materials did not arise from Boyle's own experimentation; rather, they were compendia of other people's results—items communicated to him by personal contacts, generally within the Hartlib circle. Boyle own experimental results do not show up until the late 1650s. Nor do there exist other pieces of evidence that might argue for the existence of Boyle's own original natural philosophical or medical experimentation during this early period. Thus, in the early 1650s, although Boyle had in fact begun to have an increased *interest* in experimentation and had begun to perform some rudimentary operations, he clearly was not yet well versed in experiment or its techniques.

But the situation portrayed in Starkey's contemporaneous notebooks is quite different. One fragment tells us that he had begun his chymical experimentation while still an undergraduate at Henry Dunster's Harvard in

13. BP, vol. 44, 94–112; see Principe, "Virtuous Romance," 381.

14. BP, vol. 28, fols. 309–11; see Hunter, "How Boyle Became a Scientist," 66–7; for further material on the whole chronological range of Boyle's "work-diaries," see Michael Hunter and Charles Littleton, "The Work-Diaries of Robert Boyle: A Newly Discovered Source and Its Internet Publication," *Notes and Records of the Royal Society* 55 (2001): 373–90.

15. Hartlib, *Ephemerides* 1649, HP 28/1/32A, where Boyle is noted as reading one "Clave" in French, probably Etienne de Clave, or less likely, Gaston "Claveus" Duclo, both of whom are cited in the *Sceptical Chymist*. The chrysopoetic recipe is referred to in *Ephemerides* 1649, HP 28/1/8B; no evidence is given that Boyle actually tried it in the laboratory.

the mid-1640s.[16] After starting with a bookish interest in the subject that he was able to pursue in Harvard's environment of tutorial and disputation, Starkey turned to an extended iatrochemical and technological community in Massachusetts for hands-on instruction. The same fragment also reveals that Starkey had resumed laboratory work by February 1651 at the latest—within four months of his arrival in England—and probably earlier. Starkey's letter of April/May 1651 to Boyle excitedly recounts the experimental first fruits of this new laboratory. These first productions of Starkey's London laboratory already include serious and developed approaches to the chief chymical desiderata of the day. Another notebook fragment, written in December 1651, likewise reveals Starkey busy in his laboratory at work on, among other things, the Philosophical Mercury (for making the Philosophers' Stone), the alkahest, and advanced medical arcana. Starkey's 30 May 1651 letter to the Amsterdam chymist Johann Moriaen further corroborates this view of Starkey busy at work in his new English laboratory.[17] Clearly then, at the time of Starkey and Boyle's meeting, and for some time thereafter, Starkey was by far the more active and experienced laboratory experimentalist.

The divergent portraits presented here of these two young men—a Starkey skilled in laboratory practice and single-minded in his devotion to chymistry, and a Boyle predominantly a moralist with a perhaps slightly dilettantish interest in chymistry—seem at odds with some usual perceptions of Starkey's and Boyle's work, and certainly with the divergent fates the two would receive historically. Indeed, the intervening centuries of history have tended to re-create the ocean that separated them in 1648. In their own day, both enjoyed a measure of fame, but in very different ways. Boyle's eventual fame as a "father of chemistry" (regardless of how seriously we take that title), as a founder of the Royal Society, as a proponent of experimentalism, the mechanical philosophy, and corpuscularianism hardly requires comment and endures to this day. Starkey acquired fame as well; besides his lesser degree of recognition as a London medical practitioner during his own short lifetime, he achieved renown-once-removed as the alchemical adept Eirenaeus Philalethes, a character as celebrated as he was mysterious. The chrysopoetic treatises that Starkey wrote and circulated under the name of this fictive adept achieved enormous popularity and respect and ex-

16. Text printed Newman, *Gehennical Fire,* 249–50, and *Correspondence and Notebooks,* document 15.

17. On Moriaen, see John T. Young, *Faith, Medical Alchemy, and Natural Philosophy: Johann Moriaen, Reformed Intelligencer, and the Hartlib Circle* (Brookfield, Vt.: Ashgate, 1998). The texts are published in *Notebooks and Correspondence,* documents 4 and 5.

erted significant influences on such figures as Boyle (who seems never to have discovered that Philalethes was actually Starkey), Sir Isaac Newton, Johann Joachim Becher, Georg Ernst Stahl, Hermann Boerhaave, and a host of other notables. For a time in the late seventeenth and early eighteenth centuries both Starkey (mostly as Philalethes) and Boyle garnered respect and notoriety for their respective natural philosophical and experimental endeavors; but this situation did not persist. The progressive segregation of "chemistry" and "alchemy" served to erect a partition between Boyle and Starkey/Philalethes, and with the eighteenth-century repudiation of alchemy as simply fraudulent, and later as "nonscientific," "pseudoscientific," and even "occult," Starkey (along with Philalethes) slipped into the shadows beyond the fringe of scientific respectability. Starkey/Philalethes came to be seen as the last of an alchemical line, being called by one prominent historian of science "the last great philosophical alchemist of the seventeenth century," while Boyle came to stand at the vanguard of the "New Chemistry."[18]

This division between Boyle and Starkey—and the division between an alchemical past and a chemical future of which it is emblematic—are problematic. The problem with these divisions is thrown into relief by our knowledge of the communication and collaboration that went on between Boyle and Starkey in the 1650s. Despite the very different depictions of them in the secondary literature, Boyle and Starkey were—at least for a time—interested in the same issues and pursued the same goals. The world of Starkey and the world of Boyle, of a supposedly ancient alchemy and of a modern chemistry, seem to coexist. The problem becomes yet more acute when we consider the biographical depictions presented above, wherein Starkey is clearly the more experienced and dedicated chymical experimentalist of the two, and is further intensified by the fact that, as we shall show later in this study, Starkey was Boyle's primary tutor in chymistry.

Recent historical scholarship has begun to bridge the seeming gulf between the likes of Starkey and the likes of Boyle. We now know that the clean division of alchemy from chemistry, which seemed so "obvious" at first glance a generation ago, did not exist for most in the seventeenth century.[19] Moreover, various characteristics widely attached to alchemy that

18. Betty Jo Teeter Dobbs, *The Foundations of Newton's Alchemy, or The Hunting of the Greene Lyon* (Cambridge: Cambridge University Press, 1975), 52.

19. William R. Newman and Lawrence M. Principe, "Alchemy vs. Chemistry: The Etymological Origins of a Historiographic Mistake," *Early Science and Medicine* 3 (1998), 32–65; Bernard Joly, "Alchimie et Rationalité: La Question des Critères de Démarcation entre Chimie et Alchimie au XVIIe Siècle," *Sciences et Techniques en Perspective* 31 (1995): 93–107; John C. Powers, "'Ars sine Arte': Nicholas Lemery and the End of Alchemy in Eighteenth-

seemed to render a neat division from chemistry possible have turned out to be largely exaggerations or anachronistic accretions.[20] On the other side, the roster of accepted figures of the "Scientific Revolution" who were likewise devotees of traditional alchemy continues to grow—first Newton, then Boyle, now still others.

An investigation of seventeenth-century chymical laboratory practice promises to provide much deeper insight into the development of chemistry. The appearance of rational laboratory experimentalism has long been closely associated with key junctures in the history of chemistry (and the history of science more generally). On the other hand, the "alchemical" tradition has frequently been thought lacking in important qualities that are seen as characteristic of "modern" laboratory practice—quantification and control, theory-guided practice, practice-informed theory, reproducibility, and so forth. Indeed, the most prevalent schools of interpretations of "alchemy" tended to distance "alchemical practice" from anything resembling careful laboratory operations, or at least downplayed the importance of practice in that tradition. In fact, the evidence of careful experimental methods and procedures has sometimes sufficed to identify the side of the canonical "alchemy-chemistry" divide upon which a given worker or procedure should be placed. A good example of the pervasive (and misleading) character of this view can be seen in the reasoning that led both Richard Westfall and Betty Jo Teeter Dobbs to misattribute the authorship of Starkey's *Clavis* or "Key" to Isaac Newton, who had merely transcribed it. Westfall argued that the "Key," a detailed recipe for preparing a "philosophical mercury," could not have been written by an alchemist (like Starkey/Philalethes) because of the precision of its directions. Westfall's evidence was that the "Key," "unlike virtually all other alchemical literature . . . described laboratory procedures in detailed operative terms that could be repeated today." Speaking of Newton's precise chemical practice in general, Westfall claimed that "alchemy had never known anything like this before. It was indeed more than alchemy could survive." Dobbs pronounced much the same idea, saying that the "Key" betrayed "the fine-grained quality [of] the instructions for achieving the best results" that one associates with Newton's experimentalism, as opposed to the supposedly rough-and-ready character of alchemical practice.[21]

Century France," *Ambix* 45 (1998): 163–89. For an example of their continued separation, see Marco Beretta, *The Enlightenment of Matter* (Canton, Mass.: Science History, 1993), 74–157.

20. Lawrence M. Principe and William R. Newman, "Some Problems with the Historiography of Alchemy," in *Secrets of Nature: Astrology and Alchemy in Early Modern Europe*, ed. William R. Newman and Anthony Grafton (Cambridge: MIT Press, 2001), 385–434.

21. Richard S. Westfall, *Never at Rest: A Biography of Isaac Newton* (Cambridge: Cam-

But what can actually be said about alchemical practice? Was precise practice so unknown that the discipline could not survive its advent? The notion that precise practice suddenly appeared on the chymical scene (presumably some time in the second half of the seventeenth century) naturally invokes the question: Where did it come from? Was it dependent upon any precedent traditions? In short, what was Starkey actually doing during all those hours he spent in the laboratory—the laboratory he actually left his home in the New World to build? How (if at all) is the extravagant, secretive language of the Philalethes treatises linked to laboratory operations? Do the details of laboratory practices serve to strengthen or to erode the distinctions between the activities and commitments of Starkey/Philalethes and Boyle, or between alchemy and chemistry? What were the traditions and developments of early modern chymical practice, and how did Starkey (and Boyle) draw upon these? What does Starkey's work say about the nature of alchemical laboratory practice and the status and development of chymistry and its practitioners in the seventeenth century? Given Starkey's greater laboratory expertise and experience in the early 1650s, how far can we link Boyle's own chymical development as an experimentalist with the influence of George Starkey and of other chymical practitioners? Are there discernible and distinct features of Starkey's own methodology, whence did it derive, and what was its continuity with later practices? An investigation into the chymical laboratory practice of the mid–seventeenth century thus promises to shed much new light on questions of chemistry's development and status during this important period.

The remarkable survival of several of George Starkey's laboratory notebooks from the 1650s provides an invaluable primary source for answering such questions. Starkey's status as an important and prolific writer on transmutation (i.e., as an "alchemist"), his position as a vocal Helmontian chymical physician, and the fact that he was on close terms with Boyle during the writing of these records enormously enhance their intrinsic interest and their pertinence for this study. Notably, the time when Starkey and Boyle were on closest terms falls within Boyle's crucial formative period as a natural philosopher and experimentalist, a time during which, as we have seen, Starkey was the more experienced and devoted laboratory worker of the

bridge University Press, 1980), 370; "The Role of Alchemy in Newton's Career," in M. L. Righini Bonelli and W. R. Shea, eds., *Reason, Experiment, and Mysticism in the Scientific Revolution* (New York: Science History Publications, 1975), 189–232, esp. 227, 229; Dobbs, *Foundations,* 176; see Newman, "Newton's *Clavis* as Starkey's *Key*," *Isis* 78 (1987): 564–74, esp. 564–65.

two. Questions about seventeenth-century chymical laboratory practice thus form the core of this book.

BOYLE'S PORTRAYAL OF HIS RELATIONSHIP TO CHYMISTRY

Before turning to the central issues of this book regarding chymical laboratory practices, there remains one aspect of the development of chymistry in the seventeenth century and Boyle's relationship to it that should first be discussed, namely the way in which Boyle portrayed his own complex relationship with the chymistry of his day and its practitioners. The "revolutionary" appearance of the seventeenth century did not arise solely as a retrospective historical formulation without the cooperation of seventeenth-century thinkers and their champions. In the present case, Boyle often strove to distinguish himself from his predecessors, thus aiding his subsequent deployment as a point of demarcation between an older alchemy (often characterized as "obscurantist") and a modern experimental science.[22] Indeed, this fixing upon Boyle was encouraged by what Adrian Johns has recently called the "iconic status" of Boyle, particularly as an exemplar of modesty, piety, veracity, and civility.[23] Building on the work of Steven Shapin and Simon Schaffer, Johns stresses the self-image that Boyle and other members of the early Royal Society put forth—one that painted a picture of "broad experience, modesty, moderation, freedom of action, and disinterest."[24] Boyle's presentation of himself and his scientific development thus projects both the image of a disinterested and modest natural philosopher and the sense of a thinker who owed little of substance to the foregoing traditions of "the chymists." But Boyle's self-portrayal can be taken too uncritically; indeed, his writings present a distorted image of his relationship to contemporaneous chymistry and its practitioners. Boyle's public attitude toward chymistry actually involves two interrelated aspects: first, a subordination of chymistry to natural philosophy, and second, a posture of independence from previous chymical traditions.

Boyle's publications often display a pattern of accepting the technology and empirical results of contemporaneous chymists while conspicuously rejecting their theories. This pattern has been largely recapitulated in much of the secondary literature, and in both locales it has had the effect of elevating Boyle's own status and diminishing that of the foregoing traditions.

22. On the historical development of portrayals of Boyle's relationship to alchemy/chemistry since the eighteenth century, see Principe, *Aspiring Adept*, 11–23.

23. Adrian Johns, *The Nature of the Book* (Chicago: University of Chicago Press, 1999), 504, 510.

24. Johns, *Nature of the Book*, 468.

Boyle's distancing of himself from earlier chymical theory in general appears in a famous passage from the 1666 *Origine of Formes and Qualities:*

> though I have a very good opinion of Chymistry it self, as 'tis a Practicall Art; yet as 'tis by Chymists pretended to containe a Systeme of Theoricall Principles of Philosophy, I fear it will afford but very little satisfaction to a severe enquirer, into the Nature of Qualities.[25]

Here Boyle labels chymistry a "practical art," clearly subordinate to natural philosophy; contemporaneous chymistry is useful to natural philosophy only in an artisanal capacity. Boyle displays a similar position in *The Sceptical Chymist,* saying that

> though I am a great Lover of Chymical Experiments, and though I have no mean esteem of divers Chymical Remedies, yet I distinguish these from their Notions about the causes of things and their manner of Generation.[26]

In short, chymists are good technicians, and have even discovered remarkable medical cures, but they have little to contribute to the *theoretical* advancement of natural philosophy. Boyle reiterates this position later in *The Producibleness of Chymicall Principles* (1680), where he portrays chymical doctrines as positively dangerous, because their adoption leads otherwise sober men to take up "precarious and superficial accounts of divers *Phaenomena* of Nature" and to forsake "the investigation of the true and fundamentall causes" of things.[27] Hence Boyle affirms that he has "a very differing esteem of the *Notionall* and the *Practicall* part of Chymistry."[28] The operations of the chymical laboratory, such as distillation, solution, sublimation, and precipitation, are "excellent tools in the hands of a natural philosopher," but the theoretical speculations of chymists are of little worth. Boyle thus marks out a subservient role for chymistry, placing it on a low epistemic level subordinate to natural philosophy and, more importantly, subordinated to the ministrations of the natural philosopher.[29]

Boyle maintains this position for chymistry even when he defends it against the supercilious deprecations of natural philosophers. For example,

---

25. Boyle, *Origine of Formes and Qualities,* in *Works,* 5:301.
26. Boyle, *Sceptical Chymist,* in *Works,* 2:208.
27. Boyle, *Producibleness of Chymicall Principles,* in *Works,* 9:23.
28. Ibid., 9:25.
29. Indeed, Boyle outlines a similar position for medicine: "Natural Philosophy being a Science of far greater Extent than Physick, and supplying it with many of its Principles and Theories. . . . Medicine being a *Part,* or an *Application* of Natural Philosophy"; Boyle, *Experimenta et Observationes Physicae,* in *Works,* 11:397–98.

in the preface to his "Essay on Nitre," Boyle famously complains about "Learned Men" of his acquaintance who chided him for spending too much time upon such an "empty and deceitful study" as chymistry, a field worthy only of "sooty Empirics." Boyle's initial response (which we now know to be disingenuous) is that the traditional goals, such as the transmutatory Elixir, are "not at all my aim."[30] For he claims that although finding medicines and improving manual trades by means of chymistry is valuable, this would provide insufficient recompense for his time and trouble; rather, his study of chymistry is directed toward showing that "Chymical Experiments might be very assistant even to the speculative Naturalist." That is to say, his goal is to show how chymistry can be a useful ancilla to natural philosophy.[31] Indeed, Boyle clearly entitled this essay, widely known simply as the "Essay on Nitre," as "an Attempt to make Chymical Experiments useful to Illustrate the Notions of the Corpuscular Philosophy." Boyle's attitude toward chymistry and its practitioners is strikingly restated in an unpublished section of *Usefulnesse* where he criticizes chymists for confining their art to medicine and transmutation and adds that its further use—which its practitioners have hitherto neglected—lies in "its ability to service natural philosophy."[32]

Boyle thus depicts chymistry as a valuable source of techniques and products but allots it a position subservient to natural philosophy. Even though he declaims against chymical textbook writers for reducing chymistry to the status of a manual art, he himself maintains its artisanal status—even though he will redirect it to serve the greater master natural philosophy rather than medicine. His "Essay on Nitre" is thus a "physico-chymical" treatise attempting to combine theoretical principles of natural philosophy with laboratory demonstrations drawn from the chymical art. His goal is to do "no unseasonable piece of service to the corpuscular philosophers, by illustrating some of their notions with sensible experiments" drawn from chymistry. His point, bluntly stated, is that the natural philosopher will supply the ideas and theoretical frameworks, while the storehouse of the chymists' experiments can be plundered to provide the empirical evidence; chymistry is not a theoretical field in its own right but an empirical bulwark for the "new" natural philosophy.[33]

Such claims are fairly constant in Boyle's works, extending from the early

---

30. Of course, Boyle was in fact *very* interested in preparing the Elixir and *did* employ many aspects of earlier chymical theory; see Principe, *Aspiring Adept*.

31. Boyle, *Certain Physiological Essays*, in *Works*, 2:85–86.

32. Boyle, "That the Empire of Man may be Advanced by the Skill of Physicists in Chemical Matters," in *Works* 13:324–25.

33. Boyle, *Certain Physiological Essays*, in *Works*, 2:91.

1660s to the 1680s, despite variations of emphasis.[34] Although the fully mature Boyle of *Producibleness* clearly leaves the door open for the rare "Chymical Philosophers" or chrysopoetic adepti to teach him a potentially higher knowledge, even here he still exempts his "corpuscularian or mechanical" principles from correction by them, even though they may be able to teach him new things about "the subordinate Theory of mixt bodies in particular."[35] It is up to the "naturalists," as Boyle likes to call genuine natural philosophers such as himself, to reduce the phenomena afforded by chymists into an acceptable natural philosophical order. Thus, Boyle's positioning of chymistry both promotes his self-identification as a natural philosopher able to substitute his own philosophical notions in place of the ill-formed theories of the chymists and implies that he is making a fresh start for chymistry by deploying it rightly and philosophically.

BOYLE AND HIS SOURCES

Let us briefly consider just how unrealistic Boyle's portrayal of his relationship to earlier chymistry, including chymical theory, actually is. A clear example is Boyle's debt to the chymical and philosophical writer Daniel Sennert, which shows how a centerpiece of Boyle's own corpuscular theory had been substantially developed among "the chymists," of whose "notional parts" Boyle was explicitly dismissive. Sennert, a well-known figure in the seventeenth century, was born in 1572 and taught natural philosophy and medicine at the University of Wittenberg for nearly forty years, until his death in 1637. Despite having begun his career as a fairly orthodox Aristotelian in the mold of Jacopus Zabarella and Benedictus Pereira, Sennert began espousing a frankly atomistic doctrine in 1619, which he developed continually throughout the remainder of his life.[36] Sennert's atomism was heavily based on examples derived from contemporary chymistry—in particular he employed "reversible" processes to demonstrate the existence and persistence of semipermanent material corpuscles—his atoms.

These processes, called "reductions into the pristine state" [*reductiones*

---

34. Boyle, *Excellency*, in *Works*, 8:110–11; see "Alcali and Acidum," in *Mechanical Origine of Qualities*, in *Works*, 8:418; Boyle, *Producibleness*, in *Works*, 9:25–27.

35. Boyle, *Producibleness*, in *Works*, 9:27.

36. For Sennert's early Aristotelianism, see William R. Newman, "Experimental Corpuscular Theory in Aristotelian Alchemy: From Geber to Sennert," in *Late Medieval and Early Modern Corpuscular Matter Theories*, ed. Christoph Lüthy, John E. Murdoch, and William R. Newman (Leiden: Brill, 2001), 291–329; and Newman, "Corpuscular Alchemy and the Tradition of Aristotle's *Meteorology*, with Special Reference to Daniel Sennert," *International Studies in the Philosophy of Science* 15 (2001): 145–53. See also Emily Michael, "Sennert's Sea Change: Atoms and Causes," in *Corpuscular Matter Theories*, ed. Lüthy, Murdoch, and Newman, 331–62.

ad pristinum statum] in the language of early modern chymistry, made a significant impression on the young Boyle. Boyle's essay "Of the Atomicall Philosophy," which was composed in the early to mid-1650s and is the earliest expression of his corpuscular theory, relies heavily on Sennert.[37] There Boyle recapitulates two processes from Sennert—one where silver is dissolved in aqua fortis (nitric acid), the solution filtered, and the silver then precipitated by means of salt of tartar, and a second where gold and silver are fused into a homogeneous alloy that is then subjected to similar treatment with nitric acid, allowing for the separation of the two metals back into the forms they had before their fusion. In the first process, since the silver was regained from the acid intact despite its temporary invisibility while in solution, it was obviously not destroyed by the highly corrosive "menstruum." Moreover, since it passed through the filter paper without leaving a residue, the bits into which the acid had broken it must be extremely small—smaller than the invisible pores of the paper. The great fineness of the precipitated powder supplies further proof of the silver's minute particulate nature. Hence two canonical criteria of atomism—the indestructibility of the corpuscles and their invisibly small size—are displayed by Sennert's experiment.

It is important to note that Boyle used not only Sennert's experiments in his own "Atomicall Philosophy," but also the entire theoretical framework in which Sennert embedded them—namely, a corpuscular theory of matter. Nonetheless, the Wittenberg iatrochemist's name appears nowhere in the text even though Boyle begins the treatise by lauding specific figures involved in atomism. There he praises the atomism now "so luckyly reviviv'd[!] & so skillfully celebrated in divers parts of Europe by the learned pens of Gassendus, Magnenus, Des Cartes & his disciples [and] our deservedly famous Countryman Sir Kenelme Digby."[38] Despite paying obeisance to these luminaries of revived atomism, their direct contribution to Boyle's treatise is slight, especially when compared with the primary role given to the opening section of the treatise, whose content is drawn directly from Sennert. There is no doubt that Boyle was well acquainted with Sennert's work; prior to writing "Of the Atomicall Philosophy," Boyle had cited the German physician in the unpublished "Essay of the Holy Scriptures," composed between 1651 and 1653, and did so again in the first part of

37. For further details on Boyle's borrowings from Sennert, see Newman, "The Alchemical Sources of Robert Boyle's Corpuscular Philosophy," *Annals of Science* 53 (1996): 567–85. On Sennert's atomism and the *reductiones,* see Christoph Meinel, "Early Seventeenth-Century Atomism: Theory, Epistemology, and the Insufficiency of Experiment," *Isis* 79 (1988): 68–103, esp. 76–80, 92–99.

38. Boyle, "Of the Atomicall Philosophy," in *Works,* 13:227.

The Usefulnesse of Experimental Naturall Philosophy, mostly written by the mid-1650s.[39]

The same silent use of Sennert's theory and experiments recurs in Boyle's later published works. *The Sceptical Chymist* sallies forth once again with Sennert's reduction of gold and silver to their pristine state, which Boyle supplements with parallel details of similar *reductiones.*[40] Here, however, the informed reader is struck by Boyle's explicit claim to originality in using these laboratory observations as confirmations of corpuscular theory. After proposing the existence of "minute Masses or Clusters" of corpuscles that are not "easily dissipable," Boyle provides evidence by

> add[ing] something out of Experience; which, *though I have not known it used to such a purpose,* seems to me more fairly to make out, that there May be Elementary Bodies, than the more questionable Experiments of Peripatetics and Chymists prove that there Are such.[41]

In fact, Boyle here merely reuses Sennert's own experiment to demonstrate what Sennert had already used it to prove—that there are *prima mista* (or clusters of elementary atoms) that are themselves indivisible in the ordinary course of nature. Boyle's protestations that he had "not known [this experiment] used to such a purpose" before, and that it is superior to the "more questionable Experiments of Peripetatics and Chymists" is rhetorical posturing rather than reality.

Perhaps Boyle had simply forgotten his source, or even that he had a source. But the passage from "Atomicall Philosophy" remains curious, for there Boyle lists potential sources for his atomism but without mentioning Sennert, as if he were intentionally writing Sennert out of the picture as a source for *both* his corpuscular theory and the chymical experiments to demonstrate it. This explanation becomes more likely when we consider many passages where Boyle *does* cite Sennert, for these suggest that it is specifically his debt to Sennertian atomism that Boyle wishes to gloss over. Boyle is eager to cite Sennert in three capacities—as an authority in medicine, as an opponent in chymistry, and as a Scholastic. Almost nowhere does he appear as an atomist, despite the prominence of atomism in the very works of Sennert that Boyle read and cited.

---

39. Boyle, *Usefulnesse,* in *Works,* 3:254. For the date of composition of the final essays in part 1 of *Usefulnesse,* see *Works,* 3:xix–xxiv. For the date of Boyle's "Essay of the Holy Scriptures" see Hunter, "How Boyle Became a Scientist," 67; for Boyle's reference to Sennert therein, see 77. Boyle's earliest reference to Sennert (his "Institutions of Physicke") occurs in 1647 in the context of one of his unpublished moral epistles; see Boyle, *Works,* 13:70.

40. Boyle, *Sceptical Chymist,* in *Works,* 2:230–31. See Newman, "Alchemical Sources," 580–84.

41. Boyle, *Sceptical Chymist,* in *Works,* 2:230 (italics added).

Of the fifty-two references to Sennert by name that we have found in Boyle's published works, twenty-three, slightly less than half, represent him as a medical writer. In this capacity Sennert receives high accolades as a "learned physician" and is used as a source for illustrative anecdotes about such subjects as a man who lived by eating dung, the dangerous effluvia emitted by hot spiders and orpiment, and unusual medical cases.[42] In a second group of eleven references, Boyle refers to Sennert as a "chymist," and here he mixes praise with criticism. For example, the *Sceptical Chymist* portrays Sennert as "the learnedest champion for the hypostatical principles," meaning the three Paracelsian principles Mercury, Sulphur, and Salt, but the very fact that he is the "great champion" of these material principles of mixed bodies makes him philosophically unacceptable since the doctrine of the *tria prima* is Boyle's particular target there.[43] Finally, in a third group of eighteen references, Boyle portrays Sennert as a Scholastic, and here Boyle's explicit purpose is to debunk Sennert's natural philosophy.[44] In only one of these fifty-two references does Boyle even obliquely acknowledge that Sennert was in fact an atomist or a corpuscularian, and there it is to misrepresent Sennert's theory as incapable of accommodating the transmutation of metals.[45] Thus the *open* use of Sennert conforms exactly to Boyle's relative positioning of chymistry and natural philosophy. Sennert's bare experiential anecdotes are useful materials for deployment by the natural philosopher, but his theories and attempts at natural philosophy are not. Yet this apparent use is belied by the clear evidence in "Atomicall Philosophy," *The Sceptical Chymist,* and *The Origine of Formes and Qualities* that Boyle actually borrowed not only experiments from Sennert but also the theoretical conceptions for which they had already been deployed as evidence by Sennert himself. Boyle read Sennert's atomistic treatises with extraordinary care, even reprising whole passages, yet without acknowledgment.[46]

---

42. Boyle, *Works,* 1:71; 1:279; 3:340; 3:342–43; 3:343 (three times); 3:372; 3:373; 3:453; 3:470; 7:252; 7:293; 7:294; 7:295; 7:296; 7:351; 8:108; 10:127; 10:129; 10:337; 10:381. There is an additional reference that might be construed either as chymical or medical—3:346 (marginal note in a manuscript draft occurs at BP, vol. 16, fol. 216), where Boyle cites Sennert's recounting of the events related by Paracelsus's onetime follower Oporinus.

43. Boyle, *Works,* 2:281; 2:303; 2:315; 2:328; 2:330; 2:331; 2:332; 2:336; 2:337; 2:338; 3:254.

44. Boyle, *Works,* 2:267; 5:339; 5:449; 5:450; 5:451; 5:452; 5:455; 5:458 (five times); 5:459; 5:464; 5:466; 5:473; 6:269; 8:392. The largest group of such references occurs in Boyle's discourse on subordinate forms, which provided an appendix to the second edition of *The Origine of Formes and Qualities.*

45. Boyle, in *Works,* 2:267. The reference to Sennert, found on 267, is followed by Boyle's critique of his theory on 271–76.

46. For example, Boyle's discussion of "immanent" and "transient" materials on 273–74 of *The Sceptical Chymist* is based very closely on the following passage from Sennert, *De chymi-*

The evidence seems clear then that Boyle intentionally suppressed acknowledgment of the corpuscular theoretical principles of Daniel Sennert, even while praising or blaming Sennert by name in other contexts. Boyle's debt was not insignificant, since Sennert's chymistry provided Boyle with *both* theory and illustrative experiments. Thus Boyle drew a ready-made example of his Baconian attempt to ground the corpuscular philosophy on experimental evidence from the chymical writings of Sennert. The important point for our story is that Boyle's lack of explicit acknowledgment of such origins for his corpuscular theory gives the appearance of a greater break between him and the foregoing chymical tradition than really was the case.

The case of Sennert is paralleled by examples drawn from Boyle's relationship with George Starkey. Although the full study of Starkey's influence on the young Boyle is reserved for chapter 5, three examples here will suffice to indicate Boyle's similar use of Starkey's chymistry, again with the effect of obscuring the real linkage between Boyle and his chymical predecessors.

The first example comes from the *Usefulnesse of Experimental Naturall Philosophy*, where Boyle deals with the issue of "correcting" the toxic properties of opium. He writes that

> I never knew *Opium* so much Corrected by Saffron, Cinnamom, and other Aromatical and Cordial Drugs (wherewith 'tis wont to be made up into *Laudanum*) nor by the most tedious tortures of *Vulcan*, as I have known it by being a while Digested in Wine, impregnated with nothing but the weight of the *Opium* of pure Salt of *Tartar*; as we elsewhere more fully declare.[47]

Later in *Usefulnesse*, Boyle adds that he has found that opium "if duly corrected and prepared, proves sometimes a great resolver, and commonly a sudorifick." While Boyle implies that these are his own observations, they are in fact Starkey's. In a letter to Boyle of 3 January 1652, Starkey refers to a "laudanum" that he is making for Boyle, and Starkey's laboratory notebooks from the early 1650s describe a "theriac laudanum" made by correcting opium with salt of tartar (just as is described in *Usefulnesse*) a process for

---

corum (Wittenberg, 1619), 286: "Immanens est, quae in re materiata & effecta permanet, suam naturam retinens: ut lapides, ligna sunt materia domus; elementa corporis misti. Transiens est, quae non manet in re effecta, sed mutatur & formam accipit aliam. Ita Chylus est materia sanguinis; sanguis corporis humani."

47. Boyle, *Usefulnesse*, in *Works*, 3:405; the "elsewhere" is probably Boyle's lost early essay "Of Turning Poisons into Medicines" (Henry Oldenburg's excerpt of which is published in *Works*, 13:239–57); this work shows significant debts to Starkey; see below, chapter 5.

which Starkey would eventually gain a measure of fame.[48] After describing this process in his private notebooks, Starkey goes on to detail his observations on the remarkable sudorific property of the drug, just as Boyle would later do in print. But the clinching evidence that Boyle collected this knowledge from Starkey appears in Boyle's own notebooks from the early 1650s. These documents record the same processes for correcting opium with salt of tartar as described in *Usefulnesse*, but here they are openly attributed to Starkey.[49]

It must be pointed out, however, that Boyle offers an excuse for not naming informants in the area of chymical medicines. He writes that he believes it more harmful than beneficial to practitioners to have their names revealed along with their recipes. Patients will continue to purchase drugs from a given chymist even if the recipe is published, so long as the published version of the recipe is not connected explicitly with the chymist's name. Hence Boyle declines "to annex in his life time [the chymist's] name to . . . his Receipts or Processes," for fear of damaging his trade.[50] But Boyle also appropriated many nonmedical observations from Starkey, where this explanation would not work. A very clear example occurs in Boyle's celebrated observations on the production of cold by freezing mixtures.

The first experiment in Boyle's "Mechanical Origine of Heat and Cold" refers to the cold produced upon the dissolution of sal ammoniac (mostly ammonium chloride, in modern terms) in water. As Boyle notes, the resulting cold is so intense that it can freeze water that has been placed on the exterior of the flask. Significantly, at the beginning of the passage Boyle expresses his wonder at the fact that no chymists have yet "taken notice of" this strange phenomenon.[51] Yet we know from Starkey's letter to Boyle of

48. Starkey to Boyle, 3 January 1652; printed in *Notebooks and Correspondence*, document 6, and in Boyle, *Correspondence*, 1:107–11; *Notebooks and Correspondence*, document 10; Sloane MS 3711, fol. 2.

49. BP, vol. 25, 341: "Laudanum St[irkii] & Mor[iani]. Also 344, "Opii vera Correctio," which is sandwiched between two recipes openly attributed to Starkey and seems to be copied or adapted from one of his notebook entries; see chapter 5, below.

50. Boyle, *Usefulnesse*, in *Works*, 3:486.

51. Boyle, "Mechanical Origine of Heat and Cold," in *Works*, 8:332: "My first Experiment is afforded me by the Dissolution of Sal Armoniac, which I have somewhat wonder'd, that Chymists having often occasion to purifie that Salt by the help of Water, should not have, long since, and publickly, taken notice of . . . there will be produced in the mixture a very intense degree of Coldness. . . . Nay, I more than once by wetting the outside of the Glass, where the dissolution was making, and nimbly stirring the Mixture, turn'd that externally adhering water into real Ice, (that was scrap'd off with a knife) in less than a minute of an hour." Actually, Starkey's own account of his discovery was published in *Liquor Alchahest* (London, 1675), 27–28; this book was dedicated to Boyle by its publisher, Starkey's friend Jeremiah Astell, and probably was publicly available before Boyle's *Mechanical Origine of Qualities*, which was partly published in 1675 but not available until 1676 (see Boyle, *Works*, 8:xxxv–xxxvi).

16 January 1652 that the young colonial had himself observed the freezing power of sal ammoniac and told Boyle about it. Starkey says there that "in the blink of an eye" sal ammoniac that had been sublimed with antimony and then dissolved would "(through glass) freeze into true ice the water in which a glass containing it is immersed, even if near the fire."[52] Boyle's own *Philosophical Diary* of 1655 records the same phenomenon, where Boyle notes that water sprinkled on the flask in which the sal ammoniac is being dissolved will freeze. Crucially, Boyle clearly tags the observation there as *Experimentum Stirkii*—"Starkey's Experiment."[53] Hartlib's *Ephemerides* shows that Starkey had even described his observation of the sal ammoniac phenomenon to the German intelligencer. Hartlib reported in the summer of 1652 that Starkey's "Experiment of making Ice in the hottest roome or Summer would bee of great worth in Italy," where the cardinals paid large amounts to have their drinks cooled.[54] In spite of this clear provenance recorded even in Boyle's own private papers, in print Boyle claims the discovery of this phenomenon for himself, and even expresses his wonder that "no chymist" had ever noted it previously. Once again, Boyle is consciously and publicly severing (or concealing) his links to earlier chymistry.

A third and similar example occurs in Boyle's work on a specially prepared mercury that would grow hot when mixed with gold, a topic—central to chrysopoetic endeavors—on which Boyle labored for nearly forty years. Boyle published an account of this mercury in the *Philosophical Transactions* in 1676. As has recently been demonstrated, however, the recipe for this substance came from none other than George Starkey, who entrusted it to Boyle in a spring 1651 letter. In his *Philosophical Transactions* paper Boyle states that he first acquired the mysterious substance "about the year 1652," correlating strikingly with the period of his correspondence with Starkey.[55] Yet Boyle gives no indication whatsoever of any debt to Starkey or to anyone else and instead attributes his discovery of this prized alchemical product

This freezing process is also mentioned amid an assortment of Boyle's unpublished tracts and experiments forwarded by Henry Oldenburg to Joseph Glanvill, and published in the latter's *Plus Ultra* (London, 1668), 103–7, on 107: "A *New Experiment*, shewing how a considerable degree of *Cold* may be suddenly produced without the help of *Snow, Ice, Hail, Wind,* or *Nitre,* and that at any time of the year, *viz.* by *Sal Armoniack.*"

52. Starkey to Boyle, 16 January 1652, in *Notebooks and Correspondence,* document 6, and Boyle, *Correspondence,* 1:111–18, on 114.

53. BP, vol. 8, fol. 144r, no. 54.

54. Hartlib, *Ephemerides* 1652 (summer), HP 28/2/22A.

55. Boyle, "Of the Incalescence of Quicksilver with Gold," in *Works,* 8:557; see Principe, *Aspiring Adept,* 159–62.

merely to his own trials and "God's blessing."[56] Once again, Starkey's contributions are erased in Boyle's public presentation of his material but, more importantly for our present purpose, in all these cases Boyle presents himself as independent of earlier chymical practitioners.

There is also a more general example of Boyle's elision of his relationship to chymists like Starkey. We have already noted that when Starkey and Boyle met, Starkey was clearly the more able experimenter. Moreover, it is clear that Boyle was in contact with a number of proficient chymical workers in the Hartlib circle as he developed from a moralist and devotional writer into an experimental natural philosopher. In view of Boyle's early affiliation with these chymists, his autobiographical comments in the preface to the *Sceptical Chymist* are, to say the least, jarring. Directly after lamenting the "deficiencies of [the] theory" of "the chymists," Boyle notes that

> whereas Beginners in Chymistry are commonly at once imbu'd with the Theory and Operations of their professions, I who had the good Fortune to Learn the Operations from illiterate Persons, upon whose credit I was not Tempted to take up any opinion about them, should consider things with lesse prejudice, and consequently with other Eyes than the Generality of Learners; And should be more dispos'd to accommodate the *Phaenomena* that occurr'd to me to other Notions than to those of the Spagyrists.[57]

Thus Boyle claims—like a good Baconian—that he is freer from prepossessed theoretical principles than the average chymist; the reason for this is that he was taught chymical operations not by educated chymists but by "illiterate persons." Here Boyle's chymical teachers, among whom the Harvard-educated Starkey occupies a key position, are referred to as "illiterate"![58] This remarkable claim gives the appearance of a Boyle who never accepted any principles drawn from the "spagyrists," that is, the followers of Paracelsian or Helmontian chymistry, because he was never exposed to

56. Principe, *Aspiring Adept,* 162; Boyle, "Incalescence," 557.
57. Boyle, *Sceptical Chymist,* in *Works,* 2:213–14.
58. Benjamin Worsley too was college educated, having studied at Trinity College Dublin; see Webster, *Great Instauration,* 64. Interestingly, The "illiteracy" of premodern chemical practitioners became an idée fixe by the eighteenth century; see for example, Hieronymus David Gaubius, *Oratio inauguralis, qua ostenditur, chemiam artibus academicis jure esse inserendam* (Leiden, 1732), 7, who states that chemistry was "ab illiterato hoc rudique hominum genere primum exercita." Hermann Boerhaave even claimed that Van Helmont himself was "instructed in chemistry by a certain illiterate person"; see Tenney L. Davis, "Boerhaave's Account of Paracelsus and Van Helmont," *Journal of Chemical Education* 5 (1928): 679. Further such perspectives are highlighted in Christoph Meinel, "Theory or Practice? The Eighteenth-Century Debate on the Scientific Status of Chemistry," *Ambix* 30 (1983): 121–32.

them. Of course, Boyle's claim to such independence is overstated in the extreme. The young Boyle, like Starkey, was a committed follower of Helmontian chymistry, and his reliance upon Van Helmont for many of the central arguments in the *Sceptical Chymist* has long been recognized (his further adoption of Helmontian principles will be outlined in chapters 5 and 6).[59] So again we find Boyle denying his indebtedness not only to specific individuals, but to the whole of antecedent chymical tradition in general. The impression Boyle leaves is that of greater independence than was actually the case—that his chymistry was generated de novo.

There is thus clear evidence that, despite claims to the contrary, Boyle did adopt scientific theories and processes from the preexisting chymical tradition, and that he did so without acknowledgment.[60] Such practice was, however, not unusual in the period, particularly in chymistry itself. As an example, we have only to think of the history of the incalescent, or sophic, mercury. As noted above, Boyle obtained the method of preparing it from Starkey in 1651 but published an account of it as his own in 1674. Yet Starkey's process, which he in turn presented as his own discovery to Boyle, was actually taken by him almost directly from the sixteenth-century Prussian iatrochemist Alexander von Suchten. Further examples of the nearly ubiquitous silent appropriation of processes and ideas in early modern chymistry will appear in subsequent chapters. Even outside of chymistry, Boyle was again in good company as regards his tacit borrowings. Two celebrated humanists of the sixteenth century, Joseph Justus Scaliger and Justus Lipsius, have both been shown by modern scholarship to have lifted substantial parts of their work from previous, unnamed authors, even while

59. Allen G. Debus, "Fire Analysis and the Elements in the Sixteenth and Seventeenth Centuries," *Annals of Science* 23 (1967): 127–47; Webster, "Water as the Ultimate Principle of Nature: The Background to Boyle's *Sceptical Chymist,*" *Ambix* 13 (1966): 96–107; Michael T. Walton, "Boyle and Newton on the Transmutation of Water and Air, from the Root of Helmont's Tree," *Ambix* 27 (1980): 11–18.

60. Boyle's continued adherence to traditional chrysopoetic theories and practices was one thrust of Principe, *Aspiring Adept;* here we extend the argument to other branches of early modern chymistry and relate this to Boyle's self-presentation. Additionally, similar unacknowledged borrowings for the sake of the appearance of novelty extended to the purely literary realm of Boyle's early work as well. For example, the argument and characters of Boyle's early moralizing romance *The Martyrdom of Theodora*, composed around 1648, are almost certainly borrowed from Pierre de Corneille's play on the same subject, *Théodore, vierge et martyre*, written in 1646. All the same, Boyle claims in the preface to the work that he actually came upon the story of the martyred Theodora in original sources; see Principe, "Virtuous Romance," 386. For a treatment and interpretation of Boyle's similar remarks regarding his own *Occasional Reflections* in relation to Bishop Joseph Hall's meditations see Michael Hunter, "Self-Definition through Self-Defense: Interpreting the Apologies of Robert Boyle," in *Robert Boyle (1627–1691): Scrupulosity and Science* (Woodbridge: Boydell, 2000), 135–56, on 145.

they trumpeted their own originality.[61] Thus it must be clear that our goal here is not a diminishment of Boyle's stature in the history of science; rather, it is to correct the prevailing historiography that has built somewhat too uncritically upon his own presentation of himself. In order to do this, we must transcend Boyle's self-presentation and determine his real relationship to foregoing traditions. Unearthing these suppressed sources belies the "iconic" picture of Boyle and reveals his actual debt and connection to earlier chymistry, which can only have a salutary effect on our understanding both of Boyle and of the grander sweep of the history of chemistry. It is crucial in all of this, however, to ensure that Boyle's silence in regard to sources is not merely an artifact of a culture with a different view of intellectual property, and so we need to place Boyle's views in context.

BOYLE AND INTELLECTUAL PROPERTY

Early modern England had its own standards for the recognition of intellectual property. Writers remained heavily influenced by the ancient notion of *imitatio,* which viewed the business of literature as that of adopting and adapting existing themes and stylistic models. *Imitatio,* however, was not mere copying. In a classic study, Harold Ogden White outlined several key features of the Greek and Roman theory of literary production. First, ancient authors encouraged imitation of the best authors and viewed the subject matter of literary works as common property. Virgil was obviously not plagiarizing when he "copied" the Homeric epics, and the fact that he transformed Homer's work into Latin verse was viewed as the creation of a new genre. This differed from outright theft, on the other hand, which was condemned. Hence the Latin poet Martial transformed the word *plagiarius* ("kidnapper") into a term of abuse for one who abducted another's literary work wholesale by affixing his own name to it. This leads to White's second major point—that the classical theory put great emphasis on *improving* borrowed models. The business of improvement was often viewed as a sort of rivalry between the imitator and his source; hence, the imitator was encouraged to name his source openly so that the reader could make an accurate comparison. To do otherwise would have diminished the recogni-

---

61. For Scaliger's unacknowledged use of Paul Crusius, see Anthony Grafton, *Joseph Scaliger: A Study in the History of Classical Scholarship,* vol. 2 (Oxford: Clarendon, 1993), 110–15, 237, 259, 270–76; for Scaliger's indignation at his own work being used without acknowledgment, see Grafton, *Joseph Scaliger,* vol. 1 (1983), 106. For Lipsius's appropriation of ideas from Claude Chifflet and Marc-Antoine Muret, see Jose Ruysschaert, *Juste Lipse et Les annales de Tacite* (Turnhout: Brepols, 1949), 144–63, and Arnaldo Momigliano, "The First Political Commentary on Tacitus," in *Contributo alla storia degli studi classici* (Rome: Storia e Letteratura, 1955), 37–59, esp. 55. We thank Anthony Grafton for these references.

tion to be gained by improving on an acknowledged classic. Secrecy on the part of the imitator was therefore viewed as a perversity, since it worked against the very reward that one hoped to attain. The net result of this theory was that classical plagiarism consisted mainly in the failure to outdo one's model.[62]

The classical theory of *imitatio* was itself imitated in England during the Renaissance. Hence plagiarism in the sense of unacknowledged borrowing was little criticized, although base imitation and wholesale theft of books and poems continued to be held in disrepute. Around the beginning of the seventeenth century, however, a linguistic change occurred; "plagiarism" and "plagiarist" began to enter the English language. Ben Jonson used "plagiary" in public quarrels with the "poetasters" John Marston and Thomas Dekker, where it was again servile copying and wholesale theft that were condemned.[63] Bishop Joseph Hall repeatedly used "plagiary" in his *Honor of the Married Clergy Maintained* to describe those who stole passages as well as whole texts.[64] Historians of literature agree that by the time of Boyle's literary production, the use of the English term was beginning to approximate the modern one.[65] Yet even John Milton's *Eikonoklastes* (1649), which made devastating use of the fact that Charles I's gallows prayer was plagiarized from Philip Sidney, still condoned imitation that "seem'd to vie with the Original."[66] Charles's twofold sin was not that he copied without acknowledgment, but that the borrowing was taken from a "vain amatorious Poem" rather than a worthy one, and that his copying was "not better'd by the borrower."[67]

Such accusations (and the related interest in priority) were rife in the environs of the early Royal Society.[68] And upon returning to Boyle, we can

62. Harold Ogden White, *Plagiarism and Imitation during the English Renaissance* (Cambridge: Harvard University Press, 1935), 16–19. White's conclusions, though more than half a century old, have been reaffirmed recently by Laura J. Rosenthal, *Playwrights and Plagiarists in Early Modern England* (Ithaca: Cornell University Press, 1996), 7.

63. White, *Plagiarism*, 135–36.

64. Ibid., 121.

65. H. M. Paull, *Literary Ethics: A Study in the Growth of the Literary Conscience* (Port Washington, N.Y.: Kennikat, 1968; reissue of 1928 ed.), 106–11. See also Alexander Lindey, *Plagiarism and Originality* (New York: Harper, 1952), 62–94; and Thomas Mallon, *Stolen Words: Forays into the Origins and Ravages of Plagiarism* (New York: Ticknor and Fields, 1989), 1–12.

66. *Complete Prose Works of John Milton,* ed. Douglas Bush et al. (New Haven: Yale University Press, 1962), 3:361.

67. Elisabeth M. Magnus, "Originality and Plagiarism in *Areopagitica* and *Eikonoklastes,*" *English Literary Renaissance* 21 (1991): 98.

68. White, *Plagiarism*, 120–202. White links the widespread abuse of plagiarism in the period up to 1625 to classical theories of imitation. For contemporary accusations of scientific pla-

see that his silent appropriation of sources is in fact not contrary to early modern practices. What *is* unusual, however, is that even while Boyle engages in this common practice, he also develops an alternative view of intellectual property—one that is surprisingly similar to modern standards. In particular, he displays an almost surprising consciousness of the need for accreditation of sources, and the prefaces to his works are deeply concerned with the issue of plagiarism.[69] Remarks on intellectual property occur in the publisher's forward to *The Origine of Formes and Qualities* (1666), where the publisher (Richard Davis) recounts Boyle's dismay when he found one of his experiments in the work of "a very recent *Chymical Writer*." Davis (or perhaps Boyle himself) insists that *Origine of Formes* had been finished several years before its publication, implying that Boyle's experiment was written down prior to the publication of the unnamed chymist. Here, however, there is no accusation that Boyle is being plagiarized. Rather, the publisher's preface aims to defuse the possibility that Boyle himself may be accused of lifting the experiment from the chymist.[70] What is really quite striking is the great emphasis that Boyle and his publisher put on the simple presentation of an experiment—a premium is placed on priority and ownership in the discovery of a simple "matter of fact."

The text of *Certain Physiological Essays* contains further comments on intellectual property that are unequivocally Boyle's own. Boyle first excuses his loquacity in naming what he has borrowed from other authors, citing a phrase from Pliny to the effect that it is noble to acknowledge one's sources. Then he begins a lamentation that sounds very similar to what one finds in many of the "publishers' prefaces" to other works.

> Though I have seen divers Modern Writers that so boldly usurp the Observations and Experiments of others, that I might justly apply to them what the same *Pliny* annexes, [namely] "Know that I, while comparing authors, have found old texts copied literally and without acknowledgement by the most recent and attested authors." If other Writers should not prove more equi-

---

giarism, see Johns, *Nature of the Book,* 461 and accompanying notes, and on Boyle in particular, Michael Hunter, "The Reluctant Philanthropist: Robert Boyle and the 'Communication of Secrets and Receits in Physick,'" 202–22 in *Robert Boyle, 1627–1691: Scrupulosity and Science,* esp. 219–21.

69. Although much of this material occurs in so-called publishers' notes or prefaces, Boyle clearly was involved in their composition, and in the case of his 1678 *Degradation of Gold,* a draft of the "publisher's preface" occurs among Boyle's own papers, in the same hand and on the same paper as the body of the text; see Principe, *Aspiring Adept,* 226–27, 288–89.

70. Boyle, *Formes and Qualities,* in *Works,* 5:284. The editors of the *Works* suggest that the unnamed author is Johann Rudolph Glauber.

table (for I will not say more thankful) than such as these, they would quickly discourage those whose aims are not very noble and sincere, from gratifying the Publick with Inventions, whose Praise and Thanks would be usurp'd by such as will not name them.[71]

Here Boyle insists on the explicit identification of sources by name; otherwise authors will lack an incentive for publishing their work and the public will suffer. Boyle even goes on to quote Pliny's extremely derogatory comment about such ungrateful plagiarists: "Surely it is the mark of a base and miserable soul to prefer to be caught in theft than to return a debt, especially when a profit will be made from the borrowing."[72]

The sense of this passage from *Certain Physiological Essays* is reprised in the preface to *The Mechanical Origine of Qualities* (1675), where an accusation of plagiarism is directed at a specific individual. The publisher's note to this work complains about William Salmon's collection entitled *Polygraphice*, which pirated fifty experiments from Boyle's *Experiments touching Colours*.[73] What is especially interesting to us is the fact that the preface admits that Salmon claimed no originality for the processes he published: "Nor did I think this practice justified by the confession made in the *Preface*, importing, that the Compiler had taken the particulars he deliver'd from the Writings of others." Thus for Boyle's publisher (and presumably Boyle himself), it was not enough for the author of *Polygraphice* to acknowledge his own unoriginality; he must name the original authors explicitly. Otherwise, the preface continues, such derivative authors will not do "right to *particular* authors." A mere "general and perfunctory acknowledgement" will discourage the publication of experiments and should therefore be discountenanced by the commonwealth of learning.[74] While these are agreeable words to the modern reader, they lead to a certain dissonance when juxtaposed with Boyle's own practices in regard to Sennert, Starkey, and others; this dissonance must be addressed.

CONCLUSIONS

The passages noted here show that the view of intellectual property promoted by Boyle was more similar to standard modern practice than one

71. Boyle, *Certain Physiological Essays,* in *Works,* 2:29: "Scito enim, conferentem authores me deprehendisse a juratissimis & proximis veteres transcriptos ad verbum neque nominatos, &c." The translation is our own.

72. Ibid.: "Obnoxii profecto animi & infelicis ingenii est, deprehendi in furto malle, quam mutuum reddere, cum praesertim sors fiet ex usura." We prefer our nonliteral translation of this passage to the rather cryptic ending found on that page: "especially since capital arises from interest."

73. Hunter, "Self-Definition through Self-Defence," 137–38, 148–49.

74. Boyle, *Mechanical Origine of Qualities,* in *Works,* 8:317.

might expect.[75] The public Boyle, unlike Milton, is not willing to excuse borrowed work even if it has been "better'd by the borrower." The appropriation of "matters of fact" based on another's experimentation suffices to brand one a plagiarist, regardless of the elegance in which they are clothed or the purposes to which they are put. How, then, are we to harmonize the "iconic" portrait of Boyle with the obvious clash between word and practice revealed by his silent appropriation of materials noted above? There may well be a general and coherent psychological explanation—Boyle, after all, was the young seventh son of a great family and thus anxious to distinguish himself independently in the world.[76] But at least in the case of Boyle's chymistry it is possible to offer a further, more specific explanation.

Boyle's treatment of his chymical debts hinges on the way he wished to present himself to the world. In regard to Sennert, it seems that already by the mid-1650s Boyle was intent on distinguishing himself as an experimentally based, corpuscularian natural philosopher. His lack of acknowledgment of the German physician and natural philosopher, coupled with his explicit allegiance to Gassendi, Descartes, Magnenus, and Digby, suggests that he was keen to link himself to the corpuscularian avant-garde of the "New Science," as opposed to the Scholastic Aristotelianism of Sennert. Sennert's compendium of natural philosophy had been in use by Oxford students since the early 1630s, and he was in fact one of the authors ridiculed (by Starkey and others) as an archetypical "school-man."[77] Already by the time of the "Essay on Nitre," Boyle made it clear to the public that his scientific niche would consist of experimental demonstrations of the mechanical philosophy, conceived along Baconian lines.[78] The Wittenberg iatrochemist was not an intellectual companion with whom Boyle wanted to be seen publicly, in spite of the benefit he had reaped from reading his work. Boyle thus reinforced his own novelty and status by suppressing his sources—downplaying the atomism of previous writers on chymistry such

75. See also Boyle's concerns about the possibility of being accused of having plagiarized Glauber, in his "Essay on Nitre," in Boyle, *Works,* 2:89; see also Hunter, "Self-Definition," 146–48.

76. For several such analyses see "Psychoanalyzing Robert Boyle," a special number of *British Journal for the History of Science,* edited by Michael Hunter (vol. 32 [1999]: 257–324). One is also reminded of Isaac Newton's observation that Boyle was "too desirous of fame"; Newton to Fatio de Duillier, 10 October 1689, in *The Correspondence of Isaac Newton,* ed. H. W. Turnbull, 7 vols. (Cambridge: Cambridge University Press, 1960), 3:45.

77. Robert G. Frank Jr., *Harvey and the Oxford Physiologists: A Study of Scientific Ideas and Social Interaction* (Berkeley and Los Angeles: University of California Press, 1980), 120. See Starkey, *Natures Explication* (London, 1657), [xxi], 53, 155.

78. Boyle, "Proemial Essay" to *Certain Physiological Essays,* in *Works,* 2:17, where Boyle promises a work that will spell out his Baconianism.

as Sennert—while simultaneously aligning himself with the "right" class of thinkers. With this goal in mind, Boyle's works (and by extension, those of others among the "New Philosophers") would naturally tend, with a "rhetoric of novelty," to exaggerate their break with the predecessors from whom they chose to distinguish themselves.

Boyle's appropriation of material from Starkey can be partly explained on the same grounds, although the details are somewhat more complex. In his general claim to have been tutored only by illiterates, Boyle was again distancing himself from what he wished to cast as an incoherent world of chymical theory, another realm of thought—like Sennert's Scholasticism—from which he wished to dissociate himself. Boyle's more specific excision of Starkey's name from the opium correction process can be explained both in terms of his policy of omitting proprietary names from drugs while their discoverers were still alive and as a matter of emphasizing his own inventive originality. In the case of his claim to priority in observing the cooling power of sal ammoniac, however, no medical trade was involved. Here Boyle was laying claim to the first observation of a strange chymical effect and hence emphasizing his own expertise and originality in the case of a matter of fact. Similarly, the incalescent mercury paper not only announces a strange phenomenon, but broadcasts Boyle's prowess in the realm of chrysopoeia to the world of adepts, where Boyle presumably wished to stress his own expertise, not that of his actual alchemical master. The underlying thread in all these examples is Boyle's desire to present himself as an original thinker and practitioner, untrammeled by theoretical obligations to a preexisting discipline and owing his practical observations to no one but himself and God's beneficence. Again, just as Boyle did not wish to be seen in the company of Sennert, he presumably did not wish to appear too indebted to one of those practitioners of chymistry whom he had classed as "merely artisanal." This failure to acknowledge his debts cannot be explained away solely as part of a Baconian program aimed at presenting himself as one who was not "prepossess'd with any theory or principles," since these borrowings from Starkey were of a practical rather than a theoretical nature, and analogous borrowing occurred in nonscientific areas as well.[79]

The dissonance between Boyle's words on intellectual property and his actions in regard to his chymical forebears may thus result, at least in part, from a dilemma imposed by Boyle's overriding commitment to the advancement of natural philosophy. On the one hand, Boyle wished to deploy chymistry in the service of natural philosophy and to free it from its am-

79. Boyle, *Certain Physiological Essays,* in *Works,* 2:13.

biguous reputation, emphasizing that this was a fresh start for chymistry. In this regard, linking himself publicly to preexisting traditions would be counterproductive. On the other hand, Boyle's same commitment to natural philosophy required that natural philosophical authors be able to publish their observations without fear of losing their priority or authority through theft, and this required a reconceptualization of early modern ideas of intellectual property. In the end, these two results of a single commitment to natural philosophy turned out to conflict.[80]

The result of Boyle's suppression of his sources is an enduring self-portrait of a virtuoso seemingly uninfluenced by older ideas, and Boyle's apparent novelty receives further emphasis from his expressed views on intellectual property. This self-presentation led in Boyle's own lifetime to an "iconic" status reinforced by his partisans and hagiographers and then widely adopted and endorsed by subsequent secondary literature. The supposed novelty and originality of Boyle have deeply influenced the wider historiography of chymistry, for they create the impression of a break or discontinuity that begins with the appearance of Boyle's works. Thus a rather facile dismissal of Boyle's predecessors and their influence upon him is facilitated by Boyle's self-presentation. But our reason for exploring this topic is to show how little one can rely upon the image of the iconic Boyle for grounding claims about seventeenth-century chymistry and Boyle's place therein. One of our previous publications has already explored this issue specifically in terms of Boyle's continued interest in the traditional goals, methods, and theories of transmutational alchemy, thereby blurring the lines of demarcation for his chymistry.[81] The topic of laboratory practice and experiment—whether in chrysopoetic, iatrochemical, technological, or yet other subdivisions of chymistry—is another place where the reality of a putative demarcation needs to be explored. Indeed, ascertaining Boyle's exact place in the history of chymistry is secondary to the larger issue of seeing what seventeenth-century chymistry was actually like in practice, freed from the presumption of an iconic Boyle standing at the cusp of a "New Science."

In the next three chapters we will therefore turn our attention to the content, status, and bases of chymical laboratory practice in the mid–seventeenth century. Indeed, in spite of the recent interest in issues of experi-

---

80. This dilemma is analogous to other dilemmas Boyle faced when his commitments encountered the real world, for example, in medical reform; see Hunter, "The Reluctant Philanthropist," and Michael Hunter, "Boyle versus the Galenists: A Suppressed Critique of Seventeenth-Century Medical Practice and Its Significance," *Medical History* 47 (1997): 322–61.

81. Principe, *Aspiring Adept*.

mentalism, the details of seventeenth-century chymical experimental practice remain little known. In order to explore these issues, we will give special attention to two characters. The first is Joan Baptista Van Helmont, whose works dominated chymistry during the second half of the seventeenth century. Without a clear understanding of Van Helmont, neither the chymistry of Starkey nor that of Boyle can be fully understood. Yet in spite of his manifest importance in areas as diverse as medicine, chymistry, and mineralogy, the obscurity of Van Helmont's writing style has left him a rather shadowy figure in the existing secondary literature. Our second target is Van Helmont's great champion, George Starkey, and his beautifully detailed and explicit laboratory notebooks, which comprise a chief locus for this study. We will thereafter return to the issue of Boyle's early education—for at that point we will be better able to judge the young Boyle's reliance upon these two figures for his own chymical thought and practice—and finally attempt to define the position of the mature Boyle with respect to the chymistry of his youth.

TWO

Number, Weight, Measure, and Experiment in Chymistry

FROM THE MEDIEVALS TO VAN HELMONT

A lone figure sits in his cell, hunched over a table covered with half-opened books. The books are filled with strange symbols, barely perceptible in the semidarkness of the chamber, illuminated only by the flickering of a candle. In the corner of the subterranean room are monstrous wavering shapes, the shadows cast by primitive glass and earthenware vessels, filled, or partly filled, with colored liquids. A stuffed owl hangs overhead providing, along with the skull upon which the candle rests, the room's only conscious attempt at ornamentation. Suddenly the alchemist, for that is the profession of our strangely indolent figure, rouses himself from his reverie. He has been experiencing a waking dream, a vision, in which he saw a mysterious play unfolding within the polychrome contents of his apparatus. First there were reptilian shapes crawling back and forth, which suddenly died and transformed themselves into weird plantlike branching figures. The twisted, spiky branches of these growths then fused together to reveal the figure of a man, radiant with light and sitting on a throne. Reflecting on the beauty and mystery of this scene, the alchemist asks himself what it can possibly mean. He is at a loss, unable to express the ineffable sense of harmony and wholeness that he now feels, but sure that he has progressed along the path of the "great work." Although he has discovered nothing about the physical world, the goal of his "experimentation"—the perfection of his own inner self—has been partially attained. Our alchemist is on the way to becoming an adept.[1]

Who has not encountered such an image of the alchemist? This is the view of alchemy as a spiritual discipline, first popularized by the occultists of the nineteenth century—"mystical" and theosophical writers such as Eliphas Lévi, Mary Anne Atwood, and Arthur Edward Waite—then adopted and

---

1. This fictitious scene is based loosely on the interpretation of Theobald de Hoghelande's *De alchimiae difficultatibus,* presented in C. G. Jung, "Die Erlösungsvorstellungen in der Alchemie," *Eranos-Jahrbuch 1936: Gestaltung der Erlösungsidee in Ost und West* (Zurich: Rhein-Verlag, 1937), 13–III, on 23–24.

clothed in "scientific" language by writers on psychology and comparative religion, such as Carl Gustav Jung and Mircea Eliade.[2] The alchemist is not engaged in chemical experimentation as such—instead the vague matter within his flask serves as the focal point for nonmaterial processes. Within the psychologizing view, Jung came to call this process "active imagination," a form of meditation that could result in the visual projection of the unconscious in the form of vivid hallucinations.[3] For those who have adopted a more spiritual or religious interpretation of alchemy, the material manifestations of alchemical labors are viewed as predominantly or entirely symbolic of internal spiritual transformations. The clutter of alembics, crucibles, and furnaces is busywork compared with the chief alchemical goals of spiritual or psychic self-perfection. In either case, it matters little whether the ingredients within the alchemical vessels are metallic, mineral, or vegetable, nor are their quantities of particular importance. What is really significant is the alchemist's state of mind or meditation. It is such internal processes that will allow the alchemist to attain his true goal—the perfection of his own soul.

This concept of a spiritual alchemy, so appealing in the early twentieth century to both a Jung fresh from seances with his cousin Hélène Preiswerk and to an Eliade immersed in the anthroposophy of Rudolph Steiner, became a standard part of twentieth-century perceptions of the discipline,

2. See Principe and Newman, "Some Problems with the Historiography of Alchemy," in *Secrets of Nature: Astrology and Alchemy in Early Modern Europe,* ed. William R. Newman and Anthony Grafton (Cambridge: MIT Press, 2001), 385–434. See also William R. Newman, "Decknamen or 'Pseudochemical Language'? Eirenaeus Philalethes and Carl Jung," *Revue d'histoire des sciences* 49 (1996): 159–88; and Lawrence M. Principe, "Apparatus and Reproducibility in Alchemy," in *Instruments and Experimentation in the History of Chemistry,* ed. Trevor Levere and Frederic L. Holmes (Cambridge: MIT Press, 2000), 55–74. For further criticism of the Jungian approach to alchemy, see Barbara Obrist, *Les débuts de l'imagerie alchimique (XIVe–XVe siècles)* (Paris: Le Sycomore, 1982), 15–21, 183–245; and Robert Halleux, *Les textes alchimiques* (Turnhout: Brepols, 1979), 55–58.

3. Jung, "Erlösungsvorstellungen," 74. Jung's earliest writing on alchemy is found in *The Secret of the Golden Flower,* published with Richard Wilhelm in 1929 (see Luther Martin, "A History of the Psychological Interpretation of Alchemy," *Ambix* 22 (1975): 16). Sustained treatments of alchemy by Jung are found in the following of his texts: *Aion, Collected Works,* Volume IX, Part II (London: Routledge, 1959); *Psychology and Alchemy, Collected Works,* Volume XII (London: Routledge, 1953); *Alchemical Studies, Collected Works,* Volume XIII (London: Routledge, 1967); *Mysterium Conjunctionis, Collected Works,* Volume XIV (London: Routledge, 1963). See also C. G. Jung, "Die Erlösungsvorstellungen in der Alchemie," *Eranos-Jahrbuch 1936* (Zurich: Rhein-Verlag, 1937), 13–111 (in English, "The Idea of Redemption in Alchemy," in Stanley Dell, ed. *The Integration of the Personality* [New York: Farrar and Rinehart, 1939], 205–80). A retranslated and much expanded version of the original Eranos lecture appears in *Psychology and Alchemy,* 227–471.

and such views continue to have many followers to this day.[4] Even while specialists in the history of chemistry are dismantling these images, they continue to survive and to have influence in many quarters. Indeed, the idea that alchemy was primarily a meditative pursuit remains a preconception that any speaker or writer on the subject must assume of an educated audience. The more popular views of alchemy are thoroughly imbued with this idea, as it has been promulgated by literary figures such as Northrop Frye, Joseph Campbell, and Gaston Bachelard. The pervasiveness of this cliché has meant that even some historians of chemistry have adopted it in varying degrees.[5] The key result of its prevalence is that the spiritual interpretation has served—sometimes almost unconsciously—to set "alchemy" radically apart from "chemistry" in the modern sense.

Yet as we have elsewhere shown, the foundations of this interpretation of alchemy—whether spiritual or overtly Jungian—are strikingly weak, as they are based ultimately upon Victorian occultist views with very little reference to the historical reality of the subject.[6] While one cannot (and would not wish to) deny that alchemy is replete with a singular lushness of symbolism and overlapping levels of meaning or that it presents important resonances with religious speculations, it does not follow that this arises from hallucination, unbridled imagination, or a predominant focus on the spiritual to the exclusion or diminution of the kind of laboratory operations we have come to view as a property of "chemistry." Nor does it follow that

4. For Jung's involvement with spiritualism, see Richard Noll, *The Jung Cult* (Princeton: Princeton University Press, 1994), 144, and *The Aryan Christ* (New York: Random House, 1997), 25–30, 37–41. For Eliade's youthful interest in anthroposophy and other popular forms of occultism, see Mac Linscott Ricketts, *Mircea Eliade: The Romanian Roots,* 1907–1945 (Boulder: East European Monographs, 1988), 141–53, 313–25, 804–8, 835–42.

5. Joseph Campbell, *The Flight of the Wild Gander: Explorations in the Mythological Dimension* (New York: Harper-Perennial, 1990; 1st ed., 1951), 86–87, 218–19; Campbell, *The Masks of God: Primitive Mythology* (Harmondsworth: Penguin, 1982; 1st ed., 1959), 72; and Northrop Frye, *Anatomy of Criticism* (Princeton: Princeton University Press, 1957). For Bachelard, see Obrist, *Les débuts,* 22–23. From the many historians of alchemy and chemistry who adopt (wholly or partly) the spiritual or the Jungian perspective, we mention only the following: Gareth Roberts, *The Mirror of Alchemy* (London: British Library, 1994), 7, 66; Marco Beretta, *The Enlightenment of Matter* (Canton, Mass.: Science History, 1993), 77 n. 6, 330–47; William H. Brock, *The Norton History of Chemistry* (New York: Norton, 1993), 17, 678; B. J. T. Dobbs, "From the Secrecy of Alchemy to the Openness of Chemistry," in *Solomon's House Revisited,* ed. Tore Frängsmyr (Canton, Mass.: Science History, 1990), 75–94, cf. 76; Dobbs, *The Foundations of Newton's Alchemy* (Cambridge: Cambridge University Press, 1975), 26–35; Pierre Laszlo, *Qu'est-ce que l'alchimie?* (Paris: Hachette, 1996); Allison Coudert, *Alchemy: The Philosopher's Stone* (London: Wildwood House, 1980), 148–60; F. Sherwood Taylor, *The Alchemists: Founders of Modern Chemistry* (New York: Schuman, 1949), 159, 228; and E. J. Holmyard, *Alchemy* (New York: Dover, 1990; 1st ed., 1957), 163–64, 176.

6. Principe and Newman, "Some Problems with the Historiography of Alchemy."

alchemy is nothing but the manipulation of such symbolism or texts without reference to laboratory activities. Yet the widespread stress on the "otherness" of alchemy tends to support the view that alchemists in their laboratories were not focused on material substances and their actual transformations and even that these alchemists acted more or less haphazardly or randomly in their operations.

For it is clear that there did exist a strong experimental tradition in the alchemy of Western Europe from the High Middle Ages through the seventeenth century.[7] Not only did alchemy involve considerable laboratory experience and practice in a general sense, but many alchemists were even deeply concerned with testing; some employed a kind of quantitative method that is both akin to the later traditions of chemistry and quite alien to the image of alchemists as primarily seekers of an *unio mystica*.

In the present chapter, we will survey the evidence for an experimental tradition in alchemy, beginning with some alchemists of the Middle Ages. As we will see, the medieval alchemists under consideration were far from ignoring quantitative methods in the laboratory. Indeed, they employed weight measurement as one component of the testing methods they used to determine the identity and composition of mineral substances. We will argue, furthermore, that in the sixteenth century this tradition of medieval alchemy merged with the iatrochemistry of Paracelsus and his followers. The full fruit of this union appears when the new Paracelsian emphasis on analysis and synthesis combines with a remarkable emphasis on weight determination in the works of Joan Baptista Van Helmont, in many ways the archetype of George Starkey and Robert Boyle.

TESTING, ANALYSIS, AND ASSAYING IN LATE MEDIEVAL ALCHEMY

The image of alchemists as insouciant empirics who cobbled their mixtures together with little regard to purity, quantity, or even identity of the ingredients long influenced the historiography of science. Hence technical processes concerned with the refining of metals from ores and their subsequent testing for purity are often automatically consigned by modern historians to the realm of "mineralogical" and "metallurgical" literature, in distinction to alchemy.[8] Although it is certainly true that the sixteenth cen-

7. There is little doubt that this tradition could also be found in the province of earlier alchemy as well, but it is not our present brief to present a history of the discipline from its origins.

8. A good example may be found in Beretta, *The Enlightenment of Matter*, 74–93, 134–36, 330–67. But see the much-needed correction to this view by Robert Halleux, "L'alchimiste et l'essayeur," in *Die Alchemie in der europäischen Kultur- und Wissenschaftsgeschichte*, ed.

tury witnessed the birth of an autonomous literature on mining and metallurgy, as evinced by the works of Vannoccio Biringuccio, Georg Agricola, Lazarus Ercker, and others, it is not the case that alchemists were unconcerned with the purification, testing, and exact measurement of their own materials. These concerns were not exclusively qualitative in character, but often had a gravimetric focus as well; indeed, the precision balance was associated with alchemical laboratories long before the early modern development of the mining and metallurgy genre. As Robert Halleux has noted, by the Middle Ages the techniques of mineral testing and analysis had already evolved into tools for the experimental investigation of nature and the testing of products and were not used merely as empirical means of dealing with precious metals.[9] We will thus begin by presenting some of this little-known medieval literature to witness the use of practical laboratory tests to investigate nature, starting with alchemy as it stood when first appropriated by the West from the Arab world.

The alchemical concern with the testing of materials is very evident in the extensive medieval literature on salts and alums, which then formed one of the main genres of alchemy. The origin of this literature is clearly Arabic; indeed, the founder of the genre seems to have been Muhammad ibn Zakariyya al-Rāzī (c. 854–925), a well-known physician and philosopher who also wrote on alchemy. Rāzī's genuine *Kitāb al-asrār,* or *Book of Secrets,* was translated as the *Liber secretorum* and became an influential Latin text. Yet it was by means of the pseudonymous *Liber de aluminibus et salibus*—written by a much later follower—that Rāzī acquired his greatest fame among Western alchemists.[10] Rāzī and his successors were keenly interested in the classification of salts, alums, and atraments (or vitriols). The *De aluminibus et salibus* describes rock salt *(sal gemmae),* table salt *(sal panis),* a "bitter salt" *(sal amarus),* a "Nabatean salt" *(sal Nabataeus),* alkali salt *(sal alkali),*

Christoph Meinel (Wiesbaden: Herzog August Bibliothek in Kommission bei Otto Harrasowitz, 1986), 277–92. The same kind of division leads Dietlinde Goltz to the absurd conclusion that the medieval Geber, author of perhaps the most influential alchemical treatise of the Latin Middle Ages, was not an alchemist. Goltz argues that "Alchemie ist unabdingbar mit irgendeiner Art von Weltanschauung verknüpft und stellt eine Naturphilosophie dar." When she fails to find the weltanschauung of the alchemists in the highly mineralogical work of Geber, Goltz concludes that he writes in a fashion that is "nicht alchemistisch." See Dietlinde Goltz, "Versuch einer Grenzziehung zwischen 'Chemie' und 'Alchemie,'" *Sudhoffs Archiv* 52 (1968): 30–47, esp. 34, 39–40.

    9. Halleux, "L'alchimiste et l'essayeur," 290. See also Newman, "The Place of Alchemy in the Current Literature on Experiment," in *Experimental Essays: Versuche zum Experiment,* ed. Michael Heidelberger and Friedrich Steinle (Baden-Baden: Nomos, 1998), 9–33.

    10. Julius Ruska, *Das Buch der Alaune und Salze: Ein Grundwerk der spätlateinischen Alchemie* (Berlin: Chemie, 1935).

sal ammoniac *(sal armoniacus),* and others and gives detailed instructions for their purification.[11]

Another text in this genre is the *Ars alchemie* attributed to Michael Scotus, which, despite its general title, is largely a treatise on salts and alums in the tradition of Rāzī. Scotus was a thirteenth-century philosopher in the court of Frederick II von Hohenstaufen, the great Holy Roman Emperor whose wide-ranging interests earned him the title of *stupor mundi.*[12] In the first few folios the author describes the salts that are indispensable to alchemy. There the author gives, for example, three different types of *sal nitrum.* These three species are called niter *de puncta,* "leaved" niter, and "depilated" niter. The *Ars alchemie* then provides a test with glowing coals (used some ten times in all) as a means of distinguishing these three forms of niter as well as identifying other salts. The salt is to be placed on a hot coal and the observer is told to examine it for fusion, smoke or vapor, hopping about, crackling, and any residual ash if the salt burns. In the case of the different types of niter, it is clear that the *Ars alchemie* is trying to distinguish what we would call saltpeter from soda; the two were routinely confused in premodern sources.[13] The glowing coal provided an early form of testing that has been largely overlooked by historians of chemistry and deserves further examination.[14] But what is clear is that Michael Scotus is deeply interested in correctly identifying and classifying salts using an empirical test. Were he uninterested in the true identity and composition of substances, such a test would be otiose.

The testing and discrimination of different salts by medieval alchemists shows not only that they were interested in systematic experimental means for distinguishing substances on qualitative grounds, but also that they had developed effective methods for doing so. This medieval literature of salts reveals an emphasis on the testing of minerals as well for their proper iden-

11. Ruska, *Buch der Alaune und Salze,* 80–83.

12. S. Harrison Thomson, "The Text of Michael Scot's *Ars Alchemie,*" *Osiris* 5 (1938): 523–59.

13. See J. R. Partington, *A History of Greek Fire and Gunpowder* (Cambridge: Heiffer, 1960), 87–89, where Partington discusses the *Ars alchemie.* It seems that Partington has rather missed the point, however, when he says that the flame test was employed primarily to distinguish pure saltpeter from its adulteration with common salt. The main goal of the test was clearly to distinguish soda from saltpeter. See also Robert Multhauf, *The Origins of Chemistry* (London: Oldbourne, 1966), 33, and Marcelin Berthelot, *La chimie au moyen âge* (Paris, 1893; reprint, Osnabrück: Otto Zeller, 1967), 1:98.

14. The test with burning coals appears also in the *Breve breviarium* of pseudo–Roger Bacon, written around the end of the thirteenth century. The author of the *Breve breviarium* interestingly refers to his niter as *sal petrae; Sanioris medicinae magistri D. Rogeris Baconi* (Frankfurt, 1603), 251: "Talis autem naturae est, quod si immediate ignitos carbones tangat, statim accensum cum impetu evolat."

tification and quality. If we turn to other traditions in medieval alchemy, we will find a multitude of additional tests, often employing the very techniques of assaying that have been used by historians to distinguish alchemy from metallurgy. One of the most influential alchemical works available to the Latins of the High Middle Ages was the *De anima in arte alkimiae,* attributed falsely to the Persian philosopher Avicenna (d. 1037).[15] The *De anima,* actually the work of an anonymous Arabic author, was translated at an early date into Latin, whereupon it achieved immediate success, becoming the main alchemical source for Roger Bacon, among others.[16] While much of this pseudo-Avicenna's *De anima* is devoted to the decomposition of animal, vegetable, and mineral substances by means of fractional distillation, the text also relates a full complement of assaying tests for precious metals. Seven tests are given for gold; these include attempting its dissolution in "salts" (if it dissolves it is artificial gold), the use of the touchstone, weight (if the gold is heavier or lighter in specie than normal gold it is fake), loss of its color when fired, ability to sublime, boiling upon fusion, and taste.[17] In all of these, the goal is to distinguish natural gold from artificial gold in order to measure the success of the alchemist.

Although several of the *De anima*'s assaying tests are rather unorthodox, it is not at all unusual to find the more mainstream processes of cupellation and cementation given prominent descriptions in other alchemical texts. An excellent example of the former is found in the *Theorica et practica* of Paul of Taranto, probably written at the end of the thirteenth century. The recipe that he gives is so clear and precise that it deserves to be quoted in full.

---

15. Julius Ruska, "Die Alchemie des Avicenna," *Isis* 21 (1934): 14–51; Pseudo-Avicenna, *De anima,* in *Artis chemicae principes, Avicenna atque Geber* (Basel: Petrus Perna, 1572), 1–147.

16. Newman, "The Philosophers' Egg: Theory and Practice in the Alchemy of Roger Bacon," *Micrologus* 3 (1995): 75–101.

17. Pseudo-Avicenna, *De anima,* 125–26:

Modo dicam tibi, ut cognoscas aurum cuiusmodi sit. Primum in solutione, secundum in lapide, tertium in pondere, quartum in ore, ut gustes, quintum in igne, sextum in sublimatione, septimum in fusione. Et in unoquoque eorum est magisterium temtandi. Si vis scire cuius naturae sit aurum, funde cum salibus, & si solvetur, est de magisterio: Et si vis eum tentare in lapide si est de quinto, aut de sexto, aut de octavo, aut de medietate, aut de tertio: aut de quarto, secundum quod est judicabis. Et si vis temtare ad pondus, vide si est leve aut ponderatum magis alio: & si est, est de lapide, si sustineat omnes alias tentationes. Si vis tentare ad ignem, iacta in ignem. Si permenebit in colore suo, est aurum: sin autem, est falsum aurum. Et tenta in sublimatione pulverizatum cum aliis speciebus: Si sublimatur, ita quod ante non fiat calx, nec lavetur. Et si facias eum pulverem, & projicias in aludel, & sublimetur: scies quod est de nostro auro. Et in fusione est magna scientia: quia quando fundis debes videre si ferveat, aut si non ferveat. Si ferveat, est de nostro lapide, & in gustu potes cognoscere salsum, & extrahi.

Let a very well sieved cinder be taken and mixed with water of salt; let a vessel be made from it, in which silver or whatever metal that you seek to test in the cupel be put on a very violent fire. With the metal fused, let a sixth part of lead be thrown on; this is especially done in the case of silver. Let a pipe of iron or reed be had, through which one can blow on the surface of the fused metal. The lead fused on the metal will be seen smoking due to this—that it has volatile flight as well as the loss of its substance owing to its badly fixed principles. Thence it is that, passing into smoke, it will draw with it all that is imperfect in the metal to be purged. The purged metal with the lead added to it will be recognized not to be vaporized, but it will seem to be boiling, and to eject froth—as it were flying forth; then let no more lead be added.[18]

Not only does Paul relate the weight of lead that must be added to make the process work, he even describes the blowpipe that is used to remove the litharge (lead oxides) from the surface of the molten metal. As the author points out in another passage, this process was capable of separating the noble metals from the base but could not be used to isolate gold from silver.[19] Before the advent of the mineral acids it was necessary to employ another dry process, cementation, in order to remove the silver from gold. This was done by placing leaves of the silver-gold alloy in a crucible in alternating layers with the "cement," often a mixture of brick dust and salt. The sealed crucible was heated to a high temperature, but one beneath the melting point of the alloy, whereupon the silver was corroded but the gold left unscathed. The *De perfecto magisterio,* an important early thirteenth-century work misattributed to either Aristotle or Rāzī, gives a clear description of the process.

Separation of gold. Make leaves of it as thick as your fingernail, and cement them with this powder. Take two parts of common salt, one part of old brick found on the banks of rivers, or on the seashore, which is better, grind well

18. William R. Newman, "The *Summa Perfectionis* and Late Medieval Alchemy" (Ph.D. diss., Harvard University, 1986), 4:165, 3:222–24:

Sumatur cinis optime cribratus et cum salis aqua commixta fiat vas in quo recipi possit ad ignem impiisimum argentum sive quodcunque metallum quod in cineritio ponere et examinare quesieris. Et fuso metallo, eiiciatur ibi plumbi pars sexta; et hoc maxime fit in argento. Et habeatur cannolum vel de ferro vel canna, per quem sufflari possit super faciem fusi metalli. Plumbum super metallum fusum videbitur fumans ex eo— quod a principiis suis male fixis habeat fugam, volatilitatem, et sue deperditionem substantie. Inde est quod in fumum resiliens, omne quod in metallo purgando imperfectum extiterit secum trahet. Tunc enim bene purgatum metallum noscetur cum, addito plumbo in eo, non fumari videbitur, sed ebulliri et quasi spumas evolantes eiicere: et tunc plumbum ulterius non addatur.

19. Newman, "The *Summa Perfectionis* and Late Medieval Alchemy," 4:225.

and sieve through a thick cloth. Then leave the leaves thus cemented on a tri-
pod in the middle of an athanor for a day and a night, filling the athanor, that
is, the furnace, with live coals. When they are diminished, add more. When
the fire is out, open the cool crucible and you will find the gold very well separated.[20]

As in the *De anima,* the goal of the cementations and cupellations here
in the *De perfecto magisterio* may well have included the testing of alchemically produced gold and silver. But it is important to note that other alchemical authors used such assaying tests for the investigation of natural
substances. The *Summa perfectionis* of Geber (also called pseudo-Geber),
written around the end of the thirteenth century probably by Paul of
Taranto himself, was arguably the most influential alchemical text of the
Middle Ages. As Ferenc Szabadváry has noted, the *Summa* is keenly concerned with the specific weights of the different metals, presenting them in
relative form—a method of presentation that would still be found centuries
later in the *Dictionary of Chemistry* of Pierre Joseph Macquer.[21] The same
emphasis on gravimetrics appears in Geber's determination that mineral
sulphur is 97 percent volatile; only 3 percent of a given sample is left behind
after calcination.[22] The *Summa's* concern with testing culminates at the
end of the text with a battery of assays including cupellation and cementation along with a number of others. Indeed, the author distinguishes two
types of examination. First he refers to "manifest tests . . . which are known
to all"; these include "the practices of determining weight, color, and extension by the hammer."[23] Since such tests are commonplace, the *Summa*
does not describe them as such, though they underlie much of the practice

20. Pseudo-Aristotle, *De perfecto magisterio,* in *Bibliotheca chemica curiosa,* ed. J. J. Manget (Geneva, 1702), 1:644:

> *Auri separatio.* Fac de eo laminas ad modum tuae unguis, & eas cementa cum hoc pul-
> vere: Recipe salis communis separati partes duas, lateris antiqui in ripis fluviorum, vel in
> littore maris reperti, quod melius est, partem unam, tere optime & cribretur per se-
> taceum spissum: tunc laminas dictas sic cementatas fac morari in medio athnor super
> tripodem per diem & noctem unam, & implendo athanor, id est furnellum carbonibus
> vivis, & quum minuuntur addendo semper de aliis, tunc igne remoto & infrigidato
> aperi crucibulum, & invenies aurum optime separatum.

21. Ferenc Szabadváry, *History of Analytical Chemistry* (Oxford: Pergamon, 1966), 15. The
Summa notes, for example, that lead has a specific gravity closer to gold than that of any other
metal, and that silver has a specific gravity less than that of gold. William R. Newman, *The
Summa Perfectionis of Pseudo-Geber* (Leiden: Brill, 1991), 672, 727. For Macquer's ranking of
the metals by specific gravity, see Pierre Joseph Macquer, *A Dictionary of Chemistry* (London,
1771), 1:429.

22. Newman, *Summa Perfectionis of Pseudo-Geber,* 666.

23. Ibid., 769.

behind the text. Geber then launches into a discussion of tests that employ other methods; these include cupellation and cementation, but also firing or "ignition" to see if a metal incandesces before fusion, inspection of the fused metal after cooling to see if it has blackened, exposure to acid vapors with subsequent inspection of any efflorescence, extinction of the hot metal in salty or aluminous water or with sulphur to produce color changes, burning with sulphur again to induce color changes or alterations of weight, the repetition of calcination and reduction for changes in color, weight, or volume, and the easy or difficult amalgamation with quicksilver.[24] It is worth reiterating that all of these tests occur within an alchemical text geared substantially toward the goal of the transmutation of metals.

TESTING AS A PROBE OF NATURE IN THE SUMMA PERFECTIONIS

The reader may well wonder why the Summa spends so much time on these multifarious tests when cupellation and cementation would suffice to determine the genuineness of alchemically produced silver and gold. The answer is, in part, that the Summa uses these tests not only for the practical goals of assaying, but also to determine something about nature itself, that is, the fundamental composition of the metals. Unlike the Ars alchemie or the De perfecto magisterio, the Summa's practice does not consist solely of indicator tests. Here the author molds his tests into experimental tools for revealing the nature of matter.[25]

The Summa's experimental use of assaying tests to determine the nature of the metals is evident already in the definitions that Geber gives to them. Tin, for example, is

a metallic body, white but not purely, slightly bluish, little participating in earthiness, sounding a small creak, soft, possessing in its root a rapid liquefaction without firing, not waiting through cupellation and cementation.[26]

Here failure to withstand cupellation and cementation is viewed as a defining characteristic of tin, along with such manifest properties as its impure whiteness or the "creak," that is, the peculiar sound tin makes when bent. In addition, Geber refers to the test of firing, that is, pointing out that tin melts at a temperature below that of red heat. In another passage, the Summa expands considerably on these observations. Geber claims that tin is composed of a fixed (i.e., nonvolatile) white sulphur, an unfixed (i.e., volatile) white sulphur, a fixed quicksilver, and an unfixed quicksilver. He goes on to prove this composition by means of experiment.

24. Ibid. 769–83.
25. See Halleux, "L'alchimiste et l'essayeur," 289–90.
26. Newman, Summa Perfectionis of Pseudo-Geber, 675.

You will find the evidence of these if you calcine tin, since you will smell the stench of sulfur going forth from it, which is a sign of unfixed sulfur. . . . One sulfur is made known to reason by the first test *[experientiam]*. The other is proven by the persistence of it in its calx which it has upon the fire, since a more fixed sulfureity does not stink. But a two fold substance of quicksilver is also proven to be in tin, one of which is not fixed. This is because it creaks before its calcination, but after its double calcination it does not creak, which is due to the fact that the substance of the fugitive quicksilver causing the creak has escaped. But that the substance of fugitive quicksilver adduces the creak is proven by the washing of lead with quicksilver. Since if you melt lead with quicksilver after its washing with the same, and the fire does not exceed that of its fusion, part of the quicksilver will remain with it, which will adduce a creak from the lead, and convert it to tin.[27]

This elaborate series of experiments begins with repeated calcinations of tin. The author says the metal emits a smell of sulphur only in the initial calcination, revealing the presence of unfixed sulphur. After begin calcined twice, the tin also loses its well-known creak, showing that something has escaped, which the author shows to have been unfixed mercury. He comes to this conclusion because lead (which he has independently shown to contain less quicksilver than tin), when mixed with quicksilver, gains the very creak that the calcined tin lost. This is quite an interesting example of Geber's experimental procedure. The initial loss of tin's creak upon calcination told him that it must have been due to the escape of a volatile component. How does he know the lost component was unfixed quicksilver rather than unfixed sulphur? Because he can induce the same creak by adding normal volatile quicksilver to a metal that lacks it, namely lead.

As we can see, Geber's determination of the constituents of tin is based on repeated calcination of that metal—one of the assaying tests prescribed at the end of the *Summa*. But the author is not satisfied with having demonstrated the mere presence of fixed and unfixed principles in tin; he wants to determine the relative quantity of mercury and sulphur. For this he must employ yet another test, the ability of the metal to amalgamate with quicksilver:

in [tin] is equality of fixation of the two components quicksilver and sulfur, but not equality of quantity, since quicksilver predominates in their mixture, the sign of which is the easy penetration of quicksilver in its own nature into it. Therefore, if the quicksilver in tin were not of greater quantity, it would not—having been taken up in its own nature—have adhered to that easily.

27. Ibid., 733.

For this reason quicksilver does not adhere to mars [i.e., iron] or venus [i.e., copper] except by means of the subtlest craft, due to the paucity of quicksilver in them in their intermixture.[28]

The ready amalgamation of tin with quicksilver allows Geber to conclude that it, like gold, is composed primarily of that principle. Copper and iron, on the other hand, because they amalgamate only with difficulty, must contain less quicksilver than tin. One could go on to relate other tests of this sort, for Geber determines the principles of all the metals by employing the same battery of assaying techniques. He even goes so far as to compile comparative lists of the metals' ability to amalgamate and of their components' ability to sublime.[29] In short, the very core of the *Summa perfectionis* lies in its attempt to arrive at the nature of the metals by exposing them to a variety of tests that are extensions of an age-old tradition of assaying. Embedded in these alchemical assaying tests throughout is a concern with specific gravity and increase or decrease of weight. Although these gravimetric concerns are not privileged over qualitative indicators such as melting point and resistance to corrosion, they form an important part of the medieval alchemist's experimental armory.

THE BALANCE IN ALCHEMY

From the experimental determination of the metallic composition given in the *Summa perfectionis,* we should be able to see the difficulties in dissociating alchemy from the technology and practice of assaying. At the same time, Geber's determination of the relative specific gravities of the known metals, his use of changes in weight during testing to distinguish metals from one another, his measurement of the volatile component in mineral sulphur, and his attempt to determine the relative quantities of fixed and volatile principles in the metals, belie any belief that alchemy was fundamentally nonquantitative. More specifically, these aspects of alchemical practice throw any notion that alchemists did not make much use of the balance into doubt.[30] Not only is this presumption contradicted by the ev-

---

28. Ibid., 734.

29. Ibid., 656, 722.

30. This viewpoint finds succinct expression in a recent article by Anders Lundgren:

The chemical (as opposed to the hydrostatic) balance does not appear in illustrations of laboratories in the 17th and 18th centuries. Chemists did not use it in their daily work. Only in commercial mining, which typically involved amounts of material far larger than anything of interest to the chemist or the assayer, was the balance at home.

Lundgren, "The Changing Role of Numbers in 18th-Century Chemistry," in *The Quantifying Spirit in the 18th Century,* ed. Tore Frängsmyr et al. (Berkeley and Los Angeles: University of California Press), 247–48.

idence of Geber and the countless alchemical recipes that express their ingredients in apothecary measurements, but it runs counter to what we know about the history of the analytical balance itself.

The earliest positively identified illustration of an encased analytical balance is found in an alchemical text—specifically Thomas Norton's (1433?–c. 1513) *Ordinall of Alchimy*, a treatise dealing with the production of the transmutatory Philosophers' Stone.[31] This illustration, which gives a detailed picture of an alchemical laboratory with furnaces, an alembic, a pelican, and an analytical balance was printed in the *Theatrum chemicum britannicum*, a collection of English verse treatises on chrysopoetic alchemy edited by Elias Ashmole F.R.S. in 1652 (figure 1). The printed illustration is in fact a close copy of Norton's manuscripts, for at least one fifteenth-century manuscript, an important presentation copy prepared in the 1480s or 1490s, survives complete with the balance illustration. Since this manuscript was copied during Norton's lifetime, and probably under his direct supervision, the manuscript illuminations command more authority than is common with many other alchemical illustrations.[32] Beneath Norton's balance one sees a trunk containing chemical vessels. Their blue and red contents may indicate that they are crucibles rather than cupels, for one would expect only the yellow of litharge or the color of the metal being tested if the vessels were the latter. The curious upper object in the middle of the trunk is very likely the "monk" or plunger that was to be driven into clay to make a crucible. On the table before him the master alchemist has a piece of silver, represented by the conventional crescent moon, a ball, perhaps of gold, and a gold-colored vessel. The latter may be a cupel, which appears to be emitting the spume of litharge characteristically produced when a metal is assayed by cupellation with lead.

The appearance of an enclosed analytical balance in the Norton illustration raises an interesting question about the accuracy of the instrument. Here it is appropriate to consider an *ordonnance* of Philip VI of France, represented as having been issued in 1343.

The general or particular assayer must have good, light balances, faithful and exact, that do not decline to either side. . . . When one carries out the assay, it

---

31. John T. Stock, *Development of the Chemical Balance* (London: Her Majesty's Stationery Office, 1969), 2.

32. The manuscript is British Library Additional MS 10302; see John Reidy, ed., *Thomas Norton's Ordinal of Alchemy* (London: Oxford University Press, 1975), x–xiv, for dating. The illumination is reproduced in William R. Newman, "Alchemy, Assaying, and Experiment," in *Instruments and Experimentation in the History of Chemistry*, ed. Frederic L. Holmes and Trevor H. Levere (Cambridge: MIT Press, 2000), 35–54; see color plate after p. 42.

Figure 1. An illustration of an alchemical laboratory from Thomas Norton's *Ordinall of Alchimy,* as printed in Elias Ashmole's *Theatrum chemicum britannicum* (1652). Note the enclosed balance on the table. Ashmole's illustration is closely based on older manuscript illuminations of Norton's *Ordinall,* such as the one found in a late fifteenth-century copy in British Library Additional MS 10302, fol. 37v. Reproduced in color in William R. Newman, "Alchemy, Assaying, and Experiment," in *Instruments and Experimentation in the History of Chemistry,* ed. Frederic L. Holmes and Trevor H. Levere (Cambridge: MIT Press, 2000), 43.

must be done in a place with neither wind nor cold, and one must see that his breath does not affect the balance.[33]

From this one can see that even in the mid–fourteenth century, it was recognized that a precise balance must be relatively light, that is, have a beam of low mass.[34] Although we cannot give an absolute measure either to the weight of the balance or its precision, the fact that its accuracy could be affected by one's breath suggests a fair degree of precision, as Szabadváry has noted.[35] Norton's encased alchemical balance could have been of similar accuracy to that cited in the *ordonnance*.

The surprising degree of precision implied by the *ordonnance* of Philip VI and by Norton's illustration is less remarkable if we accept the evidence that medieval alchemy and assaying were closely linked, since assayers are necessarily concerned with careful determinations of quantity. This linkage did not end with the decline of the Middle Ages, but continued into the early modern period. Even in the sixteenth and seventeenth centuries, alchemy was closely conjoined with the technology of purifying and testing metals. The *Bergwerck-* and *Probierbüchlein* tradition of early sixteenth-century Germany illustrates this readily, for here one finds handbooks of mineral technology with such titles as *Rechter Gebrauch d'Alchimei* (1531), *Bergwerck und Probirbuechlin fuer die Bergk und Feuerwercker/Goldschmid/Alchimisten und Kuenstner* (1533), and *Alchimi und Bergwerck: Wie alle Farben/Wasser/Olea/salia und alumina/damit mann alle corpora/spiritus und calces preparirt/sublimirt/und fixiert/gemacht sollen werden* (1534).[36] These works predate the far more famous *Pirotechnia* of Vannoccio Biringuccio, which was published in Venice in 1540. Any attempt to dissociate alchemy from this tradition would prove fruitless, for even the earliest works in this genre, such as the *Nutzlich Bergbuchley(n)* of Ruelein von Kalbe (1505), openly acknowledge their debt to alchemical sources.[37]

---

33. Jean Boizard, *Traite des monoyes* (Paris, 1692), 166–67: "Le General Essayeur, ou l'Essayeur particulier doit avoir ses balances bonnes & legieres, loyaux & justes, qui ne jaugent d'un coste ne d'autre. . . . Quand on poise les essays, il doit estre en lieu, ou il n'y ait vent ne froidure, & garder que son halaigne ne charge la balance."

34. Stock, *Development*, 7.

35. Szabadváry, *History*, 17.

36. Paul Walden, *Mass, Zahl und Gewicht in der Chemie der Vergangenheit*, in *Sammlung chemischer und chemisch-technischer Vorträge*, Neue Folge, Heft 8 (Stuttgart: Ferdinand Enke, 1931), 3. For a short overview of the *Bergwerck-* and *Probierbüchlein* tradition, see Ernst Darmstaedter, "Berg-, Probir- und Kunstbüchlein," in *Münchener Beiträge zur Geschichte und Literatur der Naturwissenschaften und Medizin* 2/3 (1926): 101–206.

37. Darmstaedter, "Kunstbüchlein," 118–19.

ALEXANDER VON SUCHTEN AND THE SIXTEENTH-CENTURY SYNTHESIS OF CHYMICAL TRADITIONS

The coexistence of alchemical theory and practice, qualitative and quantitative assaying, and experiment in general is elegantly revealed by the remarkable treatises on antimony written by Alexander von Suchten, a Prussian nobleman of the mid–sixteenth century.[38] The *Tractatus secundus de antimonio vulgari* first published in 1604 but composed before 1579, is written in the form of a letter addressed to Johann Baptista von Seebach.[39] Suchten's main interest in this treatise is the preparation of a potent medicine from antimony. He states that this is done by reducing the native ore of antimony, stibnite, to produce the "regulus of antimony" (in our terms, metallic antimony), then alloying the regulus with silver and using the resultant alloy to "acuate" or "quicken" common quicksilver. The goal is to isolate the "volatile gold" within the regulus (originating from the iron used in its production), which will eventually be turned into "potable gold"— the desired medicinal arcanum. The acuated mercury itself has the power to penetrate the metals and to separate their Mercury and Sulphur from one another.[40] This was a crucial desideratum of chrysopoeians, since it was commonly believed that the process for making the Philosophers' Stone must begin with such a dissolution of gold into its principles. Indeed, Suchten's antimonial mercury was one starting basis of George Starkey's own alchemical practice, and we shall have cause to return to this theme later.[41] But Suchten's main interest lies in the realm of Paracelsian iatrochemistry, not metallic transmutation; although a section of his second treatise deals with chrysopoeia, he ultimately rejects it in favor of medicinal preparations. Interestingly, we will see that Suchten's eventual rejection of the artificial metals he prepared from antimony originates from careful qualitative and quantitative tests.

38. On Suchten, see Wilhelm Haberling, "Alexander von Suchten, ein Danziger Arzt und Dichter des 16. Jahrhunderts," *Zeitschrift des Westpreussischen Geschichtsverein* 69 (1929): 177–230; Włodzimierz Hubicki, "Alexander von Suchten," *Sudhoffs Archiv* 44 (1960): 54–63; and Carl Molitor, "Alexander von Suchten, ein Arzt und Dichter aus der Zeit Herzogs Albrecht," *Altpreussische Monatschrift* 19 (1882): 480.

39. Alexander von Suchten, *Tractatus secundus de antimonio vulgari Alexandri von Suchten an den Erbarn und Ehesten Johan Baptista von Seebach geschrieben,* in Suchten, *Mysteria gemina antimonii* (Leipzig, 1604). Włodzimierz Hubicki mentions another edition of the same year, edited by Jakob Foillet. Hubicki, "Alexander von Suchten," *Sudhoffs Archiv* 44 (1960): 58. For the terminus ante quem of composition, see Hubicki, 59.

40. Suchten, *Tractatus secundus,* 422; see also Newman, *Gehennical Fire,* 138.

41. Newman, *Gehennical Fire,* 135–41. On this process, see also Principe, "Chacun à Son Goût: Theory and Experiment in Sixteenth- to Eighteenth-Century Chymistry," *Sudhoffs Archiv,* forthcoming; and chapter 3 below.

The section of Suchten's text that deals with metallic transmutation describes the attempted fabrication of other metals directly from antimony regulus—a process that he distinguishes from his directions for making a Philosophical Mercury. As Suchten says, "I have made these four metals [lead, tin, copper, and iron] myself out of regulus. The other two, silver and gold, I have seen my good friend make."[42] Of the artificial silver, Suchten says that it can be fused, hammered, and cupelled just like "natural silver." Therefore,

> I thought nothing else for a long time than that it was the best silver, but when my comrade said that it was heavier than other silver, I grew distrustful. I took the same silver, wishing to dissolve it in aqua fortis [i.e., nitric acid] made of vitriol and saltpeter. When I found that it did not attack the silver at all, I grew suspicious [again]. I deliberated for a while, and then put it in an aqua regia [mixture of nitric and hydrochloric acids]. When it dissolved totally, then I conceived that it should be reducible into gold.[43]

Suchten here recounts his growing disaffection with the artificial silver—first, it did not dissolve in aqua fortis, as silver should do, then it dissolved fully in aqua regia, which silver should not do. He then hypothesized that his silver was really a form of gold, for it was denser than common silver and dissolved in aqua regia like gold. So he precipitated it from the aqua regia solution the way one would to isolate dissolved gold and recorded that he obtained a white powder, which he then fused, hoping to reduce it into metallic gold. The reduction did not yield gold, but only a "milky glass." Needless to say, the Prussian alchemist was disappointed with the failure of his antimonial "silver" to pass the assay that he performed, for it acted like neither gold nor silver. But he did not give up straightaway. Instead, he took more of his artificial silver and amalgamated it with his Philosophical Mercury, distilling off the volatile component between amalgamations. In the end he allowed the amalgam to digest in the fire for over a month and then distilled it a final time. Again to his disappointment, the artificial silver itself had become volatile and distilled with the mercury.

> Thus I learned that the silver made from regulus is nothing but a coagulated mercury, which does not remain permanently in the form of a metal [in specie

42. Suchten, *Tractatus secundus,* 438.
43. Ibid., 439.

metallica] but returns and becomes mercury again, which might well turn an alchemist into a fool.[44]

Here we see Suchten's rigorous attempt to assay the artificial silver made from antimony and his acceptance of the negative result. He describes the tests with aqua fortis and aqua regia in a manner that any educated reader could understand. His conclusion, moreover, that antimonial silver is really nothing but a "mercury" that has temporarily acquired the appearance of a solid metal will lead him to denounce transmutational alchemy in general.

Suchten's denunciation of *alchemia transmutatoria* emerges from the failure of his alchemical metals to pass the complete battery of assaying tests known to him. After testing the antimonial silver, Suchten moves on to the antimonial gold. What is truly remarkable, however, is that according to Suchten, the antimonial gold received the approval of a professional goldsmith but failed the more stringent tests employed by the would-be chrysopoeian himself.

> When I told my good comrade, who believed nothing else than that he had already reached the goal, what his silver [really] was, he did not want to believe it. He undertook the operation himself, and upon discovering the truth, began at once to doubt the gold. And thus he said: "I have assayed it repeatedly, but I do not want to trust myself. So take this *Loth* [½ ounce] of gold and test it *[probier es]* at your leisure. Master Hans the goldsmith says it is good gold." So I took the gold and brought it to the goldsmith, asking him what sort of gold it was. He said it was good gold, [and that] he could use it as gold. So far as the appearance, cutting, touchstone, and hammering *[Augenschein, stich, strich, unnd Hammer]* went, it was good gold.[45]

But Suchten was not content with the judgment of Master Hans the goldsmith, who seems to have based his decision on qualitative determinations such as color, resistance to cutting, streak left on the touchstone, and malleability. Suchten then relates how he himself performed the test of quartation on the "gold," by alloying it with silver and then dissolving the alloy in aqua fortis. The supposed gold passed the test, for it did not dissolve along with the silver. But still not content, Suchten heated his gold with stibnite, another standard metallurgical test; again the gold passed. Even after these tests Suchten was still not satisfied. He now fused the purged gold with stibnite and sulphur, leaving this with the goldsmith to blast with his own bellows, "for I myself had no flue."[46] The gold once

44. Ibid., 441.
45. Ibid., 441–42.
46. Ibid., 442.

again passed the test, a result that "might overjoy every alchemist."[47] But Suchten was not able to rest easy, for the failed assay of his analogous "silver" made him anxious. So he performed the same test that he had on the silver, amalgamating it with his acuated mercury and heating the mixture for a month, and then distilling the product. He found that when all the mercury was driven off, only two *Quintleins* [one-quarter ounce] of metal were left behind, one-half of the original quantity.[48] This, Suchten points out, was exactly the amount of gold that his friend had added to the antimony regulus in the beginning in order to convert it into "gold." Suchten had recovered only the original gold that he and his friend had hoped would act as a seed to the antimony. Suchten's comrade then explained why the artificial gold failed.

> He said, "The sulphur of antimony, which coagulates its mercury, is not radically *[in radice]* united to the same, so it does not remain with it. If you try it [on your other artificial metals], your regulus will remain neither lead, tin, copper, nor iron, but will again be a mercury. Therefore neither you nor anyone will be able to coagulate [the mercury of antimony] into a good metal, as some speculate. It will therefore escape even if two *Quintleins* of gold still remain."[49]

Suchten's disillusioned friend then concluded that the transmutation of metals is "a lunatic, melancholy fantasy" entertained only by those who lack experience "in the fire."[50] Although the metals really are made of a mercury coagulated by its own intrinsic Sulphur, the artificial replication of this process cannot succeed, at least not before the alchemist himself expires of old age.[51] Thoroughly convinced of the inevitable failure of transmutation, Suchten and his friend recommend to other aspiring chrysopoeians that they too abandon the quest.

In the work of Alexander von Suchten, we encounter many of the themes treated in this chapter. Quantification and testing are readily apparent in his *Mysteria gemina antimonii,* of which the *Tractatus secundus* forms a part. If one were unfamiliar with the long history of alchemy, it might be tempting to argue that Suchten's approach to the discipline was

47. Ibid., 443.

48. The *Quintlein* was a standard measure used in Germany for silver and bullion assay. See Anneliese Sisco and Cyril S. Smith, *Lazarus Ercker's Treatise on Ores and Assaying* (Chicago: University of Chicago, 1951), 343.

49. Suchten, *Tractatus secundus,* 445–444 (the pagination of these two pages is reversed).

50. Ibid., 446.

51. Ibid., 449.

completely novel and that his application of assaying techniques was destined to sound the death knell of the field. But as we have seen, assaying was an integral part of alchemy at least from the Middle Ages, and medieval alchemical texts served as one of the primary means of disseminating such practical knowledge in the Latin West.[52] Suchten was carrying into practice the advice of such medieval sages as Geber and Rāzī, with the addition of newer techniques at his disposal, such as the use of mineral acids.

Yet there is another point at which Suchten reveals both his proximity to medieval alchemy and his distance from it. Whereas Suchten's tests of his chymical metals show how he assayed these products to determine their identity with their natural exemplars, there is another case where he employs a purely demonstrative test, in the spirit of Geber's analysis of tin. Yet here we encounter a critical difference as well—Suchten is more keenly concerned with the exact weights of his initial and final products than was Geber. Although we have argued that quantity of ingredients was an important concern of medieval alchemists and that they dealt with issues of specific gravity, there is little evidence that they made comparative weighings of the ingredients that went into their reactions and the products that came out, except for the purpose of detecting failed transmutations. Although Geber explicitly lists increase or decrease of weight upon heating as an indicator of false gold, he expresses little interest in such determinations during his analyses of metallic principles.[53] Qualitative and gravimetric tests are given equal status in the *Summa perfectionis*. But the case is quite different with Suchten, for the Prussian iatrochemist was eager to demonstrate the nature of his Philosophical Mercury by means of quantitative analysis.

The painstaking process by which Suchten prepared his Sophic Mercury involves amalgamating quicksilver with an alloy of antimony regulus and silver; this process leads to the expulsion of a combustible, black powder, which Suchten refers to as a "Sulphur." He assumes that the fusion of antimony regulus with silver and its amalgamation with quicksilver have caused a separation of the antimony's Sulphur and Mercury.[54] The Mercury of antimony unites with the common mercury, while the Sulphur of antimony is spewed out as the black powder. Suchten collects this black powder by washing the amalgam and then analyzes it.

52. Robert Halleux, "Methodes d'essai et d'affinage des alliages aurifères dans l'Antiquité et au Moyen Age," *Cahiers Ernest-Babelon* 2 (1985): 39–77.

53. One exception is Geber's determination that calcined sulphur retains only 3 percent of its substance by weight, to which we have referred above.

54. Suchten, *Tractatus secundus*, 429–30.

Suchten advises that the powder be dried in the sun, whereupon it will appear gray like lead. In order to rid the powder of any residual quicksilver, he heats it gently in a crucible until the mercury vaporizes. He then heats the powder further until it ignites and burns like a coal. When the combustion has finished, the residue in the crucible can be reduced into regulus. But Suchten advises that the ash first be weighed, and he hints obliquely that its weight should be compared to that of the black powder (see accompanying text and table). Since Suchten assumes that the sulphureous part of the black powder will have been consumed by the burning, this weight comparison will reveal how much Sulphur was actually separated from the original regulus by the fusion with silver and amalgamation with common quicksilver. Similarly, by weighing the black powder before and after the quicksilver clinging to it was evaporated off, and then comparing this difference to the weight of the original quicksilver used, one can determine how much common quicksilver entered into the Sophic Mercury. The purpose of this analysis, Suchten iterates, is simply to reveal "the knowledge of antimony fully" and to show how the quicksilver has been acuated.

Alexander von Suchten's analysis of the "Sulphur" of antimony regulus, and his use of that analysis to determine the quantity of common quicksilver in the Sophic Mercury, falls into the tradition of the medieval alchemists whom we have examined, except that Suchten puts more emphasis on initial and final weights. While this emphasis may partly reflect Suchten's own expertise in metallurgical assaying, however, as we have shown, assaying and alchemy were interwoven throughout the period under discussion. But a further, newer influence of note on Suchten is the view of chymistry as *spagyria,* the art of analysis, pioneered by Theophrastus von Hohenheim, called Paracelsus (1493–1541). Paracelsus argued resolutely that the fundamental alchemical process was *Scheidung,* or "separation." The Swiss chymist envisioned a wide variety of processes—ranging from the digestive system's separation of nutrient from excrement to the creative act of God Himself in making the cosmos—in terms of distillation and the removal of slag during the refining of metals. A major goal of Paracelsian chymistry, then, was the analysis and purification of minerals and other substances. Yet Paracelsus's own work was largely nonquantitative, in contradistinction to the medieval traditions we have discussed. The development of a quantitative chymistry that focused on the weights of starting materials and final products and eventually engaged in the reciprocal processes of analysis followed by synthesis could occur only when the *spagyria* of Paracelsus was fused with more quantitative traditions. This fusion of the two traditions becomes apparent in the antimonial treatises of Suchten, who was both an

ALEXANDER VON SUCHTEN'S GRAVIMETRIC ANALYSIS OF THE ANTIMONY REGULUS USED IN HIS SOPHIC MERCURY

Suchten first amalgamates quicksilver with an alloy of antimony regulus and silver; this process leads to the expulsion of a combustible, black powder that Suchten views as containing the Sulphur of the regulus. Suchten collects this black powder by washing the amalgam, and then he analyzes it. First he dries the powder and weighs it. Then he evaporates off the quicksilver still contained in the black powder and weighs the powder again. Having thus determined the weight of the quicksilver that was in the powder, he now compares this weight to the initial weight of the quicksilver he amalgamated with the regulus. This gives him the weight of the quicksilver that combined with the regulus and silver in the Sophic Mercury. Finally, he burns off the Sulphur in the black powder to get a lighter ash. Weighing this ash allows Suchten to determine the amount of Sulphur that was separated out of the regulus during its amalgamation and washing.

$$\frac{\text{weight of black powder before vaporizing its quicksilver} - \text{weight of black powder after vaporizing its quicksilver}}{\textit{weight of quicksilver that was in black powder}}$$

$$\frac{\text{weight of original quicksilver employed} - \text{weight of quicksilver that was in black powder}}{\textit{weight of quicksilver in Sophic Mercury}}$$

$$\frac{\text{weight of black powder before burning off its Sulphur} - \text{weight of ash left after burning off Sulphur}}{\textit{weight of Sulphur that was separated from antimony by amalgamation with Mercury.}}$$

heir to medieval alchemical/assaying techniques and a devotee of Paracelsian *spagyria*. The full flowering of these combined emphases appears clearly in the following century in the work of Joan Baptista Van Helmont.

JOAN BAPTISTA VAN HELMONT: ART, NATURE, AND EXPERIMENT

The scion of a noble Flemish line, Joan Baptista Van Helmont studied at the University of Louvain during the 1590s, but initially refused to take a degree because of his growing disillusionment. He returned to academic learning, nonetheless, and received a medical degree at Louvain in 1599. Despite this concession to the learned culture of his day, Van Helmont remained an inveterate opponent of the disputational techniques that formed the basis of academic training. He denied the primacy of "reason" as employed in logical disputations, and in good Neoplatonic fashion he tried to supplant this with instantaneous cognition through the higher faculty of "intellect."[55] Like Plotinus and his followers, Van Helmont viewed the intuitive revelation of knowledge by intellect to be a godlike, instantaneous act, as opposed to the temporal process of logical argumentation, which proceeds by means of sequential propositions and proofs.[56] Indeed, for Van Helmont, reason was not the defining characteristic or the highest faculty of human beings; even animals possessed reason. It was intellect that represented the height of human nature and the image of God in man.[57] As part of this rebellion against the university culture of disputation and proof, Van Helmont adopted the mystical ideology of Thomas à Kempis and Johannes Tauler.[58] After reading these authors, Van Helmont says that he fell into a dream in which he saw himself as an empty bubble whose diameter reached from the earth to the heavens. Above the bubble hung a tomb, while below it was the dark abyss, a vision that horrified the young Van Helmont. Upon waking, he realized that the bubble represented his own boastful, vacuous self and that he must turn from his traditional studies to the surer knowledge provided by intellect.[59]

Van Helmont's works are peppered with such dreams, from the early *Eisagoge in artem medicam a Paracelso restitutam* of 1607 to the vast *Ortus medicinae* published posthumously in 1648.[60] In addition, his writings are filled with diatribes against school mathematics.[61] Now the combination of Van Helmont's oneiric epistemology and his apparent antipathy toward mathematics and "reason" might seem to make him an exemplar for anyone who would characterize the alchemist as irrational or "pseudo-scien-

55. Alice Browne, "J. B. Helmont's Attack on Aristotle," *Annals of Science* 36 (1979): 575–591; see esp. 580.

56. Van Helmont, *Ortus*, "Venatio scientiarum," pp. 20–32. See Guido Giglioni, *Immaginazione e malattia: Saggio su Jan Baptiste van Helmont* (Milan: Francoangeli, 2000), 26–41.

57. Van Helmont, *Ortus*, "Intellectus adamicus," pp. 705–7, and "Imago dei," pp. 708–18.

58. Walter Pagel, "The Religious and Philosophical Aspects of van Helmont's Science and Medicine," *Bulletin of the History of Medicine*, supp. 2, 1944; Berthold Heinecke, "The Mysticism and Science of Johann Baptista Van Helmont (1574–1644), *Ambix* 42 (1995): 65–78, esp. 65; Heinecke, *Wissenschaft und Mystik bei J. B. van Helmont (1579–1644)* (Bern: Peter Lang, 1996), 73–135.

59. Van Helmont, *Ortus*, "Studia authoris," p. 17.

60. C. Broeckx, "Le premier ouvrage de J.-B. Van Helmont," in *Annales de l'académie d'archéologie de Belgique* 10 (1853): 327–92; 11 (1854): 119–91; on pp. 339–43; Van Helmont, *Ortus*, "Imago mentis," no. 13, pp. 269–70; Van Helmont, *Ortus*, "Potestas medicaminum," no. 3, p. 471 and passim; Van Helmont, *Opuscula, Tumulus pestis*, pp. 5–7; Giglioni, *Immaginazione*, 35–41.

61. Allen G. Debus, *The Chemical Philosophy* (New York: Science History, 1977), 2:311–17; Debus, "Mathematics and Nature in the Chemical Texts of the Renaissance," *Ambix* 15 (1968): 1–28.

tific." Devotees of the view of a predominantly spiritual alchemy and even of Jungian analytical psychology would seem to have fodder here. This "alchemist" seems to be dreaming in his laboratory, seemingly untouched by and openly hostile to key aspects of modern science—mathematics, measurement, logical methods, and reason.[62]

A closer inspection, however, shows that this is not at all the case. Van Helmont *was* deeply attached to his dreams, and they were for him a real means of acquiring knowledge. But they were not the *only* means. Van Helmont did not have to choose between instantaneous understanding through *intellectus,* revelations in dreams, and what we might recognize as rational investigations. Van Helmont's dreams and desire for the union that brings true insight did not inhibit his ability as a vigilant laboratory practitioner, any more than the spectacular dreams of Descartes provided obstacles to his geometry.[63] More importantly, Van Helmont's attack on *mathesis* was not a wholesale rejection of mathematics in the study of the world, but only as it was applied in the Scholastic medicine still taught in early modern universities, and as he surely encountered it at Louvain.

VAN HELMONT AND THE USE OF MATHEMATICS IN NATURAL PHILOSOPHY

Van Helmont's early *Eisagoge* already gives a plenary rejection of Scholastic mathematics. Since the *Ortus medicinae* merely expands upon this theme without substantially modifying it, we will focus initially on the earlier work. Van Helmont first rejects the Scholastic physicians' predilection for arguing "in the manner of geometers" (*geometrarum more*), by which he means their use of a type of deductive process that begins with axioms (*axiomata*) derived from reason rather than from experience. He gives a list of ten such axioms that together constitute the "method" of Galen. The first two of these so-called axioms begin to throw light on Van Helmont's attitude toward mathematics:

---

62. As one example, P. Nève de Mévergnies, *Jean Baptiste Van Helmont, Philosophe par le Feu,* in *Bibliothèque de la Faculté de Philosophie et Lettres, Université de Liège,* fasc. 59 (Paris: Droz, 1935), despite his valuable contributions to Van Helmont's biographical details, did in fact dismiss Van Helmont as an unworthy contributor to modern science in large part due to his dreams and interest in the mystic writers. Similarly, Herbert Butterfield, *The Origins of Modern Science, 1300–1800* (New York: Macmillan, 1951), 98, states that Van Helmont "made one or two significant chemical discoveries, but these are buried in so much fancifulness . . . that even twentieth-century commentators on Van Helmont are fabulous creatures themselves, and the strangest things in Bacon seem rationalistic and modern in comparison."

63. For a comparison of Van Helmont's dreams to those of Descartes, see the important article by Robert Halleux, "Helmontiana II: Le prologue de L'Eisagoge, la conversion de Van Helmont au Paracelsisme, et les songes de Descartes," *Academiae Analecta, Klasse der Wetenschappen* 49 (1987): 20–36.

1. That the four elements, as sensible fabric, combine in separate mixtures, in certain but unknown weights, and immeasurable measures.
2. That a "complexion" is a proper measure of mixture, and that this is the cause of health, while an improper measure is the cause of diseases and death.[64]

Already one can see the scorn that Van Helmont hopes to throw on university medicine—the "axioms" underline the contrast between Scholastic claims to quantitative certainty and the fact that the academic physicians cannot *actually measure* the quantities of the four elemental humors, blood, phlegm, yellow bile, and black bile. Van Helmont goes on to argue that despite the university doctors' claim that good health consists in a proper measure *(synmetria)* of the four humors, they cannot even demonstrate that the two biles actually exist naturally in the human body. In fact, according to Van Helmont, these physicians actually *cause* the production of bile within the body by administering poisonous purges such as scammony, which liquefy and putrefy the body's internal structure and cause the expulsion of this supposed bile as an artifact.[65] Moreover, they employ pseudomathematical rules in their doctrine of critical days and recurrent fevers that are supposed to return daily, every other day, or at other intervals. Their tables of bloodletting, governed by the phases of the moon, and their calculations of the "climacteric period," along with the foregoing, are all "incorrect borrowings of Pythagorean numbers," from whose tedium Van Helmont hopes to be excused.[66]

We can now begin to pick out several distinct themes in Van Helmont's critique of mathematics in Scholastic medicine. In general terms, university physicians have tried to ape Aristotelian natural philosophy by abstracting their practice from "axioms" based on reason rather than on experience.[67] Not content with feigning such abstractions, the Scholastics have developed fanciful and complex mathematical rules for determining the flow and balance of humors and their correctives. Their metrical determinations of the elements composing drugs and humors are likewise fictitious, and so Van Helmont rejects the long tradition of quantitative pharmacology, say-

---

64. Van Helmont, *Eisagoge,* in Broeckx, "Le premier ouvrage" (1853), 358: "I. Elementa quatuor, machinas sensibiles nempe, in singulis mistis, certis et incognitis ponderibus et mensuris immensurabilibus concurrere. II. Synmetriam mistionis, complexionem esse, hancque sanitatis productricem: ametriam vero morborum et mortis causam."

65. Van Helmont, *Eisagoge,* 161–62.

66. Ibid., 159. On "climacterics," see Tamsyn Barton, *Ancient Astrology* (London: Routledge, 1994), 187.

67. He would later claim in the *Ortus medicinae* that even the *prima materia* of the Aristotelians is a pseudomathematical entity, being a pure abstraction without body, a sort of a dimensionless being, pure quantity without any given quantity.

ing that Scholastic determinations of the intension and remission of the four primary qualities, hot, cold, wet, and dry, are worthless *(frustra)*.[68] Since Van Helmont rejects the Aristotelian notion that each element is characterized by a pairing of elemental qualities, it follows that the tactile and sensory qualities that are supposed to flow from them are also fruitless tools in the hands of the Scholastics.[69]

Yet there is more to Van Helmont's critique than a rebuff of premature abstraction and a dislike of ineffectual calculation in the fashion of Francis Bacon. At the root of his attack lies the belief that the Aristotelian approach adopted by physicians is irretrievably superficial and artificial.

> We feel far otherwise than most Peripatetics do about the properties and nature of place, for they try to fit mathematical descriptions to natural things. A sculptor skillfully feigns the shape of a man on the outside. But he does not know how to imitate the interior organs and the multiple shapes of the vessels within; still less can he emulate the vital spirit—the foundation of the body. To us the places of things are not idle, but are the very things that display life by means of what is located there, that is, their *semina*. This is a natural consideration, not a fantastic contemplation of the perimeter of the surface *[fantastica superficiei circumductionis contemplatio]*.[70]

68. Van Helmont, *Eisagoge*, 147. See Michael McVaugh's brilliant study of this pharmacological tradition: introduction to *Arnaldi de Villanova opera medica omnia*, vol. 2, *Aphorismi de gradibus*, ed. McVaugh (Granada-Barcelona: Seminarium Historiae Medicae Granatensis, 1975).

69. Van Helmont, *Eisagoge*, 364: "Pigritia enim, somnus, torpor, membrorum resolutiones, proprietatibus papeverinis, sulphureis, vitriolatis debentur: non autem frigiditati vel humiditati, non albedini vel nigredini, non magnitudini aut parvitati, non obliquitati, rectitudini, circulari figurationi." Van Helmont probably has in mind the primary, secondary, and tertiary qualities of the Scholastics, for which see Anneliese Maier, *An der Grenze von Scholastik und Naturwissenschaft* (Roma: Edizioni di Storia e Letteratura, 1952), 9, 14–15.

70. Ironically, this passage seems to owe a substantial debt to Galen, usually one of Van Helmont's favorite targets; see Galen, *On the Natural Faculties*, tr. Arthur John Brock (London: Heinemann, 1947), 129. Note that in the above passage *circumductio* also has the meaning of "a deceit," in the sense of a "wild-goose chase," possibly implying that the Peripetetics have been led completely astray by superficial appearances. Van Helmont, *Eisagoge*, 153:

> Longe secus de loci natura et proprietatibus sentimus ac plerique peripatetici: qui descriptiones mathematicas rebus naturalibus adaptare conantur. Statuarius exterius quidem figuram Hominis affabre mentitur. Interiora tamen viscera, ac multiplices vasorum ductus nescit imitari, multoque minus, corporis principium vitalem spiritum emulari potest. Apud nos igitur loca rerum non sunt otiosa: sed quae vitam exhibent suis locatis, id est seminibus. Hoc est naturalis consideratio, non autem fantastica superficiei circumductionis contemplatio.

Van Helmont's association here of Aristotelian natural philosophy with the artificiality of the plastic arts like sculpture is striking. Van Helmont is surely thinking of the famous section in book 2 of Aristotle's *Physics,* where the Stagyrite distinguishes between natural and artificial things, claiming that only the former have an innate *principle of movement (or change) [echonta en heautois archēn kineseōs],* whereas the artificial have *no inherent trend toward change [oudemian hormēn echei metabolēs emphyton].*[71] Key to understanding Van Helmont's point is the fact that throughout the *Ortus medicinae,* the Belgian philosopher links this position implicitly to the Aristotelian principle that in cases of efficient causality, the mover must be in contact with the moved (*Physics* VII 2 243a 12–17), the principle that forbids action at a distance.[72] According to Van Helmont, the Aristotelians have

---

71. Aristotle, *The Physics,* tr. Philip H. Wicksteed and Francis M. Cornford (London: Heinemann, 1929), 106–15. Despite the fact that Van Helmont criticizes this formulation in *Ortus,* "Physica Aristotelis et Galeni ignara," no. 2, p. 46, he explicitly adopts it elsewhere. See *Ortus,* "Ignotus hospes morbis," no. 86, p. 504:

> Illud demum notabile, quod in artis operibus, efficens sit semper extra: Illiusque errore deceptae Scholae, nesciverunt, in naturalibus ac substantialibus generationibus, agens esse internum. Idcirco enim naturalium causarum catalogo, efficientem, ut externam relegarunt. Imo nescitum est, ambas naturalium connexas causas (ut suo loco demonstravi) non differre ab suo effectu, nisi fluxus prioritate. Quae res decepit, quotquot per artificia naturam similitudinarie sunt contemplati.

72. Van Helmont, *Opuscula, De lithiasi,* no. 11, pp. 34–35: "Profectò, corpora non agunt in corpora, per naturalem compositionis actionem: sed quicquid corpora in invicem peragunt, id fit ratione ponderis, magnitudinis, duritiei, figurarum & motuum. Inserviuntque enim Mathesi, vix Physicae." Van Helmont, *Ortus,* "Ignota actio regiminis," no. 6, p. 331: "Ut autem ostendam, ejusmodi Matheseos respectus non habere actionem, manantem à potestatibus rerum: sed tantum relationem Matheseos (cujus omnis mera actio, quanquam per corpora fiat: non est tamen ipsius corporis, ut talis) sufficit ostendisse per praefata, determinationem motuum distare procul à motuum activitate interna, juxta quam censentur res à veteribus repati & reagere, in omni actione." Van Helmont, *Ortus,* "Ignota actio regiminis," no. 16, p. 335: After describing the proper way for Christians to philosophize, Van Helmont says, "In motu autem locali, viribus motricibus, adeoque & in exercitio Matheseos, axiomata Aristotelis inserviunt quidem, quae violento Scholae imperio, atque importunè, in naturam introduxere." More generally, Van Helmont explicitly rejects the Scholastic position that the cause and the caused must be distinct, once again attributing this view to their "mathematical" approach, as in the *Ortus,* "Ignotus hospes morbus," no. 35, p. 491: "Seductae ergo Scholae, per proprias somniorum libertates, autumarunt, quod quia consideratio causarum ac principiorum differt à consideratione rei, per illa productae; quod proinde de necessitate formaliter causante causae omnes deberent in fiendo, essendo, operando, ac permanendo, in perpetuum manere separatae à rebus causatis. Non attendentes, quod plerumque consideratio causarum, & principiorum, non differt alias à consideratione causati, quàm per relationem entis mentalis, que etsi recepta sit in Mathesi, & sermocinalibus: minimè tamen in cursu naturae. Itaque Scholae delusae per ejusmodi elenchos, credidere causam omnem, efficientem, de necessitate externam: nec proinde cum causato uniri posse. Ideoque nec generans esse partem generati. Cum alioqui id natura, ens proximè generans, semper sit moderator internus, vitalis, & assis-

overlooked the real internal principles that direct both animate and inanimate things—namely the hidden and self-moved *semina* and the *archeus* (which he calls the "internal efficient cause" of the body). As Van Helmont puts it, "Aristotle was ignorant of this [the *archeus*], and erroneously pointed to external efficient causes in a way that showed only the understanding of a country fellow or simple mechanic."[73] The *semina* and the *archeus* lurk deep within the recesses of physical bodies and are responsible for their specificity, their transmutations, and their development.[74] Unlike bodies at the macro level, the *semina* operate by means of a "radial activity" that need not involve physical, bodily contact; hence, the principle that the mover and the moved must be in mutual contact does not apply to *semina*.[75] Therefore, according to Van Helmont, Aristotelian natural philosophy, in its ignorance of the *semina* and *archeus*, reduces much of nature to the status of an artifact, that is, something that has no internal principle of motion or change, but must receive motions from the contact of external agents.

At this point Van Helmont introduces a medieval notion that such artificiality and mimicry of nature are best exemplified in the making of machines. Many medieval authors (like the twelfth-century Hugh of Saint Victor) believed that the word "machine" *(machina)* was derived from "adultery" *(moicheia);* a maker of machines feigns the appearance of a natural object or phenomenon in the same way that an adulterer pretends to be a husband.[76] Now the proper way to analyze artifacts, or rather machines, is according to the rules of simple mechanics, such as the law of the lever. These rules rely on spatial measure and proportion, such as the relation between the distance from a fulcrum to the motive agent and the distance

---

tens Architectus generationis: adeoque qui in finem dirigit cuncta ad scopos, sibi omnia patrat, & pro se egit universa."

73. This unusually clear definition of the *archeus* is found in Van Helmont's correspondence with Mersenne. See Van Helmont to Mersenne, 11 January 1631, in C. de Waard, *Correspondence du P. Marin Mersenne* (Paris: Presses Universitaires de France, 1946), 3:13: "*Archeus sive causa efficiens interna* (quam Aristoteles ignoravit; omnem causam efficientem externam indigetans, rustico ac plane mechanico intellectu)."

74. Walter Pagel, *Joan Baptista van Helmont* (Cambridge: Cambridge University Press, 1982), 24–28, 39–49, and passim.

75. Van Helmont, *Ortus,* "Ignota actio regiminis," no. 18, p. 335: "Nam quanquam haec illarum axiomata locum habent in actionibus corporalibus, quibus agens, de necessitate, fovet, & tangit suum objectum, atque hactenus suam eidem inspirat vim: Attamen id prorsus est impertinens in Agentibus, quae in supposita, loco eminus seposita, agunt."

76. Jerome Taylor, ed., *The Didascalicon of Hugh of Saint Victor* (New York: Columbia University Press, 1991), 51, 55–56; see William Newman, "Technology and Alchemical Debate in the Late Middle Ages," *Isis* 80 (1989): 424.

from the fulcrum to the body to be lifted; hence, the science governing machines and, by extension, governing artificial things in general is *geometry*.
Such geometrical techniques are employed by Aristotle in book 7 of the
Physics (249b27–250a19), where he discusses the proportionalities relating
force, resistance, distance traveled, and time required in the case of a boat
pulled by a group of men.[77] But this geometrical approach to motion, Van
Helmont claims, while quite appropriate for "examples drawn from artificial
things," does not apply to the natural actions of the *archeus*.[78] Therefore, by
Van Helmont's rather obscure logic, Aristotelian science, insisting on the
need for contact between the mover and the moved, has reduced the world
to a mere machine governed by laws of mathematics. In their ignorance of
the *semina* and the *archeus*, which are not only self-moved entities (and thus
untouched by Aristotle's geometrical laws of motion) but the true internal
efficient causes of natural things, the Aristotelians have erred.

Now we can begin to acquire a correct grasp of Van Helmont's opinion
concerning the relation of mathematics to natural philosophy. Mathematical methods are properly applicable to machines and to those aspects of natural bodies that involve spatial measurement. At the same time, however,
natural bodies have internal principles such as the *semina* and *archeus*
that—unlike machines—need not act by the principles of contact mechanics, and so mathematics alone is inadequate for their understanding. Application of mathematical methods to the natural world would work only in
those cases where one is not trying to determine the action of *semina*. To
modern ears, Van Helmont's argument may at first sound like a blanket appeal to vitalism in opposition to mechanism. Yet it can be read more com-
pellingly as an attempt to distinguish superficial physical change from
intimate chymical interactions that result in change of substantial identity.
This distinction underlies much of Van Helmont's chymistry, which distinguishes the mere mechanical grinding and spatial translation of corpuscles
from the "deep connection" and "marriage" of substances brought about

77. For an excellent introduction to the medieval commentaries on this Aristotelian locus
see Marshall Clagett, *The Science of Mechanics in the Middle Ages* (Madison: University of Wisconsin Press, 1959), 421–44.

78. Van Helmont, *Ortus*, "Butler," p. 594:

Pro more Scholarum namque ex artificialibus exempla deducens, etiam erravi cum
earundem doctrina. Seductus putavi enim, sicut duo equi fortius trahunt unico, poten-
tiusque alit panis integer quam ejus mica, putavi similiter, pro remedio restaurativo
Archei requiri unciarum & drachmarum quantitatem, quae viribus ac pondere, morborum producta superaret. Nondum scilicet deposueram contractam antiqui erroris
labem, qua metiuntur morbi dumtaxat ex causa occasionali, ejusque pondere: non
autem ex vero morborum efficiente.

by seminal interaction.[79] Van Helmont expended considerable energy in sorting out these two types of change. His distinction may be seen most clearly in the two varieties of change that water can undergo—some of these metamorphoses, such as the conversion of water into ice, vapor, and *gas* (a word of Van Helmont's own coinage), are of the superficial variety, while others involve the radical transmutation of water into another substance altogether.[80]

Van Helmont's explanation of the superficial changes that water can undergo hinges on the idea that this element consists of corpuscles made up of the three principles—Mercury, Sulphur, and Salt.[81] Driven by his concern that a genuinely elemental substance must by definition be simple, Van Helmont explicitly denied that water could be analyzed—physically separated—into its three principles. The *tria prima* do not exist in water as anterior "principles of composition," but they do exist as "principles of heterogeneity."[82] Since the three principles of water cannot be separated by analysis, their very existence must be conjectured on the basis of water's activity, Van Helmont asserts, in the same way that premodern astronomers postulated the existence of eccentrics or epicycles from the apparent motions of the planets.[83]

---

79. Newman, *Gehennical Fire*, 141–51.

80. On Van Helmont's gas, see Paulo Alves Porto, *Van Helmont e o Conceito de Gás: Química e Medicina no século XVII* (São Paulo: Educ/Edusp, 1995), and Guido Giglioni, "Per una storia del termine Gas da van Helmont a Lavoisier: costanza e variazione del significato," *Annali della Facoltà di Lettere e Filosofia dell'Università di Macerata* 25–26 (1992–93): 431–68.

81. For the origins of the theory, see Newman, *Gehennical Fire*, 92–114, and Newman, "The Occult and the Manifest Among the Alchemists," in F. Jamil Ragep and Sally P. Ragep, ed., *Tradition, Transmission, Transformation: Proceedings of two conferences on pre-modern Science held at the University of Oklahoma* (Leiden: Brill, 1996), 173–198; on 175–185.

82. Van Helmont, *Ortus*, "Tria prima chymicorum principia," no. 54, p. 407: "Sufficit enim, ibidem quoque monuisse, partes heterogeneas aquae esse in simplicissimo elementi corpore, arte, natura, omnibusque seculis indivisibiles, atque realiter impossibiles, constantes simplicitate extrema. Itaque, licet ibidem tria prima aquae vocaverim, non sunt tamen tria compositionis, quasi anteriora, aquae initia: sed tria heterogeneitatis."

83. Van Helmont, *Ortus*, "Gas aquae," no. 8, p. 74: "Considero corpus Aquae, continere elementalem sibi, atque genialem mercurium, liquidum, atque simplicissimum: salem denique insipidum, atque simplicem. Quae ambo, intra se amplectuntur uniformè, homogeneum, simplex, & inseparabile sulfur. Haec [tria principia in aqua] suppono, prout Astronomi suos excentricos, ut intelligendi imbecillitati nostrae, eatur obviam." In his recent study of chymical corpuscular theory, Antonio Clericuzio has construed Van Helmont's reference to eccentrics to mean that water, being a simple element, does not really contain the three principles or at least something like them. In reality, what Van Helmont means is that the three principles cannot be separated from water, not that the element lacks them. On the contrary, they are present in water as "principles of heterogeneity" rather than as "anterior principles of composition." The *tria prima* or something analogous to them must exist in water in order to account for its change of state, but it does not follow that water is "made out of" three principles

What is the activity of water that requires the chymist to make this conjecture? Precisely the physical processes that do not involve the intimate change induced by *semina* or *archei*, those internal principles of motion erroneously ignored by the Scholastics. In the *Ortus medicinae,* Van Helmont goes to great lengths to explain the sublimation of water vapor from ice, which he had observed on frigid days in the northern European winter.[84] He argues first that normal vaporization of water by heat occurs in the following steps: the water is first extenuated and divided into corpuscles composed of shells corresponding to the three principles; the Mercury and Salt are found in the two outer shells, while the Sulphur inhabits the core. These corpuscles, being very small, and hence light, are driven up by heat in the normal process of evaporation.[85] Van Helmont then moves to the sublimation of ice. When water is exposed to extreme cold, it crusts itself over into ice in order to avoid being consumed by the cold. This change of state occurs because the Sulphur migrates to the periphery of the water corpuscle, while the Mercury and Salt retreat within, reversing the usual order of the layers within each particle.[86] Since Sulphur is a "dry" principle, the water solidifies ("dries") into ice, and because of the extenuation of the corpuscles that occurs during this internal rearrangement, the resultant ice is less dense (specifically lighter) than the water from which it froze. This ice can now sublime when the inverted water corpuscles are divided still further into "minimal parts" that retain the inverted order of the three principles. Van Helmont's goal in postulating an inversion of the water corpuscles' principles is precisely to explain in physical rather than chemical terms the striking changes observed in the freezing, vaporization, and sublimation of water. As he explicitly states, "It is not a new substantial generation when vapor is elevated from water, since it is only an extenuation, due to the extraversion of the parts."[87]

The point of Van Helmont's analysis of vaporization and the production of *gas* is that changes of state are superficial physical processes, in which no real substantial change occurs—"no mutation of essence occurs where there is only local division and extraversion of particles."[88] Yet water, as the

that somehow existed separately before the element itself. See Van Helmont, *Ortus,* "Tria prima chymicorum principia," nos. 52–54, p. 407, and Clericuzio, *Elements, Principles, and Corpuscles: A Study of Atomism and Chemistry in the Seventeenth Century* (Dordrecht: Kluwer, 2000), 56–58.

84. Van Helmont, *Ortus,* "Progymnasma meteori," no. 1, p. 67.

85. Van Helmont, *Ortus,* "Gas aquae," nos. 8–9, p. 74.

86. Ibid., nos. 10–13, p. 75.

87. Ibid., no. 10, p. 75: "Non est itaque nova, ac substantialis generatio, dum ex aqua vapor elevatur, cum sit tantum extenuatio, propter partium extraversionem."

88. Ibid.: "Non intercedit enim essentiae mutatio, ubi sola est localis divisio, & partium extraversio."

fundamental element from which all other materials spring, can in other instances undergo true "mutation of essence." Indeed, this is how water can be the basic substratum of all other matter: the action of *semina* radically converts water into all the various substances in the world. This mutation of water into metals, woods, oils, salts, and all other substances is the second type of change—a profound transformation brought about by the action of *semina*.

Having distinguished intimate chymical changes from their superficial physical counterparts, we can now further understand Van Helmont's view of mathematics in natural philosophy. Since superficial physical changes involve only processes such as division and change of location, they are purely mechanical and are thus readily analyzed by means of mathematical methods such as geometry. While mathematical treatment is perfectly appropriate in such cases, it is not useful for explaining the radial activity of *semina* that are involved in change of substance. Thus the useful application of mathematics is restricted to the explanation of superficial change.

We must note, however, that Van Helmont's idea of a "chymical change" is considerably more restricted than our own analogous division between physical and chemical changes. For the Belgian views all processes where the initial ingredients are recoverable as examples of superficial rather than of genuine change.[89] Only when the change is so radical that the original substances cannot be recovered—owing to the intervention of new *semina* or the mortification of the old *semina*—does Van Helmont consider a true chymical change to have taken place. For example, the dissolution of gold in aqua regia into a transparent yellow liquid is not a genuine chymical change for Van Helmont, because the original gold may be recovered from the solution unchanged by precipitating it with salt of tartar (potassium carbonate). Since all laboratory processes where the initial ingredients are recoverable are to Van Helmont examples of superficial change rather than genuine mutation, it follows that his chymistry leaves considerable room for explanation in terms of atoms and corpuscles that undergo no alteration other than change of place.[90] Paradoxically, it was Van Helmont's acute ob-

89. Van Helmont's ultimate goal was the creation of completely new substances that did not result from mere mechanical grinding and apposition of "atoms," but involved a "deep connection" and "marriage" of substances. See Newman, *Gehennical Fire*, 141–51.

90. We differ here from Clericuzio, *Elements, Principles, and Corpuscles*, 56–57, who marginalizes the role of Van Helmont's corpuscular theory. Clericuzio does not note that for Van Helmont many, if not most, of the operations of ordinary chymistry involved the artificial displacement and association of corpuscles. Although Van Helmont does view the "intimate marriage" of substances as a major desideratum, this does not impede his ability to explain chymical operations in corpuscular terms.

servations of such superficial change that allowed him to debunk the old belief that iron is actually transmuted into copper in vitriol springs such as those at Goslar. Instead, he argued, the iron atoms are slowly pulled away by the corrosive menstruum (solvent) in the vitriol, and the pores left by the departing iron are filled by incoming copper "atoms."[91] Mere change of location, while relegated by Van Helmont to the realm of the superficial, was nonetheless of key importance in his chymistry cum natural philosophy.

Now finally we can see that Van Helmont's attack on *mathesis* does not entail a wholesale rejection of the mathematical approach to nature as such; rather, it involves an extraordinary critique of Aristotelianism as a *misapplication* of mathematics to nature. The problem lies in the fact that the Peripatetics have inserted local motion and mathematics into nature without proper discretion.[92] A Scholastic medicine that employs excessive "geometrical" abstraction based on fictive humors, engages in useless calculations, and ignores the action of the all-important *semina* must be avoided. Instead of studying the distorted geometrical method of the schools, students should apply themselves to practical and descriptive mathematical sciences that can be put to genuine use in their proper domains.

I would have it that in this so brief period of life, the spring of young men no longer be steeped with trifles of this sort and mendacious sophistry. In these useless three years, or even seven years, they should [instead] successively learn arithmetic, algebra, the *Elements* of Euclid, [and] geography, with the circumstances of the oceans, rivers, fountains, mountains, lands, and minerals. Also the properties and habits of nations, plants, animals, minerals, and places. Especially the use of the *annulus* and the astrolabe. Then let them pass to the study of nature; let them learn to recognize and separate the first principles of bodies.[93]

91. Van Helmont, *Ortus,* "Paradoxum tertium," no. 14, p. 692.

92. Van Helmont, *Ortus,* "Physica Aristotelis et Galeni ignara," no. 8, p. 49: "Ac postremo motum localem, quatenus Mathesi inseruit, in naturam aeque futiliter, atque indistincta indiscretione introducunt."

93. Ibid., no. 9, pp. 49–50:

Optarem certe in tam brevi vitae spatio, ver adolescentum, ejusmodi nugis posthac, & sophismate mendaci non amplius imbui. Discerent nempe inutili isto triennio, totoque septennio, Arithmeticam, Algebram, Euclidis elementa, Geographiam deinceps, cum circumstantiis marium, fluminum, fontium, montium, Provinciarum, & Mineralium. Itemque proprietates, & consuetudines nationum aquarum, plantarum, animalium, mineralium, & locorum. Insuper usum annuli & Astrolabii. Dein accedant ad naturae studium, discant prima corporum initia noscere, & separare.

As he makes clear in this passage, Van Helmont has no brief against mathematics if it is applied to nature in the proper fashion. Indeed, he wishes to replace the traditional propaedeutic curriculum of "arts" with an introduction solely based on the mathematical sciences, followed in due course by hands-on training in natural philosophy. In an almost Baconian vein, Van Helmont says that this mature natural philosophy should be promoted by means of "histories" devoted to such topics as extractions, divisions, connections, maturities, promotions, impediments, consequences, failures, and utilities. All of this, and more, should be taught by "practical demonstration of the fire" (*demonstratione Ignis mechanica*), not by "mere logical description" (*nuda Logismi descriptione*). But what role will mathematics really play within this new natural philosophy itself, beyond training the mind as a preparation? Here Van Helmont now reveals a surprising predilection for the use of a specific kind of mathematical method within natural philosophy.

> We read in our furnaces that there is no more certain genus of acquiring knowledge (*sciendi*) for the understanding of things through their root and constitutive causes than when one knows what is contained in a thing and how much of it there is.[94]

What Van Helmont has in mind here is chymical analysis, the *spagyria* of the Paracelsians, but with attention paid to *quantitas* as well as *quidditas*. It is analysis performed in conjunction with a careful quantitative measurement of weight that will provide "a mathematical demonstration stronger than any syllogism."[95]

VAN HELMONT AND THE CONCEPT OF MASS BALANCE

As we have seen, Van Helmont felt a profound repugnance for the Scholastic mathematics to which he was exposed as a university student. At the same time, however, he argued that students should be taught the practical mathematics involved in such pursuits as navigation, surveying, and mapmaking. Even more fundamentally, Van Helmont emphasized the superiority of the quantitative knowledge derived from weighing things over the Scholastic determination of their essences by means of logic. If we give closer scrutiny to the Helmontian oeuvre, we shall find that this quantitative approach to matter is deeply imbedded in his natural philosophy. Van Helmont is sometimes given credit for having "expressed clearly the law of

94. Van Helmont, *Opuscula, De lithiasi,* chap. 3, no. 1, p. 20: "In nostris furnibus legimus, non esse in natura certius sciendi genus, ad cognoscendum per causas radicales, ac constitutivas rerum; quam dum scitur quid, quantumque in re quaque, sit contentum."
95. Van Helmont, *Opuscula, De lithiasi,* chap. 1, no. 2, p. 10.

the indestructibility of matter."[96] This is indeed a genuine Helmontian doctrine, but one that actually originated with the pre-Socratics and was repeated by virtually every Greek philosopher up to Aristotle. Much more significant is Van Helmont's principle that *weight,* along with matter, is always conserved, although the significance of this distinction might be easily missed by moderns inured to the Newtonian synonymity of matter and mass. As Van Helmont puts it, "Nothing comes into being from nothing. Hence weight comes from another body weighing just as much."[97] What Van Helmont has in mind is what we would call in modern terms "mass balance," that is, the mass that goes into a reaction must also come out at the other end, regardless of any transformations that have taken place.[98] This was not a conception that a strict follower of Aristotle could have maintained.

Van Helmont argues for the conservation of weight in the context of his belief that all things are made from water. He points out that a Scholastic opponent might reply that water (which cannot be compressed and is therefore nonporous) could not possibly be the substratum of gold, because the precious metal is vastly heavier (i.e., denser) than water. How can gold acquire its great weight without violating the Aristotelian rule that "two bodies cannot occupy the same place"? Van Helmont replies first with an ad hominem argument—the Scholastic himself can have no answer to the question. The Aristotelian theory of the four elements postulates that fire and air have no weight, but always rise to their "natural places," while earth and water, being absolutely heavy, always sink to theirs (*De caelo,* book 4, 310a–312a); therefore, since the Scholastic theory of mixture asserts that the four elements are always combined in mixts, it follows that gold should be lighter than pure earth and water owing to its incorporation of fire and air. In fact, however, gold is "ten times heavier" than the two "heavy" elements.[99] Following this argument, Van Helmont then asserts

96. J. R. Partington, "Joan Baptista van Helmont," *Annals of Science* 1 (1936): 368.

97. Van Helmont, *Ortus,* "Progymnasma meteori," no. 18, p. 71: "Ex nihilo nihil fit. Pondus ergo, ex alio corpore, tantundem ponderante, fit."

98. We do not mean to imply that Van Helmont had an articulated concept of mass in the modern sense, or sophisticated Newtonian conceptions such as "inertial mass." Yet he clearly did believe that the gross weight of initial and final ingredients remained constant, and he explicitly linked this fact with the indestructibility of matter.

99. The decimal proportion adopted here probably has nothing to do with actual measurements of specific weight, but rather with the Scholastic principle that air is "ten times" denser than fire, water "ten times" denser than air, and so on. (See Aristotle, *De generatione et corruptione,* II 6 333a16–23 for the lemma behind this Scholastic discussion. We thank John Murdoch for this reference.) Since Van Helmont is employing an ad hominem argument, he adopts the Scholastic proportion. See Van Helmont, *Ortus,* "Progymnasma meteori," no. 18, p. 71.

that the Aristotelian position forbidding interpenetration of bodies is simply wrong and that water, when acted upon by *semina,* actually *does* interpenetrate itself so that "gold is produced from water when sixteen parts of the water are compressed into the place of one."[100] Hence the weight of the water is exactly preserved in the gold, despite the vast increase of density.

It is highly significant that in his imagined debate with a Scholastic, Van Helmont points to the difficulty that the Aristotelian theory of elements with "natural places" poses for laboratory chymistry. If the changes wrought on matter by chymistry result in a transmutation of elements, then an element that is absolutely heavy could be converted into one that is absolutely light within the chymist's flask. Such elemental transmutations were the daily bread of the Scholastics, for Aristotle himself envisioned even the evaporation of water as a transmutation of that element into air.[101] If we accept the reality of such transmutations, then conservation of *weight* becomes an obvious impossibility, even though no *matter* is destroyed. Under such circumstances the balance would be totally useless for comparing the initial and final products in the laboratory. But Van Helmont simply denies the Aristotelian transmutation of elements once and for all; for him, the very notion "element" implies simplicity and permanence. An element could neither be reduced into something simpler nor be transmuted into another element.[102] Moreover, Van Helmont retains only water out of the Aristotelian quaternary as a constitutive element—he asserts that fire is no element, but merely an "artificial death of things," and that air, while primordial, cannot combine genuinely with other substances, and that earth is not an element but a product of water.[103] Thus all generation and corruption in the world is really due to the action of *semina* on water. By arguing that there is no real transmutation of elements and that all substances are merely water that has undergone rarefaction or condensation, that is, "interpenetration of dimensions" by the action of *semina,* Van Helmont is able to maintain the conservation of weight and the concept of mass balance.

100. Van Helmont, *Ortus,* "Progymnasma meteori," no. 18, p. 71: "Erunt ergo sedecim aquae partes, in unius locum compressae, ubi ex aqua, aurum constituitur." The same argument is resumed in Van Helmont, *Ortus,* "Natura contrariorum nescia," no. 32, p. 172.

101. Aristotle, *De generatione et corruptione,* book 2, 338b5 ff.

102. Van Helmont, *Ortus,* "Progymnasma meteori," no. 7, p. 68: "Quod in elementis, aqua, & aere, non contingit, eo quod ob summam sui simplicitatem, & destinationis prioritatem, recuset in aliquid prius, aut simplicius migrare, aut transmutari." See also pp. 67, 108, 172–73, 689, and passim.

103. Van Helmont, *Ortus:* on fire, "Complexionum atque mistionum elementalium figmentum," no. 34, p. 109, and "Arbor vitae," p. 798; on air, "Paradoxum alterum," no. 2, pp. 688–89; on earth, "Elementa," no. 12, p. 53.

Indeed, as we shall see, throughout his works Van Helmont extensively uses the principle that the initial weight of ingredients and final weight of products must balance. Van Helmont sometimes uses this notion as an implicit principle and sometimes as a tool for probing experimental results. But before we can consider these examples, we must confront the terminological difficulties associated with the concept of weight itself. These difficulties, if left unresolved, can too easily mask Van Helmont's intent.

QUANTIFICATION AND THE PROBLEM OF WEIGHT IN VAN HELMONT

Understanding Van Helmont's discussion of weight—like various other of his discussions—is sometimes made difficult by his tendency to present traditional and fabulous hearsay as experimental evidence. Like most writers of the early modern period, he did not clearly distinguish between what we would call "experience" and "experiment," and he frequently uses untested phenomena for rhetorical effect.[104] His acceptance of bizarre facts drawn from the daily currency of marvels led him to accept the extravagant notion that the heart withdrawn from an enchanted horse could be used against the witch who had caused the injury, if it was transfixed with a nail.[105] Other examples of *experimenta* include Van Helmont's transmission of the belief that rubbing oneself with the fat of a sea-calf will give protection from thunderbolts, and his claim that if "someone defecates at your door, and you wish to keep this from happening again, apply a hot iron to the fresh excrement; the defecator will presently get scabies on his rump from the magnetism."[106] Yet, as Robert Halleux has noted, Van Helmont used other terms, notably "hands-on demonstration" *(mechanica probatio)* and "questioning by means of fire" *(quaerere per ignem)* to refer to actual experiments carried out by himself in a laboratory.[107] Yet the borderline between what Van Helmont accepts on authority and what he has discovered or tested himself is not always clear. This leads to the practical problem for historians that whenever a passage from Van Helmont's work is obscure or

---

104. Robert Halleux, "Theory and Experiment in the Early Writings of Johan Baptist Van Helmont," in *Theory and Experiment,* ed. Diderik Batens (Dordrecht: Reidel, 1988), 93–101.

105. Van Helmont, *Ortus,* "De magnetica vulnerum curatione," no. 110, p. 769.

106. Ibid., no. 21, p. 752. See Halleux, "Theory and Experiment," 96.

107. We prefer the translation "hands-on demonstration" to Halleux's "mechanical demonstration," since it is clear that *mechanica* did not necessarily refer to machines. Van Helmont uses it, rather, in the sense of the "mechanical arts"—that is to say, "techniques" or "practices" as opposed to theory. Similarly, we prefer "questioning" to Halleux's "searching," since it is probable that by using the term *quaerere,* Van Helmont wants to contrast his form of interrogation with the Scholastic *quaestio* used in university disputation. Cf. Halleux, "Theory and Experiment," 96.

difficult, some modern commentators have been too ready to dismiss it as an artifact of excessive credulity or lack of judgment. This tendency appears clearly when we consider some of Van Helmont's comments on weight.

The *Ortus medicinae* describes an experiment for weighing water and ice, which seems at first glance to be obviously erroneous. The purpose of the experiment is to throw doubt upon the Scholastic theory that water can be transformed into air, and to do so it emphasizes the permanency of the water even as it undergoes changes of state.

> Fill a large glass vessel with pieces of ice, and seal the neck hermetically, that is, by melting the glass there. Put this vessel on a balance, with a weight on the opposite side, and after the ice melts, you will see that the water will be heavier than the ice by almost its eighth part. Since this can be performed with water a thousand times without changing its weight, it cannot be said that any part of it can be turned into air.[108]

What are we to make of this seemingly wrongheaded experiment? It was "disproven" by Robert Boyle in 1665 and several modern commentators have dismissed it as an example of flagrant error.[109] Yet, as T. S. Patterson

---

108. Van Helmont, *Ortus*, "Gas aquae," nos. 34–35, p. 79:

Imple lagenam vitream & magnam, frustris glaciei, collum vero claudatur sigillo Hermetis, id est, per vitri ibidem liquationem. Ponatur haec tum lagena, in bilance, adjecto pondere, in oppositum, & videbis quod propemodum octava sui parte, aqua, post resolutam glaciem, erit ponderosior seipsa glacie. Quod cum millesies ex eadem aqua fieri possit, reservante semper idem pondus, dici non potest, quod ejus pars aliqua in aerem sit versa.

109. Boyle's "disproof" of the experiment, found in his *Experiments and Observations touching Cold* of 1665 (*Works*, 4:415–16), is discussed by T. S. Patterson, "Van Helmont's Ice and Water Experiments," *Annals of Science* 1 (1936): 463–64. Note also that Boyle attributes the "levity of ice" to the bubbles it contains rather than to the increase in its volume relative to liquid water. Modern writers who have seconded Boyle's opinion about the "unintelligibility" of Van Helmont's experiment include de Waard, *Correspondence*, 3:71, and Halleux, "Theory and Experiment," 94. Halleux bases himself partly on a passage in Van Helmont's *Dageraad* (Rotterdam: Naeranus, 1660), 217–18, which is embedded in a discussion of flatulence, and how it is generated by acid from undigested humors. Van Helmont's reason for introducing the ice experiment here is to show that increase in the volume of a substance by expansion (and hence its reduced specific weight) does not entail the generation of new matter, as made evident by the floating of witches, bodies killed by poison, and the like. Although he does seem at first to say that the vessel in which the ice and water are weighed becomes heavier, he immediately clarifies himself and says that it is only the water, rather than the whole vessel that ac-

has argued, the difficulty can be resolved if we assume that Van Helmont is not referring to an augmentation of what we would call "mass" when the ice thaws, but rather an increase in *specific weight*, which Van Helmont would have measured by the change in volume between the frozen and liquid water. Patterson's explanation appears to be the correct one for several reasons. First, there are other cases where Van Helmont indisputably discusses specific weight while using the Latin *pondus* or the French *poix* (for *poids*). In his *Opuscula medica inaudita*, for example, Van Helmont describes his careful attempts to determine the densities of urine taken respectively from an old man, a healthy woman of fifty-five, a youth of nineteen when healthy and when he was no longer healthy but suffering from a double tertian fever and had drunk little the night before, another healthy youth "abstemious of drink," and a man of thirty-six suffering from tertian fever and a cough. The experimental technique employed by Van Helmont is revealing. First he placed the vessel in which all the urines would be weighed on a balance, and found it to weigh 1,354 grains (Van Helmont tells the reader carefully that in his system of measurements, 600 grains equals one ounce). Then he filled the vessel with rainwater and found the weight of the water and the vessel to be 4,670 grains. An equal volume of the old man's urine, weighed in the same vessel, was found to come to 4,720 grains (50 grains more than the rainwater, as noted in the text). The urine of the nineteen-year-old weighed 4,766 grains when he was healthy and 4,848 when he was unhealthy, giving a difference of 82 grains, as Van Helmont notes. The woman's urine, finally, weighed 4,745 grains, the abstemious youth's 4,800, and the thirty-six-year-old man's 4,763. What

---

quires "weight," leaving open the interpretation in terms of specific gravity. We reproduce the passage as follows:

> Eyndelijck de suerte werckende, maeckt wint, sy moeten wercken op alle het gene niet volkomen versuert en is; dus de suerte, niet verwesent in de twaelf duymigen, en eersten darm, (soo namaels sal geleertworden) en vindende eenig onverduwt rot sap, maeckt den buyckwint, naer den aert van 't selve onverduwt lichaem. Dit is genoeg van soo vuyle dingen geschreven. Een glas met ys gevult, den hals door de lamp toegesmolte, en gestelt in enn weegh-schael, en sijn effen wicht in de andere schael, latende het daer soo hangen, tot dar het ys van selfs doyt, wort het glas, dat is het water, swaerder bevonden, dan doen het ys was; de reden heb ick geschreven in t'boeck, waerom alleen alle tovressen altoos naturelijck drijven, met de helft hares ronts buyten water, cap. De weegh-konst des waeters. Alle menschen, gestorven door tovereye oft gift, drijven; item, alle gift (dit dient den Rechteren) 'twelck merckelijck doet swellen, en des te hooger opdrijven. Het swellen dan en is niet een verkrijgen van nieuwe stoffe, maer wel eenen sueren heve, waer door onse sap sich verheft, als een broodt door den heve.

clearly interests Van Helmont in all these cases is the increase or decrease of weight in a fixed volume of urine—in other words, the specific gravities of different urine samples. Nowhere, however, does he use any term other than "to weigh" *(pendere)* and "weight" *(pondus)*. Nor was it necessary for him to do so, since his method of determining specific weight in each case depended on keeping the volume of liquid constant while observing the difference in mass between the urine sample and rainwater. But of course the absence of a distinguishing term for "specific weight" as opposed to "weight" in general leads to considerable confusion for the modern reader. This becomes quite clear at the end of the passage cited, where Van Helmont explicitly notices that warm urine "is always lighter by a few grains" than cold urine and immediately notes that this is because it occupies more volume *[ut & extensior]*.[110] Clearly Van Helmont is saying that the *density* of urine varies with its temperature, as he also observed in the case of ice when compared to water. In both cases, he has no distinct terminology for specific weight as opposed to absolute weight.

Van Helmont's careful determination of specific weights occurs frequently in his studies. In his extensive 1630–31 correspondence with the French Minim and natural philosopher Marin Mersenne, Van Helmont makes many comments on specific weight. At one point he even chides Mersenne, who is famous as a champion of "mechanism," for his failure to employ Simon Stevin's technique for determining specific weights "by means of water."[111] In addition, Van Helmont gives his own determinations of the specific weights of the seven metals known to him. His technique consists of adding a quantity of the least dense metal, tin, to a known volume of water, weighing it, and marking the increased level of the water in the vessel. This allows him to use the weight of that volume of tin as a standard unit against which to measure the weights of other metals that are needed to make the water attain the same level in the same vessel. By this means, the metal samples will all have the same volume, but will differ in weight according to their density. Using this method, Van Helmont ranks the seven metals in order of increasing density: tin, iron, copper, silver, lead, mercury, and gold. In fact, he even gives values for their specific weights. Hence the amount of iron required to fill up the same volume of water as the tin is $1\frac{1}{12}$ times the weight of the tin, the copper $1\frac{2}{9}$, the silver $1\frac{3}{8}$, the lead

110. Van Helmont, *Opuscula, Scholarum humoristarum passiva deceptio,* chap. 4, no. 31, p. 108: "Urina tepens, semper frigida, paucis granis levior, ut & extensior."

111. Van Helmont to Mersenne, 21 February 1631, in de Waard, *Correspondence,* 3:114–15: "Quaestio praesupponit falsum. . . . In summa ars ponderandi per aquam est certissima et longe per[i]tior quam quae in aere fit ob causas renitentiae, alias explicat[i]s. Vide super hac Stevinus *de Arte ponderandi per aquam.*"

$1\frac{5}{8}$, the mercury $1\frac{7}{8}$, and the gold $2\frac{2}{3}$.[112] If we compare these values to those given in modern manuals of physical constants, we find that Van Helmont's method allowed him to attain values that differed from the modern ones by an average of less than 2 percent.[113] Clearly he was quite competent in determining specific weights, regardless of having no term by which to distinguish specific weight from weight per se.

All of this clarifies Van Helmont's ice experiment, in which he says that "the water will be heavier [*ponderosior*] than the ice by almost its eighth part." As we have seen, Van Helmont uses *pondus* for both specific and absolute weight; thus he is probably referring to specific weight here. Modern experiments show that water's specific weight is greater than ice's; molten ice occupies only about 91.5 percent of its original volume. It is presumably this change that he saw. But instead of noting differences in volume for a constant weight, Van Helmont normally notes differing weights for a fixed volume, as in his urine and metal measurements. Thus he expressed the observed change of volume by calculating what the change of weight would have been at constant volume. He stated the relative weights of equal volumes of water and ice by saying that the water weighs $1/.915$ times as much as the ice, or that the water is *ponderosior* by about one-eleventh. The difference between the modern value of one-eleventh and Van Helmont's one-eighth may be due to his using pieces of ice; the empty spaces increased the apparent volume, making the change seem greater.[114] But a question remains: if Van Helmont's method relied on direct observation of the *dif-*

112. Van Helmont to Mersenne, 30 January 1631, in ibid., 3:56–57. De Waard or his printer has clearly erred in giving Van Helmont's value for iron as "$1\frac{7}{12}$" since this value would have made the iron almost as dense as lead and would have altered the order of the specific weights as given by Van Helmont. We conjecture that a "1" has been misread as a "7" (an easy mistake in seventeenth-century hands), and that the value meant by Van Helmont is $1\frac{1}{12}$, which differs from the modern value by less than 1 percent.

113. David R. Lide, ed., *CRC Handbook of Chemistry and Physics,* 74th ed. (Boca Raton: CRC, 1993), 4-1 to 4-34. The *CRC Handbook* gives the following specific weights: tin 7.31, iron 7.87, copper 8.96, silver 10.50, lead 11.35, mercury 13.55, and gold 19.3. By expressing these densities on a scale where tin = 1.0 (as Van Helmont did), it is easy to judge the accuracy of Van Helmont's determinations (shown in parentheses): iron 1.08 (1.08), copper 1.22 (1.23), silver 1.44 (1.38), lead 1.55 (1.63), mercury 1.85 (1.88), and gold 2.64 (2.67). The value for lead is least accurate, diverging by about 5 percent.

114. Patterson ("Ice and Water Experiments," 465) thought that Van Helmont's value of one-eighth referred directly to the observed difference in volume between a fixed weight of water and ice, but this seems unlikely given Van Helmont's usual practice of using *pondus* to refer to specific weight and not to volume. Patterson also somewhat inaccurately cites the volumetric change of ice melting into water as one-ninth rather than the accepted value of between one-eleventh and one-twelfth. Ice has a density of .915 g/ml versus water (which is the modern standard) at 1.0 g/ml; thus a liter of ice melts into 915 milliliters of water, or conversely a liter of water freezes into 1093 milliliters of ice.

ference in volume between the ice and water, why did he seal the vessel containing the ice, and why did he put it in a balance once sealed? That he specifies the use of a balance has made some commentators assume that he "observed" an increase in weight when the ice melted; Patterson does not try to explain this. But the explanation is simple: Van Helmont intended this as a *demonstration* experiment. He needed to show that nothing entered or left the flask and that no factor other than change of density was involved. This is why he sealed the vessel. Placing the vessel on a balance was simply a further control to reinforce the fact that nothing had entered or left the flask. The balance was not intended to reveal a change in weight, but the opposite; it was the *absence* of weight change that the balance showed and that was key to Van Helmont's findings.

QUANTITATIVE SPAGYRIA

This reconsideration of the ice experiment demonstrates that one must be careful not to dismiss Van Helmont's experimental results too brashly. Similar problems arise in the topic of quantitative measurements applied to analysis and synthesis, the *spagyria* of the Paracelsians. While Paracelsus occasionally provided recipes with quantitative measurements, he did not emphasize the weights of substances subjected to analysis by fire or menstrua.[115] Nor have we found much evidence that he was concerned with the resynthesis of materials that he had subjected to *Scheidung* (division, analysis). The *Archidoxis,* for example, presents example after example of substances that have been separated by the mineral acids into layers corresponding to the four elements, fire, air, water, and earth. In each case Paracelsus notes the color of the given "element," but makes no effort to weigh it; nor does he try to recombine the layers to arrive at his original material.[116] Although Paracelsus did argue elsewhere, as in his *De renovatione et restauratione,* that metals could be analyzed into their three principles and resynthesized from these components, he was far more concerned with the medical benefits to be gained from purification by *Scheidung* than he was with the demonstrative power provided by their recombination.[117] This was not the case with Van Helmont.

115. Examples of Paracelsus's recipes can be found in his *Liber praeparationum,* in *Sämtliche Werke,* ed. Karl Sudhoff (Munich: Oldenbourg, 1930), 3:309–59.

116. Paracelsus, *Archidoxis,* in *Sämtliche Werke,* 3:91–200; cf. 102–17. See also T. P. Sherlock, "The Chemical Work of Paracelsus," *Ambix* 3 (1948): 43–62.

117. Paracelsus, *De renovatione et restauratione,* in *Sämtliche Werke,* 3:203–4. See Ernst Darmstaedter, "Arznei und Alchemie: Paracelsus-Studien," *Studien zur Geschichte der Medizin* 20 (1931): 4, 56–57.

Early in the *Ortus medicinae,* Van Helmont devotes a chapter to the Aristotelian element earth. Although he is willing to accept that there is a hidden sand below the surface of the globe, called *Quellem,* Van Helmont denies categorically that this enters into genuine mixtures with other elements. Indeed, as we have noted, he dismisses altogether the Scholastic concept of mixture, according to which the four elements were supposed to combine to form a perfectly homogeneous new substance. For Van Helmont every substance is water transmuted by the power of *semina.*[118] But Van Helmont recognizes the possible objection that glass is a genuine mixture (in the Aristotelian sense) made by fusing sand with alkaline plant ashes—the resultant product has none of the properties of the starting materials and thus seems a wholly new substance. Yet he replies rather surprisingly that glass cannot be a genuine mixture since it is not really homogeneous, despite appearances to the contrary. For, "by means of art glass returns into its original ingredients *(pristina initia)* once the bond holding them together is broken: the sand can even be regained in the same number and weight [as before]."[119] Van Helmont explains his basis for this unlikely sounding claim a few lines later:

> If one melts a fine powder of glass with a large amount of alkali and exposes it in a humid place, one will presently find that all the glass dissolves into a water. If *chrysulca* [mainly nitric acid] is poured on in a quantity sufficient to saturate the alkali, one will at once find that the sand sinks to the bottom [of the vessel] in the same weight as it was before it was used in making the glass.[120]

In modern terms, what Van Helmont has observed is the production of potassium and/or sodium silicates by fusing powdered glass (made from sand, which is mainly silicon dioxide) with alkali, probably soda or salt of tartar (sodium or potassium carbonate, respectively). The hygroscopic alkali silicate is allowed to deliquesce and is then combined with nitric acid, resulting in the formation of potassium or sodium nitrate and the precipitation of silicon dioxide in a quantity identical to that of the sand employed originally.

---

118. Newman, *Gehennical Fire,* 141–46, 151–58.

119. Van Helmont, *Ortus,* "Terra," no. 14, p. 56: "ita per artem denuo, resoluto vinculo, ad pristina redit initia, adeo ut eadem numero, & pondere arena inde prorsus eliciatur."

120. Ibid, no. 16, p. 56: "si vitri pollinem, pluri alcali colliquaverit, ac humido loco exposuerit; reperiet mox totum vitrum, resolvi in aquam: cui si affundatur Chrysulca, addito, quantum saturando alcali sufficit, inveniet statim in fundo, arenam sidere, eodem pondere, quae prius, faciundo vitro aptabatur."

This is a highly workable experiment, and it served Van Helmont well in his attempt to disprove the Aristotelian theory of mixture. The fact that he was able to retrieve his sand in its original weight seemed to support his view that the sand was present in the glass all along, in the form of minutely divided bits. Since the Aristotelian theory postulated that a perfectly mixed body had to be absolutely homogeneous, a glass composed of the juxtaposition of minute particles of sand and ash could be no such mixture, despite appearances.[121]

This example also presented a second challenge to Van Helmont's Scholastic opponents; since the same sand that went into the glass came out at the end, the initial and final sand were identical—not just the same in *species,* but also in *number.* Here Van Helmont implicitly responds to a long Scholastic commentary tradition on book 2 (338b10–20) of Aristotle's *De generatione et corruptione,* which claimed that transmuted substances could not be regained in number, but only in species. By this Scholastic reasoning, air that had been transmuted into water could be transmuted back into air, but it would not be the *same* air; rather it would be *new* air that had been produced afresh from the water. If a Scholastic argued that the ingredients of the glass had lost their identity and been transmuted into glass, how could he account for the fact that Van Helmont retrieved exactly the same amount of sand as had been used in making the glass? Had the sand been regenerated de novo from the four elements? This would seem highly improbable, given that the sand used in making the glass was created naturally by geological means, whereas the sand derived from the glass was "created" by fusion with alkali and treatment with nitric acid. Once again mass balance allows Van Helmont to argue that the sand undergoes no substantial transmutation but only loses its granular appearance in the glass because of division and superficial apposition of individual bits of sand. Thus Van Helmont's quantitative synthesis and analysis of glass gave him a powerful tool for disputing the assumptions of Aristotelian mixture.

A similar attempt to dispute experimentally occurs in the one Helmontian experiment that every historian of science knows, namely the production of a willow tree "solely out of water." This experiment has received more press than all the myriad others in Van Helmont's work combined, in part because it was treated at length by Robert Boyle, and in part because it raises interesting issues of priority.[122] Its origins have been traced back to the fifteenth-century writer Nicholas of Cusa, who probably exercised a direct influence on Van Helmont, and it even has antecedents in the *Recog-*

121. See Aristotle, *De generatione et corruptione,* 327a–328b.
122. Halleux ("Theory and Experiment," 99 nn. 11–12) gives a partial list of the scholarship devoted to the willow tree experiment, but it must be pointed out that virtually every treatment of Van Helmont describes it.

nitiones of the late antique pseudo–Clement of Alexandria.[123] What is important to us is not the issue of originality, but rather the fact that the willow tree experiment, as presented by Van Helmont, gives a clear example of his quantitative technique. The author tells us that he took an earthen vessel in which he placed two hundred pounds of dried earth. He made a special collar of perforated, tinned steel that went over the top of the pot so that no dust could enter therein. He then planted a willow sprout weighing five pounds in the pot. After five years of watering this plant with rainwater and water distilled for purity, Van Helmont found that the uprooted tree weighed 169 pounds and about three ounces, not including the weight of the leaves that fell from the tree over its lifetime. Van Helmont then dried the earth in the pot and reweighed it. Since the earth still weighed almost two hundred pounds (less two ounces), Van Helmont reasonably concluded that the increase of "164 pounds of wood, bark, and roots had arisen solely from water."[124]

While Van Helmont does not claim in the *Ortus* that he then analyzed the components of the tree that had been synthesized from "water alone," the experiment is immediately followed by another where animal fat is "reduced to water." The demonstration consists of taking fish, which Van Helmont assumes to be nourished on water alone, and extracting their fat. Here, as elsewhere, Van Helmont argues that fat or oil can be converted to water by passing through the intermediate stage of soap. By reacting fats and oils with salt of tartar or another alkali, Van Helmont was of course able to produce soaps. He then distilled his soaps to separate water (and glycerin). Van Helmont does not present a detailed quantitative analysis of this soap or make the claim that all the fat becomes water: instead, "almost all" *(paene totus)* returns to water.[125] The failure to perform a complete and "aequiponderant" analysis into the original ingredient could be excused,

123. Halleux, "Theory and Experiment," 93. It would be hard to believe that Van Helmont had not at some point been exposed to Nicholas of Cusa's *Idiota de staticis experimentis.* The work not only lays out an ur-form of Van Helmont's willow tree experiment, but also (albeit in very cursory fashion) advises that the specific gravities of different urine samples be found, suggests that the Mercury and Sulfur in metals be determined by weight, and notes that ice is less dense than water. See Nicholas of Cusa, *Nicolai de Cusa opera omnia* (Hamburg: Felix Meiner, 1983), 5: 230, 222–23, 229, 234. See also Hebbel E. Hoff, "Nicolaus of Cusa, van Helmont, and Boyle: The First Experiment of the Renaissance in Quantitative Biology and Medicine," *Journal of the History of Medicine and Allied Sciences* 19 (1964): 99–117; and Pagel, *Joan Baptista Van Helmont,* 55.

124. Van Helmont, *Ortus,* "Complexionum atque mistionum elementalium figmentum," no. 30, p. 109. A similar experiment is reported in the *Dageraad,* 60, but with quite different measurements for the earth and the tree (three pounds for the sprout, forty-nine pounds six ounces for the grown tree, and thirty-seven pounds for the earth before and after the experiment).

however, by the fact that Van Helmont had been forced to add "additaments," such as the salt of tartar, in order to carry out the chain of processes terminating in the return to water. The presence of these unspecified additional ingredients would obviously throw the balance of initial ingredients and final products out of kilter, unless the additaments could be extracted from the analyzed fat before a final weight determination was made.

THE EXANTLATION THEORY AND VAN HELMONT'S MERCURIAL ALUM

While the foregoing examples of Van Helmont's emphasis on quantitative experimental method might be deemed "successful" in modern terms, some of the most emphatic examples of his faith in quantification appear, paradoxically, in cases where the results violated the logic of common sense. The most obvious of such instances occurs in his explanation of a phenomenon that he calls "exantlation" (from the Greek *exantlein*, "to pump out"). Exantlation refers primarily to the loss of activity that corrosives such as acids suffer when they act on another substance. For example, a given quantity of nitric acid can dissolve only a certain amount of silver, after which time it is used up and the residual fluid is no longer corrosive. But Van Helmont did not think of the corrosives as being neutralized by going into combination with some other substance; instead, they gradually became enfeebled in the same way that an animal might become tired after exerting itself (although the corrosive, unlike an animal, would not become reinvigorated by repose). Now what was it, we might ask, that led Van Helmont to this counterintuitive result, other than his general tendency to favor the action of internal principles of activity (such as *semina*) over external efficient causes? In part he was driven by his rejection of a prevailing chymical theory that acids acted merely by grinding substances into bits so small that they were no longer visible in a solution. He points out that this theory does not account for the lessening of the acid's strength, nor does it explain why some acids "congeal" in the very act of dissolving another substance.[126] But we are still left with the obvious question: why did Van Helmont create the theory of exantlation rather than simply arguing that the particles of acid went into combination with those of the solute? After all, he had a workable corpuscular theory, which he employed to debunk the belief that vitriol springs actually transmuted iron into copper. Part of the answer is located in an experiment that Van Helmont mentions many times

---

125. Van Helmont, *Ortus*, "Complexionum atque mistionum elementalium figmentum," no. 32, p. 109.

126. Van Helmont, *Ortus*, "Ignota actio regiminis," no. 11, p. 333.

throughout his work; indeed, his return again and again to this example indicates its importance for his thought:

> If you distill oil of vitriol from running mercury, the oil [i.e., the sulphuric acid] is coagulated with the mercury, and they both remain in the bottom in the form of a snow. Whatever you distill thence is mere water. But this snow, if it is washed, beomes a yellow powder, which is easily reduced into running mercury in just the same weight as before. But if you distill off the wash water, you have a pure alum in the bottom, from the acid salt of vitriol. Therefore dissolvents are mutated even if the dissolved lose nothing of their substance or matter.[127]

Van Helmont's experiment is easy to follow in chemical terms. He treats a quantity of mercury with sulphuric acid and then distills the mixture, leaving behind a white, snowlike solid—which we would now call mercuric sulphate. Since the acid has reacted with the mercury, the distillate is no longer corrosive and is composed only of the water present in the original sulphuric acid. Van Helmont then washes the "snow" with fresh water, and observes that it turns bright yellow. What he has observed (in modern terms) is the hydrolysis of the white mercuric sulphate ($HgSO_4$) into a yellow basic sulphate ($HgSO_4 \cdot 2HgO$, often called *turbith minerale* in early modern pharmacopoeias). This hydrolysis liberates a quantity of sulphuric acid into the wash water. Now although the white sulphate is unstable and the yellow basic sulphate insoluble in neutral water, the liberated sulphuric acid allows a portion of the mercuric sulphate to dissolve in the wash water. Thus, when this wash water is distilled off, Van Helmont finds a fine crystalline residue, which he calls "alum," but which is in reality the redissolved mercuric sulphate. (The use of the term "alum" perhaps comes from the astringent taste of this product or the appearance of its crystals.) Finally, Van Helmont reduces the yellow mercuric sulphate back into mercury (possibly by heating with charcoal or salt of tartar), and claims to have gotten all his original quicksilver back.

Obviously, Van Helmont erred at some point in his measurement of the weights involved, as part of the weight of original mercury would be in the "alum." Possibly he washed the mercuric sulphate "snow" with a very large

---

127. Van Helmont, *Opuscula, De febribus,* chap. 15, no. 20, p. 57:

Si oleum vitrioli distilles a Mercurio currente, oleum coagulatur cum Mercurio, manéntque ambo in fundo, forma nivis. Et quidquid inde stillatur, est mera aqua. Nix ista autem, si lavetur, fit citrinus pulvis, qui facilè in pristinum mercurium currentem reducitur, eodem prorsus pondere, quo ante. Si vero aquam ablutionis distillaveris, habes in fundo merum alumen, ex acido sale vitrioli. Sic nempe dissolventia mutantur, tametsi dissoluta, nequidquam de sui materia, aut substantia perdiderint.

amount of water, which would have kept the concentration of sulphuric acid in the wash water very low and thus able to dissolve only a correspondingly small quantity of the mercuric sulphate. If so, relatively little of the initial mercury would have been lost. This would also help explain Van Helmont's failure to identify the "snow" and the "alum" as being chemically identical. If the sulphuric acid is sufficiently dilute, the crystals of mercuric sulphate will be large and will actually resemble those of alum (potassium aluminum sulphate) and thus be very unlike the powdery "snow" initially formed.

Regardless of the apparent inaccuracy regarding weights, the surprising part of the process is the conclusion that the solvent (sulphuric acid) has congealed of its own accord into an alum, without going into any real combination with the mercury. This, in essence, is the doctrine of exantlation. As Van Helmont points out elsewhere, mercury can be dissolved in nitric acid *(spiritus aquae fortis)* instead of sulphuric, but in that case, no snow is formed. Hence he concludes that the formation of the snow cannot be a "proper action of the mercury, but of the spirit of vitriol, altering itself differently and according to its innate propensity toward various objects of its own accord."[128] Employing a metaphor drawn from marriage, as he is fond of doing, Van Helmont states that the acid becomes thickened and enfeebled by the "dotal unfolding" of its powers, rather then by combining with the mercury. So sure of this is Van Helmont that he exclaims that the coagulation of the oil of vitriol into "alum" could be done a thousand times with the same mercury, since it loses nothing of its substance or weight in the process.[129] All of this is accomplished without material intermixture; rather, as Van Helmont claims, there is a marvelous "radial commixture" of the mercury and the oil of vitriol, in that this immaterial communication is what stimulates the acid into coagulating. Such "radial commixture" also accounts for the fact that a given weight of mercury left in water for a time would extend its virtue of killing worms to the water and yet could still be recovered without any (apparent) loss of weight. According to Van Helmont, the mercury acts upon the water by immaterial means without losing any of its substance.[130]

Although the alum experiment is adduced many times by Van Helmont as evidence of the exantlation of acids, he has other evidence as well. He ad-

---

128. Van Helmont, *Ortus,* "Ignota actio regiminis," no. 11, pp. 333–34: "seipsum sponte propria disponentis diversimode, juxta innatam propensionem, ad varia objecta, differenter seipsum mutantem."

129. Van Helmont, *Ortus,* "In verbis, herbis, et lapidibus est magna virtus," p. 576.

130. Ibid.

vises, for example, that one make a vitriol of iron (iron sulphate) by dissolving a pound of crocus of iron (probably iron oxide) in six times that amount of oil of vitriol. Upon boiling off the liquid, one will have vitriol of iron. Van Helmont then advises the reader to "take away the iron, and you will [still] have vitriol of iron. I mean a salt like vitriol, in which there is the taste of iron, but retaining no iron."[131] Although the command to "take away the iron" is unclear, a possible explanation of this experiment may be that Van Helmont's initial dissolution of the crocus, followed by evaporation, led to the production of ferrous sulphate ($FeSO_4$), which then oxidized under the influence of atmospheric oxygen to ferric sulphate ($Fe_2(SO_4)_3$), with a corresponding precipitation of rusty iron oxide and loss of color in the solution. If the iron precipitate were separated and the ferric sulphate solution congealed by evaporation, Van Helmont would indeed still have a crystalline "vitriol," despite the clearly observable release of a rusty iron-containing powder from the solution. It is probably significant that he gives no information about the respective weights of the products, for again, as in the case of the mercurial alum, the vitriol supposedly produced by mere exantlation of the acid would in reality contain some metal.

Van Helmont's exantlation theory may seem exotic from the perspective of modern chemistry, but it illustrates the faith he held in regard to quantitative approaches to practical experimental transformations and the use to which he put them. As in the case of the synthesis and analysis of glass and the willow tree experiment, the exantlation experiments were intended to prove a point. In all these cases, Van Helmont was attacking the "artificiality" of Aristotelian physics, with its emphasis on external efficient causes and denial of *semina*. It is the latter, of course, that are directly responsible for the "radial activity" of the mercury and for the autonomous response of the acid in coagulating itself. Above all, Van Helmont wants to disprove the Aristotelian theory that such phenomena are due to a mixture of the four elements, with their qualities of hot, cold, wet, and dry acting and reacting upon one another.[132] The alum experiment, with its appeal to the initial and final weights of the mercury involved, illustrates the way in which mass balance was fully integrated into Van Helmont's chymistry.

131. Van Helmont, *Opuscula, De lithiasi,* chap. 4, no. 13, p. 35: "Libram croci Martis necte sextuplo olei vitrioli, tum distilla, quicquid fuerit aqueum. Reperies vitriolum ferri. Aufer inde ferrum: & habes vitriolum ferri. Salem inquam, instar vitrioli, cui est gustus ferri. Nil retinens tamen de Marte."

132. Van Helmont, *Ortus,* "Ignota actio regiminis", no. 11, pp. 333–34; and Van Helmont, *Ortus,* "In verbis, herbis, et lapidibus est magna virtus," p. 576.

THE HELMONTIAN THEORY OF FIXED ALKALIES

Let us now consider a final case of Helmontian chymical experiment, where, as in the example of exantlation, a particular theory is built upon laboratory results. Here our concern will be the class of substances that Van Helmont calls "alkalies." In general, alkalies were made by incinerating combustible matter (such as wood, leaves, or the tartar found in wine barrels), leaching the resulting ashes, and evaporating the extract (or "lixivium") to provide the alkali salt. Since alkalies are also salts, questions regarding their production led directly to the question of whether or not they preexist in the combustible bodies. As is well-known, Van Helmont did not uncritically accept the Paracelsian theory that a preexistent Salt, Mercury, and Sulphur could be found in all things and could be separated therefrom by the fire.[133] Indeed, Van Helmont devoted a treatise within the *Ortus* to this issue, where he explicitly denies that the resolution of a body by fire analysis (i.e., thermal decomposition) gives direct evidence of its origins—such analysis reveals instead the "heterogeneity of its ultimate matter."[134] Although Van Helmont found some utility in the *tria prima* (as exemplified by his discussion of water, ice, and *gas*), he did not view the three principles as initial constituents of mixed bodies.

> We are accustomed—when speaking generally among chymists—to speak of things under the name of the *tria prima*, namely Salt, Sulphur, and Mercury. Not that I impute these to be the principles of things, however. But because they are separated from many things by means of fire, we use that name in order to distinguish the heterogeneities of concrete bodies.[135]

Van Helmont's position then is that the *tria prima* are not elemental principles or ingredients of mixed bodies; in reality the single material principle of all things is water. Instead, the *tria prima* are "heterogeneities," that is, corpuscular combinations or alterations of matter sometimes existing within fully formed substances. As he says elsewhere, Mercury, Sulphur, and Salt are not universal substances common to all species; at best they are simply the "dissimilar particles" found in other bodies, existing in a threefold variety.[136] They are not primary constituents of matter, there-

133. Allen G. Debus, "Fire Analysis and the Elements in the Sixteenth and Seventeenth Centuries," *Annals of Science* 23 (1967): 127–47.

134. Van Helmont, *Ortus*, "Tria prima chymicorum principia," no. 52, p. 407.

135. Ibid., no. 65, p. 409: "Non quidem quod putem illa esse rerum principia: sed quia è plerisque separantur per ignem, utimur illorum etymo, ad distinguendum heterogeneitates corporum concretorum."

136. Van Helmont, *Ortus*, "Complexionum atque mistionum elementalium figmentum," no. 4, pp. 104–5: "Sunt nempe sal, sulfur, & mercurius, sive sal, liquor & pingue, in speciebus

fore, but alterations of the fundamental material of all things—water. Indeed, the so-called *tria prima* are in some cases mere artifacts of the fire, meaning that they are produced by the action of the fire used to decompose the mixed body.[137] It was this position of Van Helmont's on the artifactual nature of Mercury, Sulphur, and Salt that Robert Boyle adopted as a key argument in his *Sceptical Chymist*, and which he reiterated in maturer years in the second edition's appendix, *Producibleness of Chymicall Principles*.[138]

There is one difficulty inherent in Van Helmont's position: how can we distinguish between those instances where one of the *tria prima* really exists in a substance before its fire analysis, and those where it is merely an artifact of the fire? For Van Helmont, there is no universal answer to this question; it must be solved experimentally on a case-by-case basis. A revealing example is found, however, in the Helmontian treatment of *sal alkali*, where Van Helmont first makes the case for the existence of a volatile Salt in at least some mixed bodies and then treats the transformation of this volatile Salt into the fixed alkali salt that is actually isolated from fire analyses.

As is often the case, Van Helmont embeds his treatment of a chymical topic in a medical discussion. His concern in this case is with blood and the role of respiration; he wants to argue that venous blood becomes a volatile salt that is continuously released from the body by means of exhalation and insensible perspiration.[139] As part of this discussion, Van Helmont turns to chymical experiments. He claims that many things that are volatile can become nonvolatile if heated. This is what happens in the case of some Salts in mixed bodies. Salt does exist in some mixed bodies prior to fire analyses, but it exists as a volatile Salt, not a fixed one. To Van Helmont's contemporaries this would seem untrue, or at least surprising, for the Salt of a body was routinely isolated from its residual ashes as a "fixed," i.e., nonvolatile,

---

specialissimis: non quidem ut corpora quaedam universalia, quae cunctis speciebus sunt communia; sed partes sunt simulares *[Aufgang* corrects to *dissimilares]*, in concretis corporibus, varietate triplici, pro seminum exigentia distinctae." Our interpretation again differs from that of Clericuzio, who in an overly simplistic fashion views Van Helmont as rejecting the theory of the three principles tout court. In reality, Van Helmont's position is that certain substances, including water, contain something analogous to the Paracelsian *tria prima* acting as principles of heterogeneity, even though the *tria prima* are not primordial ingredients of composition. As we show throughout the present chapter and elsewhere, the three principles actually do considerable work for Van Helmont, for example in his theory of alkalies. See Clericuzio, *Elements, Principles, and Corpuscles*, 60; and Van Helmont, *Ortus*, "Tria prima chymicorum principia," no. 54, p. 407.

137. Van Helmont, *Ortus*, "Tria prima chymicorum principia," nos. 46–47, p. 405.

138. Principe, *Aspiring Adept*, 27–62; Debus, "Fire Analysis."

139. Van Helmont, *Ortus*, "Blas humanum," pp. 178–92; Pagel, *Joan Baptista Van Helmont*, 88–91.

material. What Van Helmont proposes is that the original volatile Salt exists in bodies before they are heated, but that this Salt becomes "alkalized," that is, fixed, upon the strong heating of the body.

> Therefore when a fixed alkali is made from a Salt that was previously volatile, this is not a new production of a thing, but only an alteration of a thing. For the alkali was materially present in the concrete before its cremation, and it flowed together with the Mercury and Sulphur. But when the fire raises the Mercury and Sulphur, the Salt, being as it were the more resistant principle, seizes to itself during the liquefaction of concremation a neighboring part of the Sulphur or fattiness. And when it cannot save that part [of the Sulphur] sufficiently from the torture of the fire, it [the Salt] partly passes off under the "mask" of Gas at the same time [as the Sulphur], and acquires the odor of empyreuma, and is partly incorporated with the Sulphur seized in the colliquation and becomes a true coal. Whence the Sulphur, now fixed by its marriage with the salt, does not quickly pass out of the coal into smoke, but only gradually, and not unless the vessel is left open; indeed, a regular weight of the volatile Salt flies off with the former Sulphur (for thence it is that the Sulphur of a thing retains the sharp taste of the volatile Salt) and finally with the Sulphur in the coal.[140]

Thus Van Helmont argues that when a combustible material—say wood—is cremated, part of its initial volatile Salt and Sulphur may distill off. This can be collected as an essential or empyreumatic oil—the oiliness and flammability indicate that it is sulphureous, and its sharp taste indicates the presence of a saline component as well. The rest of the volatile Sulphur and Salt, however, are "fixed together" into the alkali salt found in the residual ashes or coals. Part of this complicated theory rests upon old corpuscular ideas invoked from the Middle Ages onward to explain the conversion of a volatile substance into a fixed one. A larger corpuscle will have

---

140. Van Helmont, *Ortus,* "Blas humanum," no. 38, p. 187:

Dum ergo alkali, fixum fit ex sale antea volatili, non est productio rei nova: sed rei alteratio saltem. Alkali enim materialiter quidem erat in concreto, ante cremationem, fluebatque unà cum mercurio, atque sulfure. Veruntamen dum ignis tollit mercurium & sulfur: sal quidem, tanquam principium magis subsistens, in liquatione concremationis, arripit sibi vicinam sulfuris, sive pinguedinis partem, eamque, dum ab ignis tortura, non sat valet tueri, partim sub larva Gas simul evolat, atque empyreumatis odorem acquirit, partimque apprehenso sulfuri colliquando incorporatur, & verus carbo fit. Quapropter sulfur conjugio salis jam fixus, non celeriter è carbone in fuliginem tendit: sed sensim, ac non nisi aperto vase: adeoque cum sulfure priore (nam hinc sulfur rei, plerumque acutum, salis volatilis saporem retinet) ac tandem cum sulfure carbonitio, justum sui salis volatilis pondus evolat.

more difficulty being raised by fire than a small one, so the uniting of different substances inhibits their volatilization by producing a compound of larger particle size.[141] As a result, some of the volatile Sulphur becomes partly or wholly fixed (nonvolatile) by entering into combination with the volatile Salt, which undergoes fixation itself in the process of combination. The fixed and volatile, however, whether Salt or Sulphur, remain the same in species, except that "one is imprisoned, [and] the other is free."[142]

To show that this fixity is relative, however, Van Helmont deploys an experiment in which honey is destructively distilled to a coal. If this coal is subjected to a greater fire, it will remain unchanged so long as the vessel remains closed to the air; but as soon as the vessel is opened, the coal begins to be consumed and eventually vanishes entirely, indicating that its fixation was only relative. The role of the open air in Van Helmont's alkali experiments will prove to be significant, as we shall see shortly.

Van Helmont then demonstrates the ability of volatile Salt to seize upon neighboring Sulphur by a quantitative study of the calcination of tartar. When strongly heated in an open dish, tartar (our modern potassium bitartrate) blackens, melts, smokes copiously, and finally leaves behind a black coal that can be extracted with water to provide an alkali—salt of tartar (potassium carbonate). Van Helmont states that sixteen ounces of tartar provide only two and a half ounces of salt of tartar; therefore, by the use of the principle of mass balance, he knows that "thirteen and a half ounces were volatile and lost in calcining." As a comparison, however, Van Helmont subjects another sixteen ounces of tartar to strong heating, but this time in a retort, not an open dish. He collects the oily distillate and then returns this distillate to the burnt residue. Now, after redistillation and extraction of the salt in the residue, Van Helmont collects "four ounces and a third" of salt of tartar instead of only two and a half.[143] Why this increase? In the initial distillation, only part of the Salt and Sulphur united to form fixed alkali. Much of the Sulphur escaped into the distillate as did some of the volatile Salt. The Salt and Sulphur of the combustible body are thus partitioned—some of each passes off with the volatile components, while some of each combines to form fixed alkali (salt of tartar, in this case). Thus, when the distillate is poured back over the residue and resubjected to the "torture of fire," another opportunity is granted for the Salt to "seize upon" the Sulphur to form an additional quantity of alkali. Thus Van Helmont has presented a quantitative experiment—complete with a control

141. Newman, *The Summa Perfectionis of Pseudo-Geber,* 146–47, 152–53, 683–87.
142. Van Helmont, *Ortus,* "Blas humanum," no. 42, p. 188.
143. Ibid., no. 39, p. 188.

and use of the balance—to support his theory of the capture of Sulphur by Salt in order to form alkalies.[144]

So far, then, Van Helmont has argued for a volatile component that becomes fixed in incinerated materials. But how does he know what the initial ingredients of the isolable alkali salt were? Might the alkali salt be simply produced from some other substance by the fire? After presenting some further evidence, the Belgian returns once again to the principle of mass balance. Van Helmont shows that if the ingredients of an alkali are missing, no alkali can be formed upon heating. He points out that rotten wood yields almost no alkali when its ashes are extracted. This absence of fixed alkali, he argues, is due to the fact that the wood's volatile Salt all evaporated with its Sulphur "through the ferment of putrefaction," that is, during the process of rotting.

> However much Salt was volatile in a thing or a concrete, so much is found to be lacking finally in the ash. That is, [in the case of rotten wood,] all of it. Whence it follows that the volatile Salt, whether retrieved from the Sulphur or the Mercury, is materially the same as the alkalizate Salt, and consequently the volatile can be alkalized and the alkali in turn volatilized, while maintaining the formal property of the concrete.[145]

Here once again, Van Helmont appeals to his belief that however much weight goes into a reaction must come out at the other end. He knows that a quantity of normal wood will yield a predictable amount of alkali upon burning and leaching. When he finds that a similar quantity of rotten wood does not give any alkaline salt, he concludes that the necessary ingredients

144. It should be noted that by superficial chemical thinking, Van Helmont's experiment cannot be correct. However, as we have tried repeatedly to show in the chapter, it is wiser to form judgments of Van Helmont's techniques circumspectly. When tartar is thermally decomposed, it puffs up greatly and expels carbon dioxide and other volatiles. Thus calcination in an open dish unavoidably involves some loss by the expulsion of solid particles during the vigorous bubbling, and this dust is carried away by the fire. When the calcination is done in a retort, this source of loss is avoided—any mechanically expelled solid is caught by the walls of the retort or preserved in the distillate, which is poured back over the residue. Stoichiometrically, 16 ounces of potassium bitartrate can yield 5.9 ounces of potassium carbonate; therefore, the fact that open calcination gave Van Helmont only 2.5 ounces of salt implies considerable mechanical loss, and his recovery of 4.3 ounces by repeated distillation is well within reason, even taking into account that his tartar (obtained from wine lees) was far from pure potassium bitartrate.

145. Van Helmont, *Ortus,* "Blas humanum," no. 41, p. 188: "Adeoque tantundem salis fuit volatile in re, sive concreto, quantum saltem in cinere deficere reperitur. Id est, totum. Unde sequitur, quod sal volatile, tam à sulfure quam abs mercurio repetitum, idem sit materialiter cum alkalizato: ac proinde volatile alkalizatur, & vicissim alkali volatilizatur, manente formali concreti proprietate."

for the alkali had already departed from the wood before its combustion. Since the only difference between the two woods was that one was rotten and the other not, it follows that the escape of the alkali ingredients must have been due to the wood's putrefaction. This, of course, fits very nicely with Van Helmont's theory of putrefaction and fermentation in general, for he believed that such processes resulted in the division of gross substances into smaller, more active corpuscles.[146] And as we have already mentioned, smaller corpuscles were supposed to be more volatile than larger ones.

Van Helmont goes on to note that even if only one of the two ingredients necessary to form an alkali is missing, no alkali will be formed.

> If the air (let him who can, grasp an arcanum) first of all volatilizes the Sulphur of a concrete with complete separation of its Salt, this Salt (which otherwise would be fixed by the fire into an alkali in the coal) will be made entirely volatile and will ascend sometimes in a liquid form and sometimes in the form of a sublimate.[147]

With no Sulphur remaining in the body, the Salt has nothing to "seize upon" and with which to fix itself, and so it retains its volatile nature, and the applied heat drives it up as a distillate or sublimate. This is what happens in blood. According to Van Helmont, the function of respiration in warm-blooded animals is constantly to volatilize the Sulphur from venous blood. If this function were stopped, the heat of the body would soon cause the venous Salt and Sulphur to "degenerate into a dry mass" akin to what happens when alkali is produced by heating. But since the air perpetually carries away the Sulphur, the Salt itself now remains volatile since there is nothing for it to seize upon and become fixed.[148]

Note that in the passage above, Van Helmont hints at an "arcanum." The Belgian leaves this aside unexplained and says only that the presence of this now permanently volatile Salt is proven by experimental demonstration, but that the process is known to few people, and it cannot be made public.[149] This hint by Van Helmont, however, would prove to have enormous significance for George Starkey, as we shall see in the next chapter.

---

146. Newman, *Gehennical Fire,* 143–46.

147. Van Helmont, *Ortus,* "Blas humanum," no. 45, p. 188.

148. Ibid. nos. 45–46, pp. 188–89.

149. Ibid., no. 45, p. 188: "Hoc sal, per mechanicam est demonstratum; ejus autem demonstratio est paucis cognita, non licet eam tamen palàm facere." Note that the *Ortus* actually reads "nos licet," but this is either ungrammatical or nonsensical; it is likely a misprint for "non licet," as we have read it. The German translation (*Aufgang,* 1:239) reads "nicht . . . zu offenbahren."

CONCLUSIONS

Van Helmont's deep concern for determination of weight and specific gravity is in itself a logical extension of the heritage of the metallurgical alchemy of the Middle Ages. By Van Helmont's own testimony he read and digested such authors as Geber and pseudo–Ramon Lull, and their ideas reappear both in his own corpuscular considerations and in his discussion of the alkahest.[150] Yet unlike most of these authors, Van Helmont put great emphasis on the paired chymical analysis and synthesis of bodies—an idea that is an outgrowth of the Paracelsian identification of chymistry with *spagyria* or *Scheidung*—in order to understand chymical transformations. Paracelsus himself, however, as we have noted, was far more interested in the resolution of bodies for medicinal preparation than he was in their recombination or the philosophical understanding of their composition. It was, after all, such *Scheidung* that allowed metallurgists to separate gold from its dross and iatrochemists to remove the noxious components from the active ingredients of drugs. Not until the inauguration of the chymical textbook tradition in the late sixteenth and early seventeenth centuries do we find authors proposing the etymology of *spagyria* as a fusion of the Greek *span* (to pull apart) and *ageirein* (to put together), thereby embodying in the chymical art the opposed processes of analysis and synthesis.[151] It was this definition of chymistry as the art of analysis and synthesis that would provide a disciplinary identity to the field lasting well into the nineteenth century.[152]

Van Helmont's adoption of analysis and synthesis as the touchstone of the chymical art would have provided little to distinguish his work from that of the early textbook writers if he had not introduced quantitative experimental methods based upon the notion of mass balance. It was his reliance upon the quantitative input and output of reactions that led him to develop a chymical practice quite distinct from that of the earlier Paracelsians. Thus, while Van Helmont was heir to alchemical traditions inextricably allied to metallurgical assaying where quantitative issues were of great importance, he joined this not only with Paracelsian sources but also with other early modern authors on the subject of weight and density determi-

150. For Van Helmont's use of Geber, see Newman, *Gehennical Fire*, 110–13, 145–49. For his use of pseudo-Lull, see Van Helmont, *Opuscula, De febribus,* chap. 14, no. 10, p. 52, and Van Helmont, *Opuscula, De lithiasi,* chap. 3, no. 1, p. 20.

151. One can find this definition in, for example, [Andreas Libavius,] *Commentariorum alchymiae . . . pars prima,* in Libavius, *Alchymia* (Frankfurt: Joannes Saurius, 1606), 77.

152. An example may be found in J. L. Comstock, *Elements of Chemistry* (Hartford: D. F. Robinson, 1831), 2; this book went through numerous editions and became a standard textbook of chemistry.

nation, such as Simon Stevin and possibly Nicholas of Cusa. The result of this integration of quantitative sensitivity and Paracelsian analysis and synthesis was the new turn to mass balance as a tool of chymical demonstration. Even if his urge to disprove Aristotle and his followers occasionally led Van Helmont into the use of mass balance as a rhetorical device, his reliance upon it is noteworthy, and his example bore fruit among his heirs. As we will see, the Helmontian tradition of quantitative *spagyria* had powerful repercussions lasting even up to the "chemical revolution" of the eighteenth century.

We have seen then how a chymist who claimed to have had visionary dreams could still be an avid experimentalist who believed that certainty in physical matters could best be acquired by means of exact weights and measures. What he saw in his furnace was not a manifestation of the development of his own soul or a vision of his unfolding psyche. Instead, he read an edict there—that true certainty and understanding of natural causes could be revealed only "when one knows what is contained in a thing and how much of it there is." As we shall see in the following two chapters, this pronouncement found no more eager adherent than George Starkey.

THREE

Theory and Practice

STARKEY'S LABORATORY METHODOLOGY

In the previous chapter we saw that Van Helmont's chymistry made substantial use of quantitative techniques and assumptions, as well as of demonstrative experiments. But Van Helmont has always been difficult to understand owing to his love of paradox, his elliptical Latin prose, and the lack of clear organizational principles in his massive *Ortus medicinae*. Beyond these issues of style, Van Helmont's publications, like the published writings of any laboratory worker, are automatically limited in their ability to reveal to us the actual workings of a laboratory practice or an experimental mind. The reasons for this are obvious and have been well treated by other historians who have noted, above all, that published treatises are "cleaned-up" versions of a research program, prepared and organized for public consumption, and generally left with very few traces of the methodological scaffolding within which the program was constructed. Errors, false starts, and dead-ends are generally purged, and the progress of a course of investigation is made to appear far more linear and preconceived than it actually was. Sometimes this may be done for rhetorical purposes, but often it is only a means of simplifying the presentation of complex material. Raw data is often not a useful (or even comprehensible) material to transmit, and so the experimenter as a matter of course winnows and digests it into a form appropriate for his purposes in publishing his research.[1] Unfortunately, this practice, while often pedagogically helpful, leaves the historian at a loss when trying to reconstruct the actual historical course of events or the methodological practices of the investigator. Historians wishing to learn more about genuine laboratory practice—what a laboratory practitioner actually did and thought during the course of his operations—

---

1. One exception to this is Kepler's presentation of his calculations of the orbit of Mars in the *Astronomia nova*, which recounts all of his false starts and errors—this method was itself, of course, for rhetorical purposes. See James R. Voelkel, *The Composition of the Astronomia Nova: The Context and Content of Kepler's New Astronomy* (Princeton: Princeton University Press, 2001).

must penetrate beneath the veneer of public presentation. Unfortunately, in many (if not most) cases this is difficult or even impossible to do simply because the first-generation laboratory documents are unavailable, due to their being restricted or lost (if indeed they ever existed).[2]

This situation makes all the more valuable the survival of several detailed laboratory notebooks kept by George Starkey during the first half of the 1650s. These documents are rare witnesses to the daily laboratory operations and thought processes of a practicing mid-seventeenth-century chymist involved in almost all of the various pursuits classed under the broad title of chymistry—from metallic transmutation to "industrial" and pharmaceutical preparations. In addition, his manuscripts reveal a surprising degree of quantitative reasoning and gravimetric technique inherited from his mentor Van Helmont. One cannot overstate the remarkable divergence in style between the dry, deliberative prose of Starkey's notebooks and the wild tropology of his Philalethes texts. Starkey's notebooks provide an unparalleled view of the chymical laboratory, allowing us to see the way in which a celebrated practitioner of the art actually organized, set about, and carried out his work in the realms of chrysopoeia, chemical pharmacy, and other technical pursuits associated with chymistry.[3]

Starkey's surviving laboratory records date from c. 1651 to 1658 and consist of three complete autograph notebooks, one fragmentary autograph notebook, and four partial transcriptions. The range of projects described in the notebooks is extremely broad. Some entries involve fairly standard chymical medicines or the distillations of oils and perfumes. Some treat the preparation of the Philosophical Mercury or attempts at the Philosophers' Stone. Others detail the search for the preparation of rare Helmontian secrets such as the alkahest or various antimonial arcana. Yet others record at-

2. In regard to recent studies of experiment see, for example, David Gooding, Trevor Pinch, and Simon Schaffer, ed., *The Uses of Experiment: Studies in the Natural Sciences* (Cambridge: Cambridge University Press, 1989); Peter Galison, *How Experiments End* (Chicago: University of Chicago Press, 1987); Steven Shapin, "The House of Experiment in Seventeenth-Century England, *Isis* 79 (1988): 373–404. Required reading is Frederic L. Holmes, "Do We Understand Historically How Experimental Knowledge Is Acquired?" *History of Science* 30 (1992): 119–36.

3. Although other chymical laboratory notebooks have survived from Starkey's period, few if any display the straightforward integration of theory and practice found in those of the American chymist. A more typical example may be found in the "Aqua vitae: non vitis" of Thomas Vaughan, which consists primarily of recipes in the imperative mood interspersed with dream sequences that frequently concern his dead wife, Rebecca. Vaughan's notebook also contains far more *Decknamen* (cover names) than Starkey's, making it less useful as a plaintext for decoding his published works. See Donald R. Dickson, *Thomas and Rebecca Vaughan's Aqua Vitae: Non Vitis* (Tempe: Arizona Center for Medieval and Renaissance Studies, 2001).

tempts to "multiply" the metals—that is, to increase the quantity of precious metals by using small portions of them to transmute baser metals. A careful consideration of these projects allows us to document and explore Starkey's thought processes and experimental methodology as well as his broader aims for chymistry, both philosophical and intellectual. The notebooks bear witness to his commitment to such practical laboratory issues as experimental design, the interplay of theory and practice, the sources of authority and proof, quantitative methods, the analysis and assessment of experimental results, and even the value of failure in experimental work. Starkey was a highly reflective laboratory worker, and the notebooks are full of his cogitations on the nature and method of investigative laboratory science as he pursued it in the mid–seventeenth century.[4]

THE USE AND FORMAT OF STARKEY'S NOTEBOOKS

It will be useful first to reconstruct how Starkey actually used his notebooks. Starkey purchased blank, bound books at the stationer's to serve for recording his laboratory investigations, rather than using bundles of papers or sheets.[5] Starkey did not move from one notebook to another in a chronological sequence. In general, he seems to have had several notebooks in use simultaneously, and the entries in any given notebook often stretch over several years. The notebook now known as Sloane 3711, for example, contains entries dated over a three-year period, and the different inks employed give witness to the fact that the physical location of an entry does not necessarily indicate its date. Indeed, all three of the complete notebooks overlap chronologically, and some entries dated to the same week and even to the same day appear in different volumes. Far from being an indication of disorder, these observations actually attest to the highly formal and orderly manner in which Starkey managed his laboratory records. The division among volumes or parts of volumes was based largely on the contents, not on the dates. Sometimes the distinction depends upon the nature of the entry. For example, the first section of Sloane 3750 was reserved primarily for the formal analysis of written sources and the formulation of experiments therefrom. Others, such as one of the sections of RSMS 179, were devoted to sequential entries for a specific project, in this case the

---

4. Interested readers are encouraged to see our companion text, *The Laboratory Notebooks and Correspondence of George Starkey* (forthcoming), where they will find all of Starkey's presently known manuscripts, including the notebooks in their entirety, edited, annotated, and accompanied with full English translations and scholarly apparatus.

5. Only one notebook, RSMS 179, preserves remnants of the original binding, but it is clear that all the other notebooks began as blank bound volumes as well.

volatilization of alkalies. Different trials relating to one specific project are often found together, segregated from those dealing with other projects carried out at the same time. In cases where two distinct projects were under way simultaneously, Starkey would sometimes begin entries for one project, leave several blank pages for its continuation, and begin recording the other project several leaves further on in the book. It sometimes happened that he left insufficient blank space for the first project, and so successive entries must leap-frog over one another, the correct order being managed by marginalia such as "look for this sign below" or "turn over four leaves."[6] The bound volumes themselves were often divided into two sections by writing from both ends of the notebook toward the center, the writing of one section being inverted relative to the other. In Sloane 3750, both rectos and versos were used until the two texts met in the middle, while in RSMS 179, the two texts are interleaved from first to last page, one sequence written on the versos and the other on the rectos.

Frequently, Starkey would also return to earlier entries to emend or annotate them with the results of later experiments. For example, in Sloane 3711 he recorded several pharmaceutical "corrections of opium" using salt of tartar sometime before mid-1653, but when in 1655 he finally succeeded in his long-term goal of volatilizing the salt of tartar, thus making it more active, he returned to this earlier list and—in the red-brown ink he used at Bristol in late 1655 and early 1656—appended a new "best correction of opium" that incorporated the newly discovered procedure and superseded the old entry. Thus the text within a given notebook is often quite layered, consisting of both the initial writing and later additions, deletions, and annotations.

The sense of cohesiveness that must have originally existed between the notebooks is underscored by the fact that the three complete volumes now known all contain references to one another, as well as to an additional volume presently unaccounted for.[7] Indeed, we must remember that there were once many more notebooks than we now have, and so the picture that the extant volumes portray is necessarily fragmented. For ease of reference, Starkey even titled each of the notebooks, calling one the "Codex veritatis," another his "Manuale experimentorum," and so forth. Thus we are

6. These examples of marginalia are taken from the second text of Sloane 3750, where running accounts of the preparation of the Philosophical Mercury and the production of metals from antimony are interspersed one within the other.

7. An early eighteenth-century manuscript catalogue of the collection of R. Jones (presumably the bookseller Richard Jones) that survives in the British Library records the ownership of a substantial number of Starkey's laboratory records of which we have no current knowledge. This list from Sloane 2574 is printed in Newman, *Gehennical Fire*, 252–53.

left with the impression of a shelf of volumes in Starkey's laboratory, each one appropriated to a project or purpose, and each receiving various sorts of entries from Starkey's pen.

STARKEY'S LABORATORY

But in what kind of a place did this shelf of records lodge? We have already encountered, at the start of the previous chapter, the popular image of the (al)chymist in his cluttered laboratory. But the reality of a working alchemical laboratory remains somewhat elusive. There are numerous artistic renderings of laboratories, including important examples from the seventeenth century itself when the alchemist and his laboratory/workshop became a stock figure in Dutch genre paintings. Yet these are only partly descriptive in character, for they also contain numerous conventional and iconographic elements geared to the genre and appropriate for the moral message of the composition; they were never intended to be "photographic" records of chymical laboratories. Even the plan of Libavius's famous *domus chemiae* is not only imaginary, but iconographically rhetorical.[8] Recent archaeological discoveries and surviving laboratory expense accounts give some hard evidence of the layout and contents of early modern laboratories, but have told little about how these workplaces were actually used.[9] The physical appearance of Starkey's laboratory likewise remains difficult to reconstruct with certainty, as he gives us no explicit description of its whereabouts, size, or outfitting. Nonetheless, we can glean some information about it from his correspondence and notebooks and from the records of those who knew him.

It is certain that at some times Starkey maintained a special room of his lodgings as a "laboratorium." There he stored his chymical implements and conducted his experiments. It is clear from his letters, for example, that he had a separate room devoted to chymistry while he was residing at Saint

8. Bernard Joly, "Qu'est-ce qu'un laboratoire alchimique?" *Cahiers d'histoire et de philosophie des sciences* 40 (1992): 87–102; C. R. Hill, "The Iconography of the Laboratory," *Ambix* 22 (1975): 102–10; Jacques van Lennep, "L'Alchimiste," *Revue Belge d'archeologie et de l'histoire d'art* 35 (1966): 149–88; A. A. A. M. Brinkmann, *De Alchemist en de Prentkunst* (Amsterdam: Rodopi, 1982); Jane P. Davidson, *David Teniers the Younger* (London: Thames and Hudson, 1980), 38–43; Newman, "Alchemical Symbolism and Concealment: The Chemical House of Libavius," in *The Architecture of Science,* ed. Peter Galison and Emily Thompson (Cambridge: MIT Press, 1999), 59–77; cf. Owen Hannaway, "Laboratory Design and the Aim of Science: Andreas Libavius versus Tycho Brahe," *Isis* 77 (1986): 585–610.

9. For archaeological studies of chymical laboratories see Rudolf Werner Soukup and Helmut Mayer, *Alchemistisches Gold, Paracelsische Pharmaka: Chemiegeschichtliche und archaeometrische Untersuchungen am Inventar des Laboratoriums von Oberstockstall/Kirchberg am Wagram* (Vienna: Boehlau, 1997); and Jost Weyer, *Graf Wolfgang II. von Hohenlohe und die Alchemie* (Sigmaringen: Jan Thorbecke, 1992).

James' in 1652. One of his letters to Boyle refers explicitly to his nodding off to sleep in the laboratory at 1:00 AM while reading. In his subsequent dream (to which we will later return) he saw a figure "entering the laboratory," thus making it clear that this was a separate room. A few days later, when someone came to call on him, he "left the laboratory" to see his guest, because he "did not want anyone to come hither."[10] This same Saint James' laboratory was apparently ill suited to its purpose, for here Starkey fell ill in early 1652 with "very horrid and seemingly Pestilential Symptomes," presumably as a result of poor ventilation while working on antimonial, mercurial, and probably arsenical preparations.[11] At that time Boyle encouraged Starkey to remove the glass from the windows of the room to let in fresh air, which Starkey did, but with the concomitant problem that the drafts and changes of weather then made it impossible to regulate his fires accurately.[12] The choice in this laboratory thus seemed to be between controllable furnaces and breathing.

At other times Starkey probably did not have the luxury of an entire room as a dedicated chymical laboratory. He relocated often, for we know of twelve different addresses for him during fourteen years.[13] At some points, when his financial health was extremely poor, he had to forgo experimentation altogether. During the worst of these times he was confined to debtor's prison, where experimentation was impossible. At other times his laboratory may have been only one section of his lodging area. Later in his career, in 1658, he expressed delight at being "confined"—apparently under some sort of a house arrest—because it cut him off from external contacts and allowed him the leisure to spend all his hours at home involved in laboratory experimentation.[14]

Starkey seems to have been constantly outfitting his laboratory with new designs of furnaces and apparatus. Several notebooks contain references to designs for new furnaces, and his 26 January 1652 letter to Boyle mentions a furnace then newly completed and already in use. Hartlib in fact made note of Starkey's "admirable skil in making all manner of furnaces."[15] Even the chrysopoetic works published under the name of Eirenaeus Philalethes dis-

10. George Starkey to Robert Boyle, 26 January 1652; in *Notebooks and Correspondence*, document 6, and in Boyle, *Correspondence*, 1:120–21.

11. Boyle, *Usefulnesse*, in *Works*, 3:501; Starkey's friend Robert Child recounted later to Hartlib that he had often warned Starkey that "he would ruine himselfe by using charcoale in places without chimneyes, as also by the preparations of mercuriall & Antimonious medicines"; Child to Hartlib, 2 February 1653, HP 15/5/18B.

12. John Dury to Samuel Hartlib, 2 April 1652, HP 4/2/15B.

13. Newman, *Gehennical Fire*, 247.

14. George Starkey, *Pyrotechny Asserted and Illustrated* (London, 1658), 168–69.

15. Samuel Hartlib, *Ephemerides* 1651 (January), HP 28/2/6A.

play Starkey's characteristic concern for proper and precise furnace construction in order to deliver and regulate the desired level of heating; in
some places Philalethes gives the exact measurements in inches for various
parts of the furnace.[16] *The Marrow of Alchemy* (1654–55) not only claims
that "a good Furnace is the choicest thing / Next to the matter [of the
Philosophers' Stone] which a man should seek" but also suggests where
and how it should be built and how the laboratory that contains it should
be situated.[17] Perhaps Starkey learned a lesson from his illness of 1652, for
here Eirenaeus Philoponus Philalethes writes

> Nor let thy room be so . . .
> . . . that the fumes arising
> From Coals no vent may finde, for thou maist get
> (as some have done, hereof less care devising)
> Thereby such harm, which late thou wilt repent,
> Hazarding life by their most hurtful scent.[18]

One of Starkey's notebooks also includes designs for special stills of
glass, iron, and copper, accommodated to specialized purposes.

> An iron pot of four or five gallons capacity is set in a furnace with a copper
> neck adapted to it so that the neck enters the pot to the depth of one thumb,
> and secured with the best luting. To this the head of the alembic is to be at-
> tached so that the neck enters it likewise to the depth of one thumb. Distill all
> non-corrosive spirits in this apparatus.[19]

Starkey accompanies this description with a list of the benefits of this
particular apparatus and how and for what purposes it is most advanta-
geous. Similarly, Frederick Clodius informed his father-in-law Hartlib in
1653 that Starkey had invented an excellent new kind of iron retort, and in
July 1656 Starkey traveled from Bristol (where he was in charge of a mining

16. George Starkey [Eirenaeus Philalethes, pseud.], *Ripley Reviv'd* (London, 1678), 37–38;
for further study of Philalethes' concern over correct furnace design, see Lawrence M.
Principe, "Apparatus and Reproducibility in Alchemy," in *Instruments and Experimentation
in the History of Chemistry,* ed. Frederic L. Holmes and Trevor H. Levere (Cambridge: MIT
Press, 2000), 55–74.
17. George Starkey [Eirenaeus Philoponus Philalethes, pseud.], *The Marrow of Alchemy*
(London, 1654–55), pt. 1, 30–33.
18. Ibid., 32.
19. *Notebooks and Correspondence,* document 12; RSMS 179, fol. 45v; see also fol. 2, which
refers to a similar apparatus and was originally apparently accompanied by a facing-page illustration; the previous page, however, has been torn out.

operation) to London in order to apply for a patent for a new kind of "continuall blast" refining furnace.[20]

Many of the notebooks contain lists of financial accounts that shed a bit more light on the working aspects of Starkey's laboratory. Some accounts show him spending sixpence on a retort, or a shilling on aqua fortis, and so on. He also occasionally estimated his costs to run an experimental laboratory, as for example when he reckoned an expenditure of two shillings sixpence per week on coals. The all-important furnaces that were central to Starkey's workplace (as is implied by his self-assumed title of "Philosopher by Fire") show up repeatedly. For example, in August 1656, when he sets about preparations for the "great work" of making the Philosophers' Stone, he first estimates the costs for the furnace that must be devoted to the project.[21]

A FURNACE	
Bricks	0 - 1s - 6d
Iron Worke	0 - 3s - 0
Kettle	0 - 2s - 6d
Mortar & making it	0 - 11s - 0
Building furnace	0 - 2s - 6d

The last itemized cost, for "building furnace," implies that while Starkey designed his own furnaces, he sometimes had others come to do the dirty work of constructing them from bricks and mortar. Indeed, another notebook records that he had previously "payd Mason 8s 10d," presumably for similar services.[22]

This use of tradesmen in helping to outfit the laboratory brings up the related question of whether or not Starkey also used paid operators or assistants to tend the fires or to carry out operations for him. Boyle's use of operators (as well as of satellite laboratories) has already received some scholarly attention.[23] In Starkey's case it is clear that he did in fact keep laboratory assistants, at least when his finances allowed. One of these assistants is mentioned in Starkey's spring 1651 letter to Boyle. There Starkey mentions that after he had himself prepared a medicinal confection, he "left it in a furnace with a gentle fire & gave order to have a fire put under when that went out." This operator is left unnamed but is likely to be the Mr. Webbe mentioned later in the letter. Starkey had in fact been lodg-

20. Hartlib, *Ephemerides* 1653, HP 28/2/73B; ibid., 1656, HP 29/5/86B.
21. *Notebooks and Correspondence*, document 12; RSMS 179, fol. 4v.
22. *Notebooks and Correspondence*, document 11; Sloane 3750, fol. 20.
23. Steven Shapin, *A Social History of Truth* (Chicago: University of Chicago Press, 1994), 361–69; Principe, *Aspiring Adept*, 135–36, 150–51, 218.

ing with a Mr. Webbe in early 1651, as Child told Hartlib.[24] This was probably Francis Webbe, a brewer, chymist, and acquaintance of Hartlib's.[25] Two years later, in March/April 1653, Webbe's name recurs as a hired assistant for Starkey alongside yet another operator:

> Hee [Starkey] hath engaged one who goes on in the maine Worke, hee giving direction from weeke to weeke and taking an Account of all. He hath also engaged one for making of wines out of Corne; & seemed to intimate that it was Mr. Webbe.[26]

In spite of this external evidence of the employment of Webbe and others, the notebooks do not mention any assistants or any work not done by Starkey himself. The notebook entries clearly recount the operations of Starkey's own hands. This restriction to Starkey's own operations is probably due to two things. First, most of the notebooks date from a time after Starkey's resources—supplied until 1653 by his wife's dowry as well as by Boyle and others in the Hartlib circle—had ebbed, presumably rendering the hiring of assistants financially unfeasible. Second, in spite of Starkey's known pursuit of various "technological" or "chymical trade" projects, the notebooks deal almost exclusively with Helmontian and other medical arcana and with experiments on transmutation. These latter attempts at the *arcana maiora* seem to have been reserved for Starkey's hand, while the "lower" (even if potentially lucrative) projects—such as the distillation of oils and perfumes and the making of spirits from corn, peas, and beans—received the labor of assistants like Webbe and whatever others there might have been.

STARKEY'S EXPERIMENTAL METHODOLOGY:
CONJECTURAL PROCESSES AND FIERY REFUTATIONS
The greatest value of Starkey's notebooks is their ability to tell us about the precise operational and methodological aspects of his laboratory practice and thereby allow us to reconstruct the thought processes that guided his

24. Starkey to Boyle, April/May 1651, in *Notebooks and Correspondence*, document 3; Boyle, *Correspondence*, 1:90–103, on 91; Hartlib, *Ephemerides* 1651 (between 19 January and 12 February), HP 28/2/6A.

25. See the description of a new method of brewing signed by Francis Webbe and others on 16 March 1648 in Hartlib, *Ephemerides* 1648, HP 8/11/1A. See also *Ephemerides* 1656 (October–December), HP 29/5/103A, where Hartlib reports on the testimony of Clodius that "Dr Goddard in Gresham College . . . hath made a new Laboratory out of Sir K. Digby and taken Mr Webbe to himself." In addition, a mid-seventeenth-century manuscript of chymical notes and treatises (Ferguson 199) is inscribed on the flyleaf as belonging to "Francis Webbe, Merchant." The manuscript contains references to acquaintances of Starkey's.

26. *Ephemerides* 1653 (March/April), HP 28/2/55B.

laboratory work. In order to make this easier for the modern reader, we will consider several experimental sequences from Starkey's notebooks, adding material, theoretical contexts, and interpretations as necessary. We will begin with the notebook now known as British Library Sloane Manuscript 3750. This volume was in use intermittently by Starkey from about 1652–53 until early 1656. As is typical, the volume contains more than one layer of text as well as varied subject matter. The section written forward through the notebook and dating probably from 1652–53 was originally entitled "A Brief Collection out of Many Authors," but Starkey deleted this title, and rightfully so, for the contents are, as we shall see, far more than a mere collection of recipes or quotations. This section of Sloane 3750 contains two discrete parts, which Starkey entitled "Antimoniologia" and "Vitriologia," that attempt to gather pertinent information regarding antimony and vitriol respectively, and to plan out future programs of research on them. We will first consider the method and contents of the "Antimoniologia."

CHASING THE ARCANA OF ANTIMONY AND OF VITRIOL
Starkey begins the "Antimoniologia" with processes for preparing the Sulphur of antimony. According to prevailing seventeenth-century chymical theory, Sulphur was—along with Mercury (and sometimes Salt)—one of the essential ingredients of metallic substances.[27] Various "Sulphurs" of antimony and methods for their preparation are common in seventeenth-century pharmacopoeias and "courses of chymistry." The difficulty for Starkey and other workers was finding an adequate method for decomposing metals or minerals (such as antimony) in order to isolate the Sulphur from the other components. Chymical physicians believed that such separated substances could have potent medicinal operations and could be freed by the proper preparative processes from the often toxic side effects of the undecompounded substances. The various Sulphurs (and Mercuries) were also of importance to chrysopoeians; since they provided the building blocks of the metals, they could give insight into the nature of metals and tools for treating the problem of producing or improving metals artificially.

Starkey begins his treatment of the Sulphur of antimony by citing some published methods for isolating it. The first is recommended by Johannes Hartmann in his edition of Oswald Croll's *Basilica chymica* (1608), and the second by Angelus Sala in his *Anatomia antimonii* (1617).[28] Sala's method

27. For a brief introduction to the seventeenth-century variations on the Mercury-Sulphur theory, see Principe, *Aspiring Adept*, 36–42.

28. On Hartmann, see Bruce T. Moran, *Chemical Pharmacy Enters the University: Johannes Hartmann and the Didactic Care of Chymiatria in the Early 17th Century* (Madison: American Institute for the History of Pharmacy, 1991). On Sala, see Zahkar E. Gelman, "An-

involves treating antimony (that is, stibnite, the native antimony trisul-phide) with a strong acid prepared by dissolving sal ammoniac and saltpeter (ammonium chloride and potassium nitrate) in common aqua fortis (nitric acid) and distilling the mixture to produce a kind of aqua regia. Treatment of antimony with this acid produces a vigorous effervescence and leaves an antimonial residue that is then to be extracted with a boiling solution (or lixivium) of salt of tartar (potassium carbonate).[29] This extract is then evap-orated to dryness, and the antimonial Sulphur sublimed therefrom.

If Starkey's project in the "Antimoniologia" were merely the compila-tion and comparison of the numerous processes published in the seven-teenth century for making a Sulphur of antimony, his work would be of little historical interest. Starkey is not interested in producing a medicine cabinet of common medicaments, however, but rather in creating a treasury of rare and powerful arcana. Thus he immediately sets about *improving* the known methods. Significantly, Starkey does not deploy merely empirical or unguided trials toward this goal but rather turns to theoretical principles to direct his practical investigations. As a committed Helmontian, he turns for guidance primarily to Helmontian chymical theory. Moreover, in order to use these theoretical principles practically, Starkey consistently deploys a highly formalized, sequential method for assessing, developing, and testing new processes.

Immediately after giving Sala's method for isolating the Sulphur of anti-mony, Starkey writes a section of text marked off in the margin as "Obser-vations." Here he notes that "Van Helmont writes that every aquafort works upon metals by reason of their Sulphur . . . Therefore it is consistent with reason that the Sulphur in minerals . . . is extraverted by corrosives, so that it can be easily removed by boiling with lixivium."[30] Starkey cites the Helmontian notion that corrosives, like the acid in Sala's process, act upon

---

gelo Sala: An Iatrochemist of the Late Renaissance," *Ambix* 41 (1994): 121–34. In terms of Sala's influence, Starkey's title "Antimoniologia," as well as the succeeding "Vitriologia" and the lost "Urinologia" (see Newman, *Gehennical Fire*, 252) are probably allusions to Sala's sim-ilarly entitled works such as the *Tartarologia* (1632) and *Saccharologia* (1637).

29. Throughout this study we give modern equivalents for the archaic names, thus en-abling readers familiar with chemistry to follow Starkey's processes at a deeper level. It must be borne in mind, however, that Starkey's "reagents" were not pure in the modern sense, and so the chemical equivalents given here represent only the major component. It is not uncommon for the success of some processes of chymistry's early period to depend entirely upon the pres-ence of impurities: for a discussion and illustrations of this idea see Principe, "'Chemical Translation' and the Role of Impurities in Alchemy: Examples from Basil Valentine's *Tri-umph-Wagen*," *Ambix* 34 (1987): 21–30.

30. *Notebooks and Correspondence,* document 11; Sloane 3750, fol. 2v. The reference to Van Helmont is to the *Ortus,* "Progymnasma meteori," nos. 15–18, pp. 70–71.

the Sulphur present in metalline bodies and "extravert" it, that is, drive it outward, thereby freeing it from the Mercurial part with which it had been associated and allowing the lixivium then to extract it. Assuming that this aspect of Helmontian theory is correct, Starkey then asks himself: "Therefore, howsoever much stronger the aquafort is, is it not that much the better? Wouldn't it be better to strengthen aquafort with sal gemmae, or with niter and sal ammoniac without distillation? How ever much more powerfully [the corrosive] acts, that much better it performs the job."[31] Thus, rather than following Sala's distillation of the acid away from the salts, Starkey proposes to leave the salts dissolved in the acid to take maximum advantage of their corrosive properties.

Starkey thus employs the more modern Helmontian theory to amend the older practical process of Sala, and he writes out now a "Process" (so denoted in the margin) to be tried that incorporates the new ideas suggested by theoretical considerations. When writing this process, Starkey gives himself two options for isolating the Sulphur from the antimonial residue left after treatment with the now-stronger aqua fortis. He suggests that the Sulphur, after its isolation with the strong lixivium, might be either sublimed away from the salt—which is akin to Sala's original method—or be precipitated out of the lixivium using vinegar. It is clear that Starkey must have then carried out both methods in his laboratory, for he returned to this entry, and annotated his process in the margin, writing "scarcely possible" next to the direction to sublime the Sulphur, and "the better way" next to the method of precipitation with vinegar.[32]

This simple example affords a short and straightforward introduction to one of Starkey's most common methodologies, found in page after page of the notebooks. In many cases, Starkey begins with a known process or set of facts, often culled from his reading, and makes formal observations upon them. These observations are then used to compose improved processes, which are to be tested in the laboratory. After testing, Starkey returns to his original entries and either records the results, positive or negative, or uses the experience in the fire to improve the process further. Such a method argues that Starkey is carrying out a systematic investigation of the chymistry of his day and endeavoring to make rational improvements to it based upon observations, theoretical principles, and experimental testing. Starkey's methods become more clearly delineated as we continue through his "Antimoniologia."

31. *Notebooks and Correspondence,* document 11; Sloane 3750, fol. 2v.

32. The use of a different ink verifies that these annotations were made at a time after the writing of the process itself.

After the Sala process, Starkey adds another method for isolating the Sulphur of antimony, this time drawn from Van Helmont. This further process involves the sublimation of antimony (meaning the native trisulphide) with sal ammoniac to produce a multicolored sublimate. Again he follows this with several questions to himself, and again he must have put these to the test in his laboratory, for, as in the previous instance, he has returned to this page and made additions (in a different ink) outlining the success of the processes. To the process initially labeled "to be tried" *[Tentandum]* he appends the words "it [the Sulphur] is collected, tested" *[Colligitur probatum]*. Thereafter, seemingly satisfied that he can indeed prepare the Sulphur of antimony by more than one process, Starkey proceeds to a "higher arcanum"—namely, the "cinnabar of antimony" alluded to by Van Helmont.[33] Unlike the simple processes for the isolation of Sulphur of antimony commonly printed in contemporaneous sources, there exists no clear description of the preparation of this Helmontian cinnabar.[34] Van Helmont is terse and secretive in its description, and so Starkey has no clear protocols to examine or critique and must work harder mentally and practically to obtain both process and product. He begins by copying out the passage from Van Helmont that deals with this substance and its medical use.

See by what means you are able to get a Sulphur [of antimony] like common sulphur, a little inclining to green. Make cinnabar, then you will sublime it six times by itself so that the sublimation may serve for the reverberation of Lili. Take half an ounce of this cinnabar, ground, and suspend it for twenty-four hours in a large jug of wine. One spoonful of this taken for several days has a wonderful effect. And the same cinnabar is sufficient for many hundred jugs of wine, as it is of equal strength if it be sublimed again.[35]

33. Curiously, Starkey described his success in this preparation in early 1651 in letters both to Robert Boyle and to Johann Moriaen (see *Notebooks and Correspondence,* documents 3 and 4). Perhaps he decided that his first preparation was not the correct one, and the treatment here is a renewed attempt at this arcanum.

34. Note that "cinnabar of antimony" in the seventeenth century referred to two different substances. One was this arcanum of Van Helmont and is found only in Helmontian sources; the other is far more common and is found in many pharmacopoeias (for example, Christophle Glaser, *Traite de chymie* (Paris, 1673), 195–96, and Nicolas Lemery, *Cours de chymie* (Paris, 1675, 210–13). This latter "cinnabar" was prepared in the same operation as butter of antimony; corrosive sublimate and antimony (mercuric chloride and antimony trisulphide) were heated together, first providing a distillate of butter of antimony (antimony trichloride) and at a higher temperature a red sublimate of mercuric sulphide that was termed the "cinnabar of antimony."

35. Sloane 3750, fols. 3r–3v; Van Helmont's original is *Ortus,* "De verbis, herbis, et lapidibus," p. 577.

Starkey then analyzes Van Helmont's words carefully in order to draw out of them the maximum amount of practical information and sets down his textual analysis as an orderly list of four points.

> I observe in this description:
>
> 1. That a Sulphur like common sulphur is desired, except that common sulphur is less green.
> 2. That from this Sulphur the cinnabar may be made, which I do not believe can be made without mercury.
> 3. That sublimation may complete this cinnabar, which also is called "Lili."
> 4. That this cinnabar is volatile, because it can be resublimed a hundred times for a hundred jugs.[36]

Starkey notes first that he must somehow prepare a greenish Sulphur from antimony, and then that this Sulphur is to be used in combination with mercury to make the "cinnabar." This latter conclusion comes from an analogical consideration of the name that Van Helmont gives his mysterious product. Common cinnabar is made by mixing common sulphur and common mercury and subliming the mixture into a brilliant red mass (composed of mercuric sulphide); therefore, this Helmontian cinnabar of antimony, Starkey predicts, is to be made by reacting *antimonial* Sulphur with mercury. Starkey's final two conclusions deal with the preparation and physical properties of the Helmontian cinnabar and its alternate name of "Lili."

Starkey now does something that is slightly puzzling at first glance. He suddenly writes out yet another known process for separating Sulphur of antimony. This process involves throwing a ground mixture of antimony, salt of tartar, and saltpeter into a hot crucible. A vigorous deflagration takes place, the residue is then fused, and the molten material poured out to cool. The molten material separates into a metallic portion called the regulus of antimony (antimony metal) and a slag or scoria composed, as the process describes, of "the Sulphur of antimony absorbed by the alkalisate salts."[37] Why is Starkey now transcribing another known process even after he believed that he had successfully obtained Sulphur of antimony by modifying Sala's prescription? Perhaps it is because he believed—for reasons that are not elucidated—that the Sulphur granted by the previous process was not adequate for the cinnabar preparation. One supporting clue is that the immediately preceding preparation from Van Helmont (using sublimation with sal ammoniac) yields a predominantly *red* product, not the *yellow*

---

36. *Notebooks and Correspondence*, document 11; Sloane 3750, fol. 3v.
37. *Notebooks and Correspondence*, document 11; Sloane 3750, fol. 4.

green one Van Helmont seems to require for the production of the antimonial cinnabar.

Starkey immediately follows this new known process with a numbered list of "Observables in this work," just as he did with the Sala procedure. The third of these *observabilia* notes that the scoria "quickly turn green in the air" and "tinge the fingers of anyone who touches them with a golden color," thus suggesting that a Sulphur of the desired yellow green color might be lurking in them. He notes also that the "metallic part" (i.e., the regulus), which is separated from "burning and flammable" Sulphur, must "still retain a combustible Sulphur within itself" because when mixed with saltpeter the regulus burns (thus revealing the presence of an inflammable Sulphur). This idea accords with Starkey's belief, drawn from Van Helmont, that antimony, like other metallic substances, has more than one Sulphur, one external (and easier to separate) and one internal (and much more difficult to extract).[38]

After having set down these "observables," Starkey then draws a set of eight "probable conclusions" *[conclusiones probabiles]*. The first among these regards the form and color of the separated Sulphur of antimony: "If this edulcorated Sulphur be sublimed, what prevents it from exchanging its redness for the color of native sulphur?"[39] Clearly Starkey is exploiting this known process in order to fulfill the command of Van Helmont to "See by what means you are able to get a Sulphur like common sulphur." But again, Starkey does not accept this known process at face value. His two following conclusions deal with improvements to the process based upon an observation that he must have made when carrying out the process, namely that a considerable amount of smoke is produced in the deflagration. This prompts Starkey to wonder if a great deal of the desired Sulphur is not burned up and lost by this method and whether a better method might not be found. Indeed, the copious smoke (largely antimony oxides) would have contained sulphur dioxide, whose malodorous sulphureous smell would have given direct observational support to Starkey's concerns about the loss of the Sulphur into the air. He suggests that the nitre (the cause of the violent deflagration) be first mixed with tartar without the antimony, the mixture of salts deflagrated by itself, and that the resultant salts be then fused quietly with the antimony. In that case, "certainly a great part of the Sulphur which otherwise would fly off in the deflagration would be preserved."[40] Starkey then poses questions to himself about the intensity of the fire to be used,

38. See Van Helmont, *De lithiasi,* in *Opuscula,* nos. 4–8, p. 69.
39. *Notebooks and Correspondence,* document 11; Sloane 3750, fol. 4v.
40. *Notebooks and Correspondence,* document 11; Sloane 3750, fol. 4v.

and even about the relative expense and ease of working the various methods. Finally, at the end of this list of conclusions he writes out a process based upon his theoretically and observationally based improvements. This is a process to be tried in the fire, and he explicitly marks it in the margin with the tag "conjectural process" *[processus conjecturalis]*.

Such conjectural processes are the centerpiece of Starkey's experimental program, and accordingly they appear again and again throughout the notebooks. What Starkey means by "conjectural process" is not something that is intended to remain no more than a thought experiment, but rather a procedure that has been developed by joining past practical experience with a consideration of theoretical principles. Indeed, immediately adjacent to the conclusions that deal with the danger of burning up the Sulphur in the deflagration with saltpeter Starkey adds the marginal note "See the process on fol. 4," thus referring to this conjectural process as the direct solution to the problems he observed to be inherent in the known process. Starkey's new conjectural process is one that should not only work, but also be an improvement over the earlier known process. Crucially, it is also one that *must* be submitted to trial in the fire. Starkey even adds a further "conjecture" *[conjectura]* to the effect that the "alkali of Roche alum" might be useful in the process, but notes that at this point he "cannot say anything from experience." The use of Roche alum is then another thing that Starkey proposes to himself to try in the fire.[41]

ALEXANDER VON SUCHTEN'S POTABLE GOLD

Records of the experimental trials of this conjectural process do not follow immediately. For, as noted above, Starkey devoted the first section of Sloane 3750 to the development of new processes rather than to their testing. So Starkey now writes a new section on another desirable antimonial arcanum using the same methodology of discovery that he employed to design the conjectural process for isolating the Sulphur. This next arcanum is the potable gold described by Alexander von Suchten, the sixteenth-century Paracelsian iatrochemist whom we encountered in the previous chapter. Suchten's influence on Starkey was substantial; most notably, Starkey's initial process for preparing the Philosophical Mercury (to be discussed further below) has already been shown to originate in Suchten's work.[42] Here, un-

41. *Notebooks and Correspondence,* document 11; Sloane 3750, fol. 5.

42. William R. Newman, "The Authorship of the *Introitus Apertus ad Occlusum Regis Palatium*," in *Alchemy Revisited,* ed. Z. R. W. M. von Martels (Leiden: Brill, 1990), 139–44. On Suchten, see Wilhelm Haberling, "Alexander von Suchten, ein Danziger Arzt und Dichter des 16. Jahrhunderts," *Zeitschrift des Westpreussischen Geschichtsverein* 69 (1929): 177–230; Włodzimierz Hubicki, "Alexander von Suchten," *Sudhoffs Archiv* 44 (1960): 54–63; and Carl

der the title of "Suchten's arcanum of antimony," Starkey notes Suchten's authorship of a two-part work on antimony and his discussion of preparing potable gold from the "martial" or "stellate" regulus of antimony. Potable gold was a much sought-after chymical medicine, and various methods for its preparation were widespread in the seventeenth century.[43] Suchten claims that it is not made from gold at all, but rather from a secret "Philosophical Gold" or "volatile Sol or Gold" extracted from the stellate regulus of antimony. This latter material is metallic antimony reduced from its ore stibnite (antimony trisulphide) using iron (symbolized by the planet Mars, hence, "martial regulus"). This reduced metallic antimony, when properly prepared, shows a stellate (starlike) crystallization pattern on its surface.[44]

As we have seen twice before, Starkey first prepares a list of his "observations" drawn from the reading of Suchten—in this case thirteen points that summarize much of the content of the first half of Suchten's antimonial treatise.[45] These observations are then followed, as previously, with sets of conclusions, but now divided into seven "negative" conclusions and five "affirmative" conclusions. The negative conclusions indicate how the Philosophical Gold is *not* to be prepared. In this case, Starkey asserts that it is not to be prepared by first making an antimonial mercury out of the regulus, as he had done previously in preparing the Philosophical Mercury. In fact, here he explicitly notes his dissent from his earlier opinion on the matter: "I may have believed this at one time, but I now think the opposite"

Molitor, "Alexander von Suchten, ein Arzt und Dichter aus der Zeit Herzogs Albrecht," *Altpreussische Monatschrift* 19 (1882): 480.

43. On potable gold, see, for example, Francis Anthony, *Medicinae chymicae, et veri potabilis auri assertio* (Cambridge, 1610); Angelus Sala, *Processus de auro potabili* (Strasbourg, 1630); Johann Rudolph Glauber, *De auri tinctura sive auro potabili vero* (Amsterdam, 1646).

44. Most seventeenth-century chymists believed that the star pattern would appear only if the antimony metal were prepared from the ore using iron as the reducing agent. Thus "martial regulus of antimony" and "stellate regulus of antimony" are the same substance (and, in fact, no different chemically from antimony reduced in other ways, although this was not known in the seventeenth century). The need for iron in generating the star was a mistaken connection; all samples of pure antimony will provide a star- or fernlike surface so long as they solidify under a molten slag, generally composed of molten salts and called a *couverture*. The processes for reducing antimony sulphide ore with iron followed by treatment with potassium nitrate do produce a slag well suited to act as a couverture. For Newton's experiments on it, see Betty Jo Teeter Dobbs, *The Foundations of Newton's Alchemy, or The Hunting of the Greene Lyon* (Cambridge: Cambridge University Press, 1975), 146–56 (an excellent photograph of the star pattern appears on p. 149). For Boyle's investigations on the use of iron, see Boyle, "Of the Unsuccessfulness of Experiments," in *Certain Physiological Essays*, in *Works*, 2:45–46.

45. Suchten's *Tractatus secundus de antimonio vulgari* (Leipzig, 1604) contains the preparation Starkey adopted for his Philosophical Mercury, but this aspect of Suchten, though crucial for Starkey's transmutational endeavors, is not of interest to him at this point.

and Starkey then writes out a numbered list of reasons for his change of mind. Starkey had by this time, of course, a great deal of experience with the antimonial or Philosophical Mercury—the topic of which his 1651 letter to Boyle (known as the "Key") was full.[46] The affirmative conclusions then suggest how Suchten's Philosophical Sol or Gold *is* to be prepared, namely, by adding "something foreign" to the regulus of antimony that allows the extraction of "a golden sublimate from this stellate regulus . . . which sublimate will be its Sol." What Starkey must discover, however, is the identity of the correct "something foreign."

A constant problem Starkey must face in his search for the higher arcana is the silence of authoritative authors about issues or ingredients key to the success of their processes. The secrecy and concealment in chymical (and many other early modern) subjects is well-known.[47] Thus Starkey must frequently hazard a guess about the meanings of things or about the right steps to take or ingredients to use in a certain process. In this case, Suchten, though remarkably forthright when describing the preparation of the Philosophical Mercury, remains either silent or obscure when discussing the isolation of the volatile Sol directly from the regulus. Thus Starkey must seek for clues in the text to guide his practice and must frequently make conjectures based upon his textual analysis. Accordingly, he clearly labels the text immediately following in the margin as "a conjecture" and notes that

> Suchten writes that there is a mystery in the scoria of the first fusion [of the regulus]; then accordingly, why might the Sulphur not be able to be sublimed from the scoria with stinking spirit [i.e., sal ammoniac], which sublimate would embrace the stellate mass in sublimation and tinge its whiteness into a solar yellowness in their conjoined ascent?[48]

In other words, Starkey conjectures that the unspecified "mystery" that Suchten claims to exist in the scoria is actually the scoria's ability to separate the Philosophical Sol from the regulus. Therefore, if Starkey separates the Sulphur of antimony from the salts in the scoria by sublimation, then that separated Sulphur (when mixed with the regulus and submitted to subli-

---

46. We will return to the important topic of the Philosophical Mercury below.

47. For a study of early modern secrecy see William Eamon, *Science and the Secrets of Nature: Books of Secrets in Medieval and Early Modern Culture* (Princeton: Princeton University Press, 1994).

48. *Notebooks and Correspondence,* document 11; Sloane 3750, fol. 7. The original source is Alexander von Suchten. Since Starkey was using Suchten in English, our references in this chapter will be to the printed English translation—Alexander von Suchten, *Of the Secrets of Antimony in Two Treatises,* "Translated out of High-Dutch by Dr. C. a Person of Great Skill in Chymistry . . . " (London, 1670), 65.

mation) may be able to carry up with itself in a "cojoined ascent" the Philosophical Sol from the regulus, "tinging" the whiteness of the regulus into a sublimate of "solar [golden] yellowness" appropriate for preparing Suchten's potable gold.

Thus, using his negative and affirmative conclusions, a conjecture about the meaning of the scoria's mystery, and the conjectural process designed previously for obtaining Sulphur from the antimonial scoria (in the context of the Helmontian cinnabar), Starkey now produces a "best conjectural process" *[processus conjecturalis optimus]*. This process prescribes the collection of the scorias from the production of stellate regulus, the sublimation of the Sulphur out of them using "stinking spirit," the recombination of the sublimed Sulphur with the stellate regulus, and finally the attempt to sublime the desired "solar yellowness," that is, the volatile or Philosophical Sol or "Gold," out of this mixture.

In writing out this process, Starkey is both highly methodical and exhaustive. The usual production of the stellate regulus involves three or four reiterated fusions of the initially produced regulus with alkali salts. Each fusion further purifies the regulus by more completely extracting the external Sulphur and produces its own set of scoria more or less impregnated with this Sulphur. Accordingly, Starkey's conjectural process stipulates that each of the four sets of scoria (from the first, second, third, and fourth fusions) be kept separately, and then each mixed independently with stinking spirit and each mixture sublimed by itself to isolate its respective Sulphur. Each of these isolated Sulphurs is then to be ground with portions of the regulus and the several mixtures resublimed. Thus Starkey creates a sort of "combinatorial chymistry," methodically testing each possible combination of the Sulphurs isolated from the different scorias with stellate regulus by a uniform protocol to determine which, if any, contains the "mystery" able to elevate the desired Philosophical Sol from the regulus. Note also that here Starkey is hedging his bets, for Suchten had in fact claimed, as Starkey noted in his conjecture, that it was the scoria of the *first fusion* that held the mystery; perhaps Starkey suspected that Suchten might have been less than perfectly straightforward in this important matter, and so he seeks the mystery as widely as possible in the various scorias of all the different fusions. Finally, Starkey had noted in one of his affirmative conclusions that if the Philosophical Sol "allows itself to be easily separated [from the stellate regulus], why not with stinking spirit?" So he tacks one further permutation of ingredients onto his conjectural process, ordering himself to "sublime also regulus separately with stinking spirit."

So now we have in the first section of Sloane 3750 several conjectural processes—or in more modern parlance, "experimental protocols"—formulated by Starkey and based upon his reading, textual analysis, theoretical principles, and conjectures. What becomes of them? In the first cases, we have already seen that Starkey must have carried out the processes based upon Sala's and Van Helmont's procedures, for he returned to his conjectural processes and annotated them with the results, noting which parts worked and which parts did not. What of the greater arcana of Van Helmont and Suchten? In this case, the experimental trials of the conjectural processes actually survive in the second part of Sloane 3750, the section written from the back of the notebook forward, and entitled in Greek "Deuterai phrontides" ("Second Thoughts"). There Starkey records about two dozen experiments, all clearly based upon the two conjectural processes outlined previously. Judging from the substantial fraction of these that are crossed through, many were unsuccessful in practice (figure 2). On the first front, Starkey was apparently unable to make a Sulphur of antimony that looked like common sulphur from the scoria. On the second, although he could reliably prepare sublimates using stinking spirit and the various scorias, none proved able to elevate a "solar yellowness" out of stellate regulus. As an immediate result of these failures, he tried further permutations, including the substitution of simple regulus for the stellate regulus, and salt of tartar alone instead of a mixture of it with saltpeter. These further changes proved to be of no avail. The final result of this failure of his "best conjectural process" is elegantly apparent. Starkey returns now to the page of the notebook where he made the original conjecture regarding the "mystery in the scoria" upon which he based his conjectural process, and in the margin he writes simply, "Frivolum hoc" ["This is worthless"] (figure 3). The judgment of the fire condemned Starkey's conjectures, and accordingly, Starkey abandoned them.

It is clear from these examples that Starkey deploys a consistent methodology for his experimental program. He begins by assessing the state of practical knowledge of a process or the hints of better methods in cryptically written sources like Van Helmont and Suchten. He then analyses known methods, and based upon direct observation and theoretical considerations he makes conjectures (or predictions) of how to attain or improve the desired goal, and then incorporates these into conjectural processes—that is, tentative experimental protocols. Finally he puts these experimental protocols into practice and assesses the results. This sequence of evaluation-interpretation-observation-conjecture-experiment-assessment is repeated frequently in numerous projects throughout the notebooks. What we find here, clearly, is a laboratory practice guided by the methodical application of theoretical principles and direct observation, and that, not to overemphasize the point, in the laboratory work of the author of

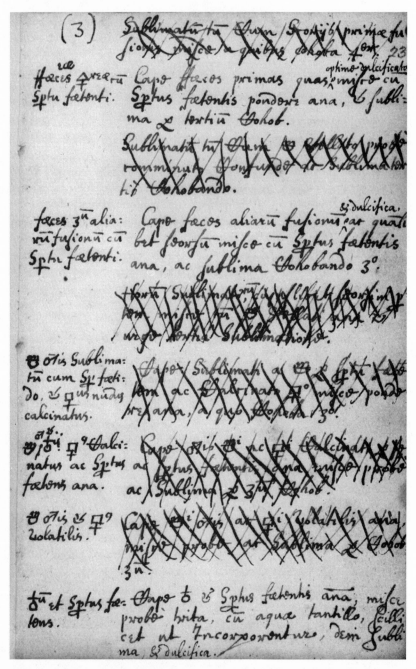

Figure 2. A page from one of Starkey's notebooks showing crossed-out unsuccessful experiments derived from a previous "conjectural process." Sloane 3750, fol. 23r. By permission of the British Library.

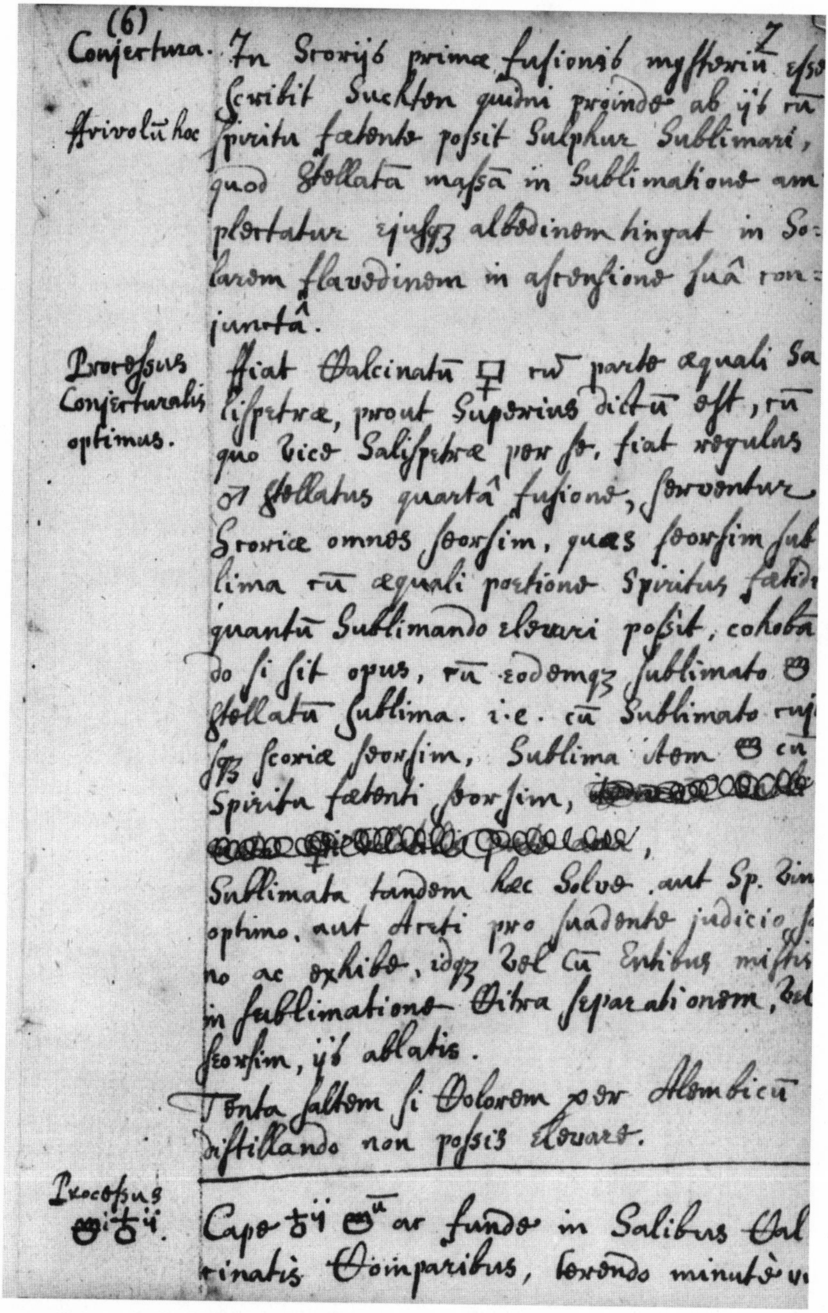

Figure 3. Practical experimentation has evaluated one of Starkey's "conjectures," and he records this with the note "this is worthless" [Frivolum hoc]. Note the "best conjectural process" indicated in the margin below. Sloane 3750, fol. 7r. By permission of the British Library.

some of the most allegorical treatises on transmutational alchemy of the
seventeenth century; one who has been called the seventeenth century's
"last great philosophical alchemist."

A RENEWAL OF EFFORT ON THE HELMONTIAN CINNABAR

Starkey was nothing if not tenacious, and so the failure of these conjectural
processes does not necessarily mean the end of the projects. While Starkey
does not seem to return (at least in this notebook) to Suchten's arcanum,
he does return to the problem of the Helmontian cinnabar. Since the trial
in the fire negated his conjectural process, and by extension, the conjectural
reading of Van Helmont upon which it was based, Starkey abandons his
earlier interpretation and approaches Van Helmont anew. Thus, further on
in the notebook Starkey gives a new interpretation of Van Helmont's terse
directions regarding the cinnabar. Now, in a section labeled "an opinion,"
he abandons the Sulphur from the scoria in favor of the Sulphur extracted
with aqua fortis (by his improvement on Sala's process) because that product "is most similar to common sulphur; why may this color not become
more green when ascending with sal ammoniac in a separate sublimation?"
He also reinterprets his earlier assumption, based upon the analogy with
common cinnabar, that the antimonial cinnabar requires the combination
of antimonial Sulphur with mercury. Now he suggests that Van Helmont's
use of the term "cinnabar" may allude only to the color of the material, not
to its composition. In conclusion, Starkey offers a new "Expositio Helmontii" by conjecturing that the greenish Sulphur and the cinnabar are actually one and the same substance, and that they merely appear in differently colored forms depending upon their modes of preparation.

> This therefore is the sense of the process: That the Sulphur which is separable
> from the kernel of Mercury, [and] sometimes appearing in a greenish form, is
> to be extracted in a cinnabarine form for this work. And then, after it has been
> freed from the added salt (by means of which it was separated in the first sub-
> limation), it is resublimed by itself six times in such fire that it may sustain the
> reverberation of Lili.[49]

We have no record of how Starkey tested his new opinion, for the record
of the pertinent experiments does not survive. These trials were presumably
recorded in a notebook that is no longer extant or that remains unidenti-
fied. Nonetheless, the tersely worded results of his trials are related clearly
enough. Immediately beneath the marginal tag labeling his new "opinion,"
Starkey has written (in a different ink) a simple phrase elegantly summariz-

---

49. *Notebooks and Correspondence*, document 11; Sloane 3750, fol. 8v.

ing a key aspect of his methodology of experiment: *Igne refutata,* that is, "refuted by the fire" (figure 4).

Starkey followed his study of antimonial processes in the "Antimoniologia" by a similar study of vitriol in a section entitled "Vitriologia." This follows the same format of citations from textual sources, textual analysis, conjectures, and the formulation of Starkey's trademark conjectural processes. There is one process, however, that adds a dimension to this investigation of Starkey's methods. After enumerating the various arcana preparable from blue vitriol (copper sulphate), Starkey turns his attention to one promised by Van Helmont in his *De lithiasi,* and called the "element of fire of Venus." This material is supposed to appear as a honey-sweet green oil, leaving the copper behind as a white metal, and represents the "external" Sulphur of copper.[50] Recall the Helmontian notion (mentioned above) that the metals contain two Sulphurs, one internal and necessary for coagulating the Mercury into a solid metal, and the other external and separable without destroying the metallic nature of the metal. We noted this previously in Starkey's studies of antimony, where he endeavored to isolate the external Sulphur of the antimony but recognized the existence of an internal Sulphur in the regulus as well. In this case, Starkey concludes from Van Helmont's words that nothing less than the alkahest is required to liberate the green external Sulphur from copper.[51] The alkahest, probably the most sought-after arcanum in seventeenth-century chymistry after the Philosophers' Stone, was a lifelong goal for Starkey, and we will examine his quest for it below. For now it suffices to note that in spite of some promising results communicated to Boyle in early 1652, Starkey was aware that he had not yet achieved the preparation of this most desirable liquor. Nonetheless, in this laboratory notebook he writes out a confident conjectural process that requires the treatment of colcothar (the residue of calcined vitriol, predominantly copper oxides) with "an equal weight of the liquor alkahest."[52] Thus Starkey here writes out a procedure for future trial, after such time as he achieves the alkahest. This entry corroborates the impression that the first part of the notebook (containing the "Antimoniologia" and "Vitriologia") was an organized phase of Starkey's program where the *development of new practical processes* was primary. The necessary testing of them occurred in a separate phase, recorded partly in the second part of the notebook ("Second Thoughts"), after which the experimental results were ex-

50. Van Helmont, *De lithiasi,* in *Opuscula,* pp. 68–73; for Boyle's interest in the same process, see Principe, *Aspiring Adept,* 88–89.

51. *Notebooks and Correspondence,* document 11; Sloane 3750, fol. 12v.

52. *Notebooks and Correspondence,* document 11; Sloane 3750, fol. 13.

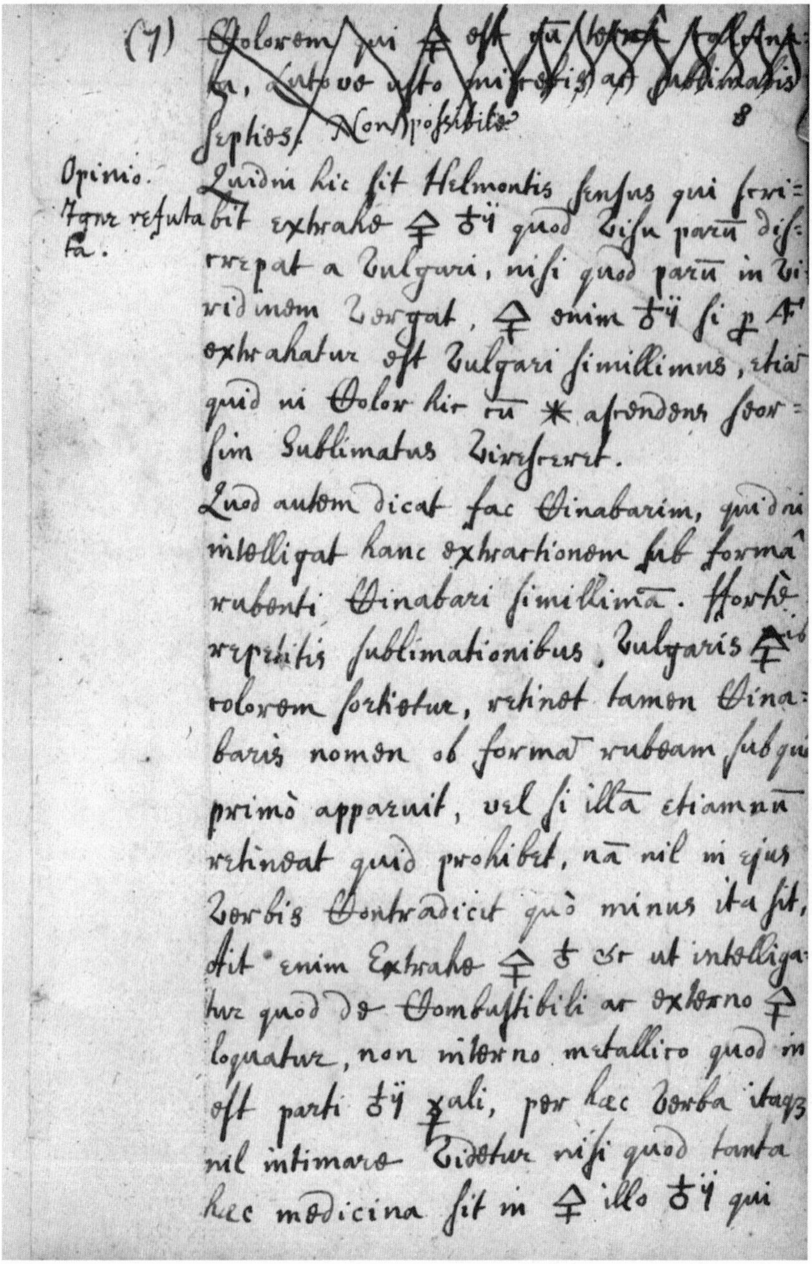

Figure 4. The fire of the chymical furnace has passed judgment on an "opinion" Starkey had formed from his reading of Van Helmont, and Starkey now records the verdict in the margin: "Refuted by the fire" *[Igne refutata]*. Note also that Starkey crossed out another unsuccessful experiment at the top of the page after he added "not possible" as the result of his practical trials. Sloane 3750, fol. 8r. By permission of the British Library.

ported back into the original conjectural materials and recorded by marginalia such as "proven" or "refuted."

These examples demonstrate some important aspects of Starkey's experimental practice. In the case of the receipt drawn from Sala, Starkey began with a simple and straightforward process, improved upon it by the application of theoretical principles to formulate a conjectural process, and then tested the process experimentally, using the results of that testing in the fire to amend the process into its final form. In the case of items drawn from Van Helmont and Suchten, Starkey endeavored to uncover higher arcana in their secretive or allusive texts by hazarding conjectural interpretations of them based upon theoretical considerations and textual analysis, drew conjectural processes from these conjectural interpretations, and then submitted these processes to the judgment of practical trials in the fires of his chymical furnaces. When the tests failed, he then rejected the conjectures upon which the experimental processes had been based and devised new conjectures that themselves had to be subjected to testing in due course. This program indicates where Starkey posited the ultimate source of authority—in the fire. *Experimental results afford final judgment on the truth of conjectures.*

Interestingly, all of the processes mentioned here begin with the authority of a written source. They were then developed through Starkey's own cogitations, but it is only through their practical success in experimental trials that real approbation is achieved. Starkey freely submits his conjectures to the fire and records in his notebooks the fire's judgment on them, both when it bears them out and when it renders them *refutatae* or *frivolae*. The title of *Philosophus per ignem* ("Philosopher by Fire") that Starkey claimed for himself was more than a clever Helmontian turn of phrase; it incapsulates the centrality he gives throughout his laboratory work to testing his thoughts and theories by accessing the phenomena of the natural world as exhibited in chymical trials.

While such insistence on the results of practical experimental trials has long been associated with the "New Science" of the seventeenth century, it has not always been widely associated with the chymistry of a Helmontian physician and transmutational writer like Starkey (much less Philalethes). And yet, if we look at Starkey's laboratory practice with a fresh eye, he seems to share more with the canonical figures of early modern science such as Galileo, Boyle, and Newton than he does with the popular image of the alchemist. In some respects, Starkey's notebooks could even be taken as models of "modern" laboratory investigation. But we have only begun to sample the richness of Starkey's laboratory notebooks, so before drawing too many conclusions, let us return to them for further illustrations of his methods.

QUANTITATIVE METHODS AND ANALYSES
IN TRANSMUTATIONAL ALCHEMY

In the previous chapter we argued that quantitative methods were by no means unknown in chymistry. Besides the simple and straightforward use of weights and measures in alchemical processes and assaying dating back to the Middle Ages, we provided some examples from sixteenth- and seventeenth-century chymistry of the use of quantitative methods as tools for monitoring chymical changes, investigating nature, and demonstrating theories. Starkey's notebooks provide—as we shall see—clear corroborating evidence by showing how quantitative methods were deployed on a regular basis in his laboratory work.

We must be careful, however, neither to overstate the case for the use of "mathematics" in chymistry nor to fail to recognize that the role of mathematics is not of equal significance in all branches of science even today. We believe that an undue degree of emphasis has sometimes been placed upon mathematics as a barometer of scientific sophistication, with the result that less mathematical sciences—such as early modern chymistry—have been marginalized and that historians of such disciplines have sometimes (unnecessarily) felt obliged to make their objects of study seem more mathematical as an argument for their dignity or importance. This situation arises partly out of the great success achieved by linear historical narratives of the Scientific Revolution that are built around the development of mathematized physics and astronomy from Copernicus through Galileo to Newton, but a full exploration of the origins of the privileged status accorded to the more mathematized sciences would extend well beyond the scope of this current study.[53] At present it suffices to point out that even modern chemistry, unlike physics, is more focused on the study of qualities than of quantities. The level of mathematical abstraction often celebrated in physics and astronomy is simply not useful (or even possible) in a discipline that focuses upon the qualities of different sorts of matter.

Moreover, the majority of chemists—from Starkey to the present—remain interested primarily in the *production of substances with specific qualities.* Practical chemistry emphasizes the making of substances and the observing of their qualities, not their reduction into numbers or mathematical relations. Later we must return to the important issue implicit in this observation regarding the experimental production of things versus the formulation of ideas. But at present we should be sensitive to the identification of the venues in which quantitative methods are likely to be useful

53. On this point, see for example, Margaret J. Osler, ed., *Rethinking the Scientific Revolution* (Cambridge University Press, 2000), esp. 3–22.

for the goals of chymistry: we can then search for the deployment of such techniques in those locales. Accordingly, in preparative chemistry quantitative methods find their most common use in the assessment or monitoring of practical processes. Assessment of experimental results often calls for quantitative measurements unable to be taken confidently by the direct application of the unaided senses. One key deployment was discussed in the last chapter, namely, the measurement of weights coupled with the notion of mass balance. In Van Helmont we saw its use predominantly in demonstrations of theoretical principles; in Suchten we saw a simple application of it to analysis. For Starkey, it is a key for his productive processes—both synthetically and analytically. In synthesis, it can be used to determine the throughput of a process, that is, how much of the starting materials are actually successfully converted to the desired product, and also to ascertain the importance of various side products. This provides the experimenter with measures of the efficiency (or yield) of a productive process, thus helping to direct improvements or to assess alternate methods. In analysis, where the relative amounts of different substances in a compound body are measured, quantitative methods are also useful, especially for identifying problems of "missing mass," that is, identifying where parts of a compound body escape invisibly or remain unaccounted for. By measuring the quantity of different substances isolated from a compound body, the experimenter can gain greater knowledge of that body—information that may itself be useful in designing and assessing productive processes. As Van Helmont put it, chymical knowledge depends upon knowing "what is contained in a body and how much of it there is."[54] Quantification of this sort—for the monitoring of specific processes of analysis and synthesis—features clearly in Starkey's chymistry. These aspects of Starkey's laboratory practice can be illustrated by two of his transmutational projects—the preparation of the Philosophical Mercury and the attempts to transmute the regulus of antimony into other metals.

THE PHILOSOPHICAL MERCURY: ANALYSIS AND OPTIMIZATION
We have already written extensively about the importance and preparation of Starkey's Philosophical Mercury. In brief, this material was believed to be the key first step in preparing the Philosophers' Stone and constituted the veiled centerpiece of the Philalethes treatises. Starkey's method for preparing this crucial substance has a long history of its own, stretching back well into the sixteenth century. Indeed, Starkey's process is based upon that of Suchten, which was outlined briefly in chapter 2. Starkey's specific contri-

54. Van Helmont, *De lithiasi*, in *Opuscula*, chap. 3, no. 1, p. 20.

butions to this avenue of chrysopoetic studies engaged the attention of
Boyle for forty years and attracted the interest of Isaac Newton and many
others as well.[55] The basis of the process lies in treating common mercury
with the martial regulus of antimony in order to "ennoble" or "acuate" it
into Philosophical Mercury, the metallic solvent capable of "radically" dissolving gold into its principles and readying it for preparation into the
Stone. In Starkey's first account of the process, candidly revealed in great
detail to Boyle in a letter of spring 1651, this treatment was carried out by
Suchten's method of alloying one part of martial regulus with two parts of
pure silver, amalgamating this alloy with mercury, laboriously grinding,
washing, and digesting this amalgam at low heat, washing away a black
powder that is emitted during the process, and finally distilling off the mercury. This lengthy operation had to be repeated seven to ten times; Starkey
termed each repetition an "eagle," and each eagle made the mercury more
noble and powerful in metallic dissolutions.[56]

Starkey's description of the process to Boyle is so specific in terms of precise quantities and detailed directions that when a Latin translation of a part
of his letter was found among Newton's papers, it was deemed to be Newton's own composition because *tanquam ex ungue leonem*—a process so
precise and clear as this could be the work only of the careful quantitative
mind of a Newton, and indeed such precision was unknown in alchemy and
"was, indeed, more than alchemy could survive."[57] This assumption says a
great deal about the view of "alchemical" operations as vague and impre-
cise; any lingering remnant of such a view should be adequately dispelled by
the present study.

The long-lost second half of Starkey's letter to Boyle (absent from the
Newton transcript and only recently rediscovered by us) records Starkey's
use of quantitative analysis to illustrate his Philosophical Mercury process.
While some of this is inspired by the vaguer analysis described by Suchten,
Starkey goes much further. Additional quantitative studies of this process
appear in *Sir George Riplye's Epistle to King Edward Unfolded* first published (without Starkey's knowledge) in 1655, but written before the sum-

55. Principe, *Aspiring Adept,* 152–180; Principe, "Chacun à Son Goût," *Sudhoffs Archiv;* Principe, "Apparatus and Reproducibility in Alchemy"; Newman, *Gehennical Fire,* 125–41, 165–68.

56. *Notebooks and Correspondence,* document 3; Boyle, *Correspondence,* 1:90–103; New-man, "Newton's *Clavis* as Starkey's *Key,*" *Isis* 78 (1987): 564–74.

57. Richard S. Westfall, *Never at Rest: A Biography of Isaac Newton* (Cambridge: Cambridge University Press, 1980), 370. See also Dobbs, *Foundations,* 175–86, and Westfall, "The Role of Alchemy in Newton's Career," in *Reason, Experiment, and Mysticism in the Scientific Revolution,* ed. M. L. Righini Bonelli and W. R. Shea (New York: Science History, 1975), 207–9, 229.

mer of 1653 and possibly as early as 1652, as part of the Philalethes corpus.[58] The emphasis on precision in Starkey's quantitative analysis and his concern with initial and final weights reflects the new style of chymistry pioneered by Van Helmont. Indeed, it is not too much to say that Starkey's gravimetric technique has much in common with the quantitative methods made famous by the chemistry of the following century.

According to Starkey's theoretical conception of the process, the reiterated incorporation of the regulus-silver alloy with common mercury has two effects on the quicksilver. First, the regulus "cleanses" the common mercury of the impurities of earthiness and saline wateriness that restrict its solvent abilities. Second, the regulus adds its own Mercurial substance to the common mercury, thus imparting a "fermental virtue" (namely the "incoagulable sulphur" that Starkey thought existed at the core of the corpuscles composing the martial regulus). This composite substance allowed the Sophic Mercury to "mercurify" other metals by separating them into their essential Mercury and Sulphur. Starkey attempted to prove these actions quantitatively with what he called an "Ocular demonstration" using weights. His instructions for washing the black powder from the amalgam are:

> make thy washings (for a tryal) with pure and clean Fountaine water; weigh first a Pint of the same water, and take the exact weight of it, then wash thy compound [the amalgam] 8 or 10 Eagles (or times,) save all the *faeces* [the expelled powder], weigh thy Body [the regulus used initially] and *Mercury* exactly, weigh thy *faeces* being very dry . . . [59]

After taking all these weights, Starkey submits the "faeces," that is, the black powder removed by washing, to gentle distillation and thereby removes "a portion of quick mercury" that the powder carried with itself from the washing.[60] Then he sets the residue in a crucible and heats it, as

58. Newman, *Gehennical Fire,* 59–60. For a transcript of Starkey's quantitative analysis of mercury by means of antimony, see *Gehennical Fire,* 323 n. 66.

59. George Starkey [Eirenaeus Philalethes, pseud.], "Epistle to King Edward Unfolded," in *Ripley Reviv'd* (London, 1678), 13; cf. *Sir George Riplye's Epistle to King Edward Unfolded,* in Samuel Hartlib, *Chymical, Medicinal, and Chyrurigical Addresses* (London, 1655), 19–47. There also exists a manuscript version of this work written by Starkey c. 1657 that is considerably different from both published versions and does not contain the precise quantitative analysis of the process; see *Notebooks and Correspondence,* document 13, for Starkey's own comments on the various editions.

60. Why Starkey first weighs the pint of wash water is not obvious from the 1655 text. Both Starkey's 1651 letter to Boyle and an unprinted version of the *Epistle* state that the black powder is collected only after it settles to the bottom of the wash water. Assuming that Starkey then decanted the liquid to get the powder, one can see how the wash water could now have been reweighed and compared to its original weight in order to reveal the presence and ap-

did Suchten, whereupon "you shall see the powder take fire & burne not as sulphur or with a flash but like a Vegitable coale"; this is the burning up of "all the faeculency of the *Mercury*," which the regulus removed from the common mercury.[61] But now Starkey reveals his debt to Van Helmont by determining exact proportions, unlike Suchten. In the *Epistle* he weighs "the remaining *faeces*," and "find[s] them to be two-thirds of thy Body [the original regulus]," wherefore he concludes that "the other third [is] in the *Mercury*," that is, that one-third of the weight of the original regulus was mercurial and that joined with the common mercury.[62] Similarly, in his letter to Boyle, Starkey recounts fusing the burnt powder and finding that it reduces back into regulus, which he weighs carefully and determines that "the pondus [weight] of this reduced Regulus will be within about $\frac{1}{4}$ or $\frac{1}{3}$ part as much as that which you added to the mercury with the silver." The fact that the majority of the regulus has been recovered verifies his assertion that the mercury gains not the entire weight of the regulus added to it, but rather only a portion of it, namely, the "fermental virtue" and not the whole body. The recovered regulus, he asserts, is now "without virtue at least not so much as it had before." Based upon the weights he has taken, he makes an analytical conclusion about the composition of the regulus, namely that the "Fermental spirit . . . is scarce a third part of the whole"; the remaining two-thirds of the weight is in the *"faeculent corporeous* parts of the Body[, which] come away with the dregs of the *Mercury"* manifested as the copious black powder removed by washing the amalgam. Thus Starkey's quantitative exercise not only works out the mass balance of the regulus through the process, but also gives him more precise quantitative information about its composition.

In the *Epistle*, Starkey goes further than in his letter to Boyle, for there he works out the mass balance not only for the regulus but for the mercury as well, an analysis not carried out at all by Suchten. To do so he compares the original weight of the common mercury he employed with the sum of the final weight of the Philosophical Mercury (minus the weight it gained from the regulus) plus the weight of the mercury recovered from the black powder, and finds that "the weight of both will not recompence thy *Mercuries* weight by far." That is, there is a problem of "missing mass." Where is the

---

proximate amount of the mercurial salts whose nature Starkey subsequently reveals to be crystalline by evaporating the water to a pellicle. See Starkey's 1651 letter to Boyle and the *Epistle* in Ferguson 85, p. 23.

61. Starkey to Boyle, April/May 1651, in *Notebooks and Correspondence*, document 3; Boyle, *Correspondence*, 1:98; "Epistle," in Starkey [Philalethes], *Ripley Reviv'd*, 13.

62. Starkey to Boyle, April/May 1651, in *Notebooks and Correspondence*, document 3; Boyle, *Correspondence*, 1:98; "Epistle," in Starkey [Philalethes], *Ripley Reviv'd*, 14.

missing mass of the mercury? Starkey has to go in search of it, so he evaporates the water used for the washings and finds it to contain "the Salt of *Mercury* Crude," which the regulus purged out of the common mercury. This Salt, along with the "faeculency", which burned away from the black powder before its reduction to regulus, accounts for the missing weight of mercury, and thus Starkey has a complete gravimetric analysis of common quicksilver. He has identified three components—pure mercury, a crude salt, and a combustible faeculency—and also found the relative weights in which each is present. Starkey then concludes happily "it is a content for the Artists to see how the Heterogeneities of *Mercury* are discovered."[63]

As we show in the accompanying text and table, Starkey not only ascertained the weights of the analyzed components of the common mercury; he also added up these weights and compared them to the total weight of the mercury before its analysis. As we will show in a later chapter, this insistence on comparing initial and final weights has more than a superficial resemblance to the "balance-sheet" methods of eighteenth-century chemistry.

This example illustrates Starkey's use of quantitative methods to analyze compound bodies; he uses similar gravimetric techniques in synthesizing products. After making the Sophic Mercury, the next step in preparing the Philosophers' Stone (according to Mercurialist principles) is the radical dissolution of gold in the prepared Mercury. Starkey apparently believed that finding exactly the correct proportion between the gold and Mercury was one key to the success of the operation. Sections in each of the Philalethes tracts witness Starkey's obsession with finding and using the correct relative weights. This "due proportion" occupies many stanzas of the *Marrow of Alchemy* and is emphasized as well in *Ripley Reviv'd:*

> for do not think it is all one . . . to put either one of the Body to two of the Water, or one to three, or two to three, or three to four; no verily, till you come to this . . . you are yet in the dark for Practise, though you may be true in Theory.[64]

Starkey's interest in determining the correct relative weights for this conjunction of Mercury and gold appears clearly in a very brief excerpt transcribed from one of his now-lost laboratory notebooks and containing experiments dated in February and March 1652. After having spent the latter half of February laboriously preparing the Philosophical Mercury (of nine eagles), Starkey turns to the digestion of this product with gold. Al-

63. Ibid.
64. Starkey [Philalethes], *Ripley Reviv'd*, 142; see also Starkey [Philalethes], *Marrow of Alchemy*, pt. 2, 23–27.

STARKEY'S GRAVIMETRIC ANALYSIS OF MARTIAL REGULUS AND COMMON MERCURY

Starkey first weighs a quantity of common mercury, martial regulus, and a pint of water. He then amalgamates the mercury with the regulus, which has been independently fused with two parts of silver, and then grinds and washes the amalgam repeatedly in the water to remove the black powder that emerges. After this treatment, he distills off the mercury and then amalgamates it again with the silver-regulus alloy, repeating the entire process of amalgamation, grinding, washing, and evaporation from seven to nine times. After the completion of this reiterated procedure, he weighs the resulting Sophic Mercury (produced from the last distillation). He then allows the black powder to collect at the bottom of the wash water and decants the water. Having separated the black powder, he now dries it and weighs it, and after that, he evaporates some residual common mercury from the powder. He weighs the black powder after the removal of this common quicksilver, then ignites the powder to a slow burn. After the powder has finished burning, Starkey reduces it to regulus (as he says in his 1651 letter to Boyle), and weighs the resultant regulus, to find that it weighs two-thirds as much as the original regulus he employed. This means, by mass balance, that one-third of the regulus is now in the Sophic Mercury.

> weight of original regulus
> −weight of regulus reduced from black powder
> _____
> *weight of regulus incorporated into the Sophic Mercury*
> *($\frac{1}{3}$ weight of original regulus)*

Starkey now adds the weight of the common mercury evaporated from the black powder to the weight of the Sophic Mercury (minus its reguline component) and finds that the combined weights do not add up to that of the initial common mercury employed.

> weight of Sophic Mercury
> −weight of regulus incorporated into the Sophic Mercury
> _____
> *weight of common mercury in Sophic Mercury*

> weight of common mercury in Sophic Mercury
> +weight of common mercury evaporated from black powder
> _____
> <*weight of original common mercury*

This missing mass leads Starkey to suspect that the wash water contains the missing mercury, a hypothesis that he confirms by evaporating the water to a pellicle to show that it now contains a crystalline salt. Since he explicitly notes that he "exactly" weighed a pint of water at the beginning, he presumably intended to weigh the decanted wash water at the end as well. We summarize Starkey's calculations as follows:

> weight of decanted wash water after washings
> −weight of initial wash water
> _____
> ≈ *weight of missing mercury*

> weight of common mercury in Sophic Mercury
> weight of common mercury evaporated from black powder
> +weight of mercurial salt from wash water
> _____
> ≈ *weight of original common mercury*

though he had received some hint from his study of the fifteenth-century writings of George Ripley regarding the relative proportions to use, Starkey nonetheless here employs an empirical approach to solving the problem definitively. Starkey sets up not one, but three digesting flasks; the first contains gold and Philosophical Mercury in a weight ratio of 4:3, the second flask uses the ratio 1:3, and a third, 1:2.[65]

The quantitative analytical procedures Starkey used to discover the composition of common mercury and antimony regulus are later used by him to improve the process of making the Philosophical Mercury. The greatest problem with the original receipt (as drawn from Suchten) was its utter tediousness. Hours of laborious grinding, washing, and digestion were required for each of the seven to ten eagles. Additionally, the use of silver was costly. In late August 1653, Starkey tried to solve both problems by replacing the cycles of grinding and gentle digestion of the regulus-silver alloy and mercury with a longer period of much more vigorous digestion of a mercury "of two eagles" with pure regulus and without any additional silver.[66] Starkey had previously followed Suchten's prescriptions, which required silver in order to allow the antimony to unite with the mercury. (Whereas mercury amalgamates readily with most metals, it will not do so with antimony; silver thus acts as a medium since it both alloys with antimony and amalgamates with mercury.) But Starkey now tries a method very different from that stipulated by Suchten.[67] Starkey is led to this new method by his own observations and his desire—evident in every project he tackled—constantly to improve upon processes in terms of yield, efficiency, cost, and ease of operation. Starkey has observed that when a strong enough fire is used, the temperature of the mercury in the digesting flask "greatly exceeds the temperature of molten lead, at which temperature however regulus is molten and is united with the mercury." Thus Sophic

---

65. *Notebooks and Correspondence,* document 7; Ferguson 322. A similar concern for proportions in the conjunction occurs in document 12 (RSMS 179, fols. 4v–2v).

66. The reader may wonder why Starkey uses a mercury "of two eagles" (meaning mercury that has undergone two cycles of "acuation") instead of common mercury since he is trying to simplify and expedite the process. After each eagle the resolving power of the Philosophical Mercury is increased, as the result of the successive additions of "fermental virtue" from the regulus and the successive purgation of the mercury from its impurities of earthiness and saline wateriness. Seven to ten eagles are required for the decomposition of the noble metal gold into its Mercury and Sulphur. Baser metals are of a looser composition and are thus easier to resolve, and so fewer eagles are required for the mercury designed to dissolve them. Starkey apparently estimated that two eagles would be sufficent for antimony regulus, an immature and weakly composed metal. Common mercury—mercury of no eagles—having no fermental virtue at all and remaining clogged with its impurities, would have no effect upon the regulus boiled with it.

67. Suchten, *Of the Secrets,* 75–76.

Mercury and antimony, both being liquids at that high temperature, could "be joined without any medium . . . except the medium of fire alone."[68]

Starkey now employs quantitative methods to monitor the success of the new process. Recall that Starkey had determined by his analyses that most of the expelled black powder was reducible back into regulus. Starkey then suggests that he can determine the ongoing progress of the permanent incorporation of the regulus with the mercury by monitoring the weight of residual regulus recoverable from the expelled powder.

> Experience repeated a hundred times taught me this, that when regulus is amalgamated with mercury through the mediation of silver . . . the greatest part of the regulus is spewed out safe and sound, and by fire may be reduced back into its pristine state after a little bit of external combustible Sulphur has burned off like beech charcoal . . . And so, I predict that this work done with a long digestion will mercurify more [of the regulus], at least as much as the mercury can take hold of.[69]

The idea here is that the regulus must be decompounded in order to release its acuating Mercury into the Sophic Mercury being formed. When this happens, the liberated antimonial Mercury unites irretrievably with the Sophic Mercury, while the Sulphur and the undecompounded regulus "is spewed out safe and sound." Therefore, by monitoring the quantity of unchanged regulus in the black powder, Starkey can determine by subtraction how much has actually been decompounded and thereby assess the progress of the incorporation of the reguline Mercury into the Philosophical Mercury. So Starkey directs himself to "determine by the usual means whether the powder collected by washing contains within itself so great a quantity of regulus as before."[70] If he finds less regulus there, it means that more has been decomposed, meaning more antimonial Mercury has been incorporated with the Sophic Mercury, and so the process is more successful.

If more regulus is mercurified in each digestion cycle, then Starkey also saves on the total number of eagles needed. Using the antimony-silver alloy limits the amount of regulus that could be used at each eagle, since regulus makes up only one-third of the total alloy. But in the newly improved method using pure regulus a greater quantity can be used at every diges-

68. *Notebooks and Correspondence*, document 11; Sloane 3750, fol. 25.
69. *Notebooks and Correspondence*, document 11; Sloane 3750, fols. 26r–26v.
70. *Notebooks and Correspondence*, document 11; Sloane 3750, fol. 26v.

tion.[71] Starkey calculates the savings by noting that "by that method [using silver], a single eagle will employ three ounces of regulus, or at most four, but by this method it will employ 16 ounces . . . thus one eagle is worth five of the other eagles."[72]

Finally, without silver, a considerable savings in cost is realized, and Starkey calculates that this translates into a substantial savings of thirty-three shillings for every two pounds of mercury.

Starkey's hands-on trial *[proba mechannica]* of the new and improved process underlines his concerns for quantifiable success. Simple observation was highly encouraging, for after a few hours Starkey saw that "the mercury had embraced the body, and in a word, the operation fulfilled expectations completely as desired."[73] But only the quantitative monitoring of the black powder could demonstrate the level of success verifiably. Unfortunately, when he tried to collect the powder for verification, misfortune struck. In Starkey's words, "though Nature may be a dear mother to me, Fortune is a step-mother"; for a considerable portion of the mercury spilled, and "fell irretrievably into cracks." Yet worse, after he set the remainder to digest further, the next morning he found the glass broken in the fire, "and I do not know with how much loss."[74] Clearly Starkey was upset not only at the loss of materials but also at the fact that the uncertainty in the weight of lost material rendered impossible his attempt at quantitative measurements.

Unfortunately, we have no further surviving records of Starkey's attempt on this improved process or information regarding how his monitoring of the weight of recoverable regulus turned out. Just at this time circumstances outside the laboratory intervened in Starkey's work, and his sequentially dated notebook entries cease abruptly at the end of August 1653. Indeed, all the extant notebooks show long gaps from September 1653 until May 1654 and again from June until November 1654. This period corresponds with Starkey's financial troubles and his confinement to debtor's prison, as recounted in a bitter letter of 28 February 1654 from Samuel Hartlib to Robert Boyle, who was then in Ireland seeing to his hereditary lands.[75] Starkey's miseries of 1653–54 are thus loudly recorded by the si-

---

71. The limiting factor is the amount of solid alloy that can be added to the mercury without rendering the amalgam itself solid and thus unable to be properly ground and washed. This factor does not play a role in the improved process of digestion at high temperatures.

72. *Notebooks and Correspondence*, document 11; Sloane 3750, fol. 25.

73. Ibid., fol. 27v.

74. Ibid., fols. 27v–28.

75. Hartlib wrote to Boyle that "Dr. *Stirk* . . . is altogether degenerated . . . I know not how many weeks he hath lain in prison for debt; but after he hath been delivered the second time, he hath secretly abandoned his house in London, and is now living obscurely, as I take it,

lence of his laboratory notebooks; indeed, in spite of the sophistication of his laboratory methodologies, and even in spite of the smiling of God and Nature upon his endeavors, stepmother Fortune did have her day.

THE ANTIMONIAL METALS PROJECT

Quantitative methods also appear in another transmutational endeavor that Starkey carried out at the same time. As noted previously, Starkey was a keen reader of Alexander von Suchten; besides the preparation of the Philosophical Mercury and of the volatile Sol, Starkey worked on a third Suchtenian project drawn from the Prussian's *Tractatus secundus*. This involved producing the six solid metals (gold, silver, iron, copper, tin, and lead) from regulus of antimony. Suchten had claimed that "out of this Regulus all Metals may be made . . . as good as the natural Metals."[76] Although these antimonial metals may be "as good as the natural Metals," Suchten notes that there are differences between them (e.g., the artificial lead is harder than the natural), and he demonstrated—as we saw in chapter 2—that they were not the same by reducing the "gold" and "silver" into mercury, even though their other observable qualities were sufficient to convince a goldsmith of their goodness.

As early as spring 1651 Starkey claimed success in producing gold and silver from regulus. John Dury apparently witnessed this operation, and several members of the Hartlib circle, especially Benjamin Worsley, strongly encouraged Starkey to devote himself wholly to this operation in order to turn a profit for both himself and the Hartlibians. Starkey, however, refused "in such a way of lucre [to] prostrate so great a secret" and "waved the motion, not willing to imbrace a life (in Exchange of a studious search of Natures mysteryes) which might be Compared with that of a Milhorse running round in a wheele today, that I may doe the same tomorrow."[77] It is difficult to determine exactly what Starkey performed before Worsley, although it is very likely that the process was related to several recipes for making *luna fixa* and tinting it to gold that are published in the *Experimenta* of Eirenaeus Philalethes and further described in Robert Boyle's work diaries.[78] The operation may have in fact involved an isolation of the

at *Rotherhith*." Hartlib to Boyle, 28 February 1654, Boyle, *Correspondence,* 1:156. We discuss this letter further in chapter 5.

76. Suchten, *Of the Secrets,* 99.

77. Starkey to Boyle, April/May 1651, in *Notebooks and Correspondence,* document 3, and in Boyle, *Correspondence,* 1:90–103.

78. George Starkey [Eirenaeus Philalethes, pseud.], *Experimenta de praeparatione mercurii sophici,* in Starkey [Philalethes], *Enarratio methodica trium Gebri medicinarum* (Lon-

trace amounts of precious metals often present in native antimony ore.[79] Nonetheless, this extraction method of 1651 differs from the transmutational experiments on antimony regulus dated from 1653 to 1655 in two of Starkey's surviving notebooks. These latter experiments use high temperatures and molten salts as fluxes, whereas Starkey explicitly states that his earlier "extractions of gold and silver" were "not done by violent heats, fluxes, waters or the like."[80]

The principle behind Starkey's experiments to prepare these artificial metals is that regulus of antimony is close to the first matter of the metals and hence "indeterminate." By treatment with portions of the particular metals it may be "specified," and thus transformed in its nature into each of them. "Regulus is a chaos as I call it," he writes, "out of which all the metals can be drawn."[81] The need for specifying the chaotic regulus toward a specific metal is made clear when Starkey asks himself the question "How are Sol [gold] and Luna [silver] made from regulus?" He answers that they are made "not from regulus alone, for in fact all metals can be drawn from regulus, which implies a certain specifying addition without which the regulus remains regulus and receives no transmutative alteration."[82] So the proper treatment of regulus with, for example, tin, would convert the regulus itself, or at least some part of it, into tin.

Starkey's experiments on this project began in late summer 1653. He was perhaps provoked to these trials as the result of information that he heard (and recorded in his notebook) about a certain Major Purling who reportedly made a "beautiful tin in London" using antimony regulus—"this work ate up almost all the antimony in the city, for the simples-sellers told me that it had all been bought up by him."[83] Besides this gossip from the sellers of simples (powdered matter of a single vegetable, usually sold for medicinal purposes), Starkey received further information (which he recorded as a "historia facti") about Purling and his tin from Sir Cheny Cul-

---

don, 1678), 183–88. For other editions, see Newman, *Gehennical Fire*, nos. 19, 22, p. 268, and no. 24, p. 269. On *luna fixa*, see below, p. 131.

79. See Newman, *Gehennical Fire*, 139–40.

80. *Notebooks and Correspondence*, document 3; Starkey to Boyle, spring 1651, and in Boyle, *Correspondence*, 1:93.

81. *Notebooks and Correspondence*, document 11; Sloane 3750, fol. 24v. The term "Chaos" is used as a *Deckname* for antimony regulus in Starkey [Philalethes], *Introitus apertus ad occlusum regis palatium*, in *Museum hermeticum* (Frankfurt, 1678; reprint, Graz: Akademische Druck, 1970), 655–56.

82. *Notebooks and Correspondence*, document 11; Sloane 3750, fol. 29v.

83. *Notebooks and Correspondence*, document 11; Sloane 3750, fol. 24v. Starkey initially spells the name "Spurling," but changes this to "Purling" on fol. 31.

peper in May 1654.[84] This person is likely to have been one Major Erasmus Purling, originally of the Isle of Jersey, who had commanded troops for Cromwell. He was imprisoned for unknown reasons in January 1654, and a letter of petition from him, dated 31 July 1654, requests not only his release but also certain "ready moneyes." He adds that if these cannot be paid, then he would be content with "part moneys for my present releife & the rest out of such discoveryes as shall be by mee made." What these "discoveryes" might be is unspecified, but if, as is likely, this is the same Major Purling as the one mentioned by Starkey, then they may have been related to transmutational endeavors.[85]

Starkey's first recorded attempt to produce silver from antimony occurred on 16 August 1653. He fused four ounces of pure silver with 29 ounces of regulus of antimony, added various salts, and evaporated off the regulus in a hot fire "with wearisome labor," but upon weighing the final product he found no increase of weight in the silver. Undaunted, Starkey tried again with the same procedure on 18 August, and a third time on 19 August, but finally concluded that "I do not see that I have gained a single grain [of silver]."[86] Therefore, on 19 August Starkey took stock of affairs in order to devise a new method, and accordingly he wrote out a series of six points regarding "how Suchten probably made metals out of this regulus." The key observation upon Suchten's text that Starkey makes here is that "the fermental odor of copper (which [Suchten] calls vegetating copper) is required."[87] Accordingly, in an experiment dated 20 August 1653, Starkey tries to produce not salable silver but lowly lead—a process Suchten claimed would be much easier. Starkey records that he first prepared a special regulus by fusing one ounce of stellate regulus with one ounce of col-

84. *Notebooks and Correspondence,* document 11; Sloane 3750, fol. 30v. On Culpeper see M. J. Braddick and M. Greengrass, *The Letters of Sir Cheney Culpeper (1641–1657),* Camden Miscellany 33 (Cambridge: Cambridge University Press, 1996), and Stephen Clucas, "The Correspondence of a XVII-Century 'Chymicall Gentleman': Sir Cheney Culpeper and the Chemical Interests of the Hartlib Circle," *Ambix* 40 (1993): 147–70.

85. British Library Additional Manuscripts 24861, f. 97; letter of Purling to Richard Major, 31 July 1654. The document also indicates that Purling was sent for out of France on 14 October 1650 by the Council of State who had received and approved "some proposals from him," although these remain unspecified. There are many references to Purling in the Parliamentary records from 1651 to 1655; on 25 April 1655, for example, Parliament did in fact order a payment "for consideration of losses of £3000 in reducing Jersey," *Calendar of State Papers* 1655, 143. Furthermore, a letter (c. 1674) from Purling to the governor of Tangiers survives in which Purling entitles himself "Engineare"; Sloane 3511, f. 258.

86. *Notebooks and Correspondence,* document 11; Sloane 3750, fol. 24.

87. *Notebooks and Correspondence,* document 11; Sloane 3750, fol. 24v (the reference is to Suchten, *of the Secrets,* 100).

cothar, the copper-containing residue left from the commercial production of aqua fortis from vitriol and niter. He then separately prepared another special regulus by fusing stellate regulus with an equal weight of minium (red lead, a lead oxide). Finally, Starkey fused these two reguluses together and "Thus the whole was made into Saturn [i.e., lead]."[88] He immediately describes the properties of this "Saturn of antimony" noting its observable differences from "natural" Saturn, or lead.[89]

Encouraged by this success, Starkey writes out a "conjectural conclusion" that for transmutation to occur, the regulus must be first fused with copper. By "conjectural conclusion" Starkey seems to mean a conclusion based upon his interpretation of an experimental result that nevertheless requires further verification. He then returns to "the making of Luna [silver] by the mediation of Venus [copper]" and writes out a series of questions that need to be answered; these include the appropriate and exact weights of starting materials to be employed, how many fusions are required, and what salts, if any, are to be used in the process. Just as we saw above in the case of the improved method of making Philosophical Mercury, Starkey's entries come to an abrupt end at the end of August 1653 owing to his financial hardships. Only in May 1654 is Starkey able to return to the practical pursuit of the conjectural conclusion he penned nine months earlier. Thereupon he draws up some further observations and conclusions from Suchten to the effect that Suchten's "Luna" is not the same as common silver, for its properties are too widely different, and in fact it contains no silver at all. These distinguishing properties include resistance to aqua fortis, solubility in aqua regia, and a greater density than common silver, all of which identify Suchten's Luna as *luna fixa,* a white metal with the properties of gold but lacking its color.[90]

On Saturday, 18 May 1654, Starkey finally begins a series of experiments on fusing regulus with copper and silver. He fuses together $1\frac{1}{2}$ ounces of silver, 3 ounces of copper, and 3 ounces of regulus and keeps the mixture in a "most fervent fire" for a total of seventeen hours. At the end of this time "the weight was just about what it was originally, that is, $4\frac{1}{2}$ ounces." Thus, relying upon the concept of mass balance, Starkey determines that only the fixed metals silver and copper remained and that none of the volatile regu-

88. *Notebooks and Correspondence,* document 10; Sloane 3711, fol. 3.

89. One of these is that "its smoke ascends vigorously in a glowing fire," which is not surprising since it must contain about 50 percent antimony, which in a molten state would be oxidized to the volatile antimony trioxide, which would pass off in white (toxic) clouds.

90. See above, chapter 2; on *luna fixa,* see Newman, *Gehennical Fire,* 140; Principe, *Aspiring Adept,* 81 incl. n. 58.

lus had been fixed into silver.[91] He tried the process again, now adding corrosive salts, but on the fourth fusion the crucible broke, and Starkey was unable to follow the process quantitatively because "I don't know how much I lost," even though he finally "obtained $2\frac{1}{2}$ ounces of mixed metal the appearance of which upon the touchstone was pale yellow, more like Sol [gold] than Venus [copper]," which he nonetheless dismissed as "a Sophistic Sol [gold]."

In the following weeks Starkey made other attempts at producing Suchten's silver, and in these cases he carefully marked down not only the exact quantities of materials used, but even marked off in the margin a running account of the cumulative amount of time the matter remained in the fire and determined the total residual weight after each of these specified periods of time (figure 5). But the recurrent problem of laboratory accidents—cracking glasses, breaking crucibles, spilling vessels—plagued Starkey's attempt to keep a quantitative eye on his work. In one case the crucible suddenly tipped over and spilled "a sizeable part" of its contents (Starkey marks this occurrence in the margin of the notebook as *casus*, "an accident"). Starkey, disheartened at the difficulty this would now present to keeping a quantitative account of the material, actually took the remarkable step of disassembling his furnace, removing the ashes from the grate and "recovered what I could by sifting the ashes with the greatest diligence; nonetheless I do not believe that this spillage occurred without loss."[92] He also notes that some of the weight is lost "by some grains flying off" when the salts are added, and this will skew his measured results, for even if the "loss is produced insensibly, it is real in the end." Again it is clear that Starkey is obsessed with getting accurate weight measurements as guides to his practice and checks to his processes. Starkey records similar problems when he tries an experiment where the regulus is added in small portions at two-hour intervals; indeed, his candid account of this new *casus infortunatus* (as it is listed in the margin) is almost enough to bring one to tears:

> I added about five and a quarter ounces of new regulus, but there was a misfortune, for by throwing a certain part in pieces into the greater fused part, it made a certain part of the uppermost liquid splash out, and scattered many little grains upon the walls of the furnace. Indeed, I diligently collected what I could, but while melting it a wild boiling carried off some part of it into the fire. Then while pouring it out into an iron dish, the crucible fell over and some spilled out, of which I collected as much as I could, and there remained

---

91. *Notebooks and Correspondence*, document 11; Sloane 3750, fols. 28v–29. Starkey also added an ounce of zinc—the reasons for this are unclear—but he noted in the margin that it was "an error in adding zinc" and that it took nine hours for the zinc to evaporate entirely.

92. *Notebooks and Correspondence*, document 11; Sloane 3750, fol. 30.

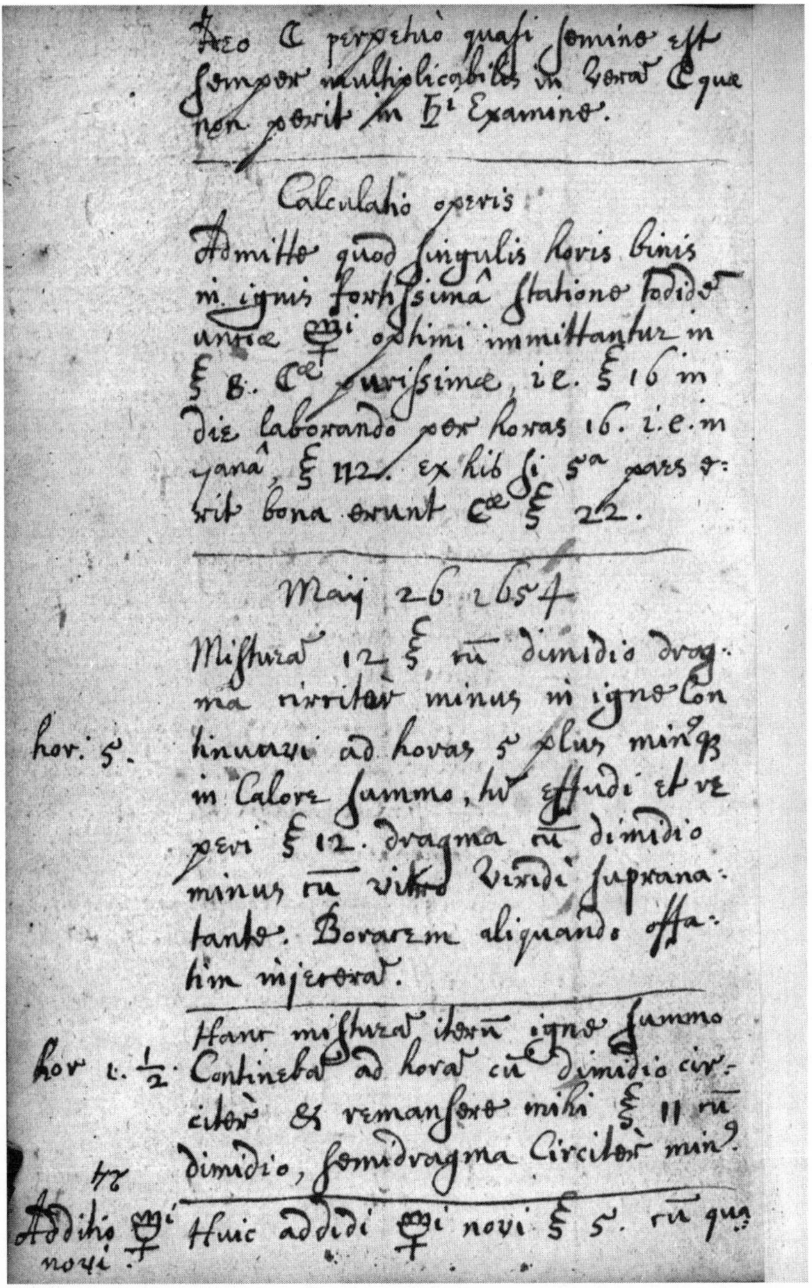

Figure 5. A page from one of Starkey's notebooks where he is trying to transmute antimony regulus into silver. Note the dated entry, the number of hours of heating recorded in the margin, and his calculation of the quantities of material needed for the long-term operation. Sloane 3750, fol. 31v. By permission of the British Library.

one and a half drachms less than 16 ounces, and so I was missing about five or six drachms of the quantity I had put in today, of which some part, in fact I think the greater part, was lost in burning off.[93]

Such candid accounts of Starkey's laboratory mishaps provide a vivid sense of the difficulty and frustration inherent in laboratory work, especially in a period when the quality of the equipment and materials was unreliable. Many of Starkey's experiments in fact—as is common in research projects of any epoch—pushed the limits set by the material technology of his time. It is all too easy for historians reliant upon the polished versions of events found in published sources to overlook the actual obstacles and inevitable accidents that occur whenever new ideas and processes are introduced to the real, physical world of laboratory apparatus. This is not rhetorical posturing or self-exculpation but rather an unvarnished private account of what Starkey faced in the laboratory, and his account will not seem at all unfamiliar to any laboratory worker of the present. For Starkey, however, the real world obstacles he faced were not restricted to the laboratory; they also included severe financial pressures. Starkey's trials on this project break off suddenly again in the middle of a process on 27 May 1654, and this is likely due to a return of his financial problems. None of the notebooks now known shows any entries at all until late November 1654, and Starkey did not return to continuous experimental activity until March 1655.

Although it will not tell us any more about Starkey's use of quantitative methods, it is worth finishing the story of his search for the secret of the Suchtenian metals from antimony for it serves further to illustrate the pathways of his laboratory work. The next entry dates from over a year later, on 23 August 1655. By that time Starkey had decided that neither fire alone nor salts were able to bring about the transmutation of regulus, and he refers to his many trials recorded in "The Book of Chymical, Medical, and Physical Miscellanies," which is presumably another notebook that is not presently known. He also recounts how he had returned to an idea he had rejected in 1653, namely, of using mercury as a "mediator" between the regulus and the silver; "but by experiment I was taught," he writes, "that that possibility hardly contains truth in its foundation. For in fact I made the experiment, . . . yet having well weighed its failings . . . I conclude that this is not the way of attaining this antimonial Luna, at least not that of Suchten."[94] On 10 November 1655, he summarizes further failed attempts and even notes that he had tried to prevent the earlier problems with the physical limita-

93. *Notebooks and Correspondence,* document 11; Sloane 3750, fols. 31v–32.
94. *Notebooks and Correspondence,* document 11; Sloane 3750, fol. 32v.

tions of his apparatus by using imported Hessian crucibles rather than the common sort, but even these "were not able to withstand the corrosive force of the salts." Nonetheless, he produces yet another conjectural process that stipulates keeping one part of silver and two parts of regulus molten in a sealed crucible for six or seven weeks.[95]

In January 1656, Starkey returned to Sloane 3711 (where the project began on 20 August 1653) to write the final entry on this project, inserting it immediately after the entry recording his production of Suchtenian lead. The impression given by this entry is that the conjectural process written on 10 November showed some kind of success: "I have learned from diverse experiments that only continued digestion in a well sealed crucible or little vessel is necessary, without the addition of anything except the body that is sought and the multiplicative air."[96]

What this success was we will probably never know, but Starkey (like Suchten before him) had previously had some success in selling what were presumably "artificial" precious metals to goldsmiths. Clodius reported to Hartlib in summer 1653 that "Mr Stirke's silver" was being bought at forty shillings an ounce; this price is more than eight times the rate for silver that Starkey himself paid in 1653, and so this metal was clearly a special kind of "silver."[97] Presumably it was the "luna fixa" that Starkey told Hartlib about on 2 March 1653—a kind of "silver equivalent to gold wanting nothing but the colour" and that "did undergoe all the trials of the Goldsmith."[98] Indeed, a recipe (though with the proportions missing) for "Mr. St. Luna-fixa" exists in Hartlib's handwriting, and the process described is very similar to those Starkey explored in Sloane 3711, involving extended fusions of regulus of antimony with silver and the copper residues from aqua fortis production.[99]

95. *Notebooks and Correspondence*, document 11; Sloane 3750, fol. 34v. It is worth noting that the expense and trouble of maintaining for nearly two months a charcoal fire hot enough to keep metals molten would have been immense.

96. *Notebooks and Correspondence*, document 10; Sloane 3711, fol. 3v. The entry is undated, but the red-brown ink in which it is written is characteristic of Starkey's time in Bristol in late 1655 and early 1656, and the immediately following entry (on a different topic, but written in the same ink) is dated 25 January 1656. Note also that if the entry was in fact written in January 1656, that would be six or seven weeks after the formulation of the 10 November 1655 conjectural process—exactly the amount of time stipulated there.

97. Starkey notes in a 1653 notebook that six ounces of silver cost 33 s, which works out to 4 s 6 d per ounce. *Notebooks and Correspondence*, document 11; Sloane 3750, fol. 2s.

98. Hartlib, *Ephemerides* 1653 (2 March), HP 28/2/54A.

99. HP 16/1/70A–B. The fact that all the proportions are left blank might suggest that the receipt was jotted down by someone who watched Starkey carry out the process, rather than actually given freely to Hartlib by Starkey himself.

THE VOLATILIZATION OF ALKALIES AND STARKEY'S
GRAND DESIGN FOR MEDICINE

The final example for this chapter comes from what Starkey himself may have viewed as the most important project of his entire career—the volatilization of alkalies. This project bears special significance for several reasons. Along with the search for the Philosophers' Stone and the alkahest, the project is one of Starkey's longest-running pursuits and probably the one to which he devoted the most experimental trials. As such, it allows us to follow the evolution and reconceptualizations of a chymical research project over a long period. After some ten years of work, Starkey achieved success in the mid-1650s, a fact that is first adumbrated in Boyle's *Philosophicall Diary* between 1 and 12 January 1655. There Boyle records that Starkey had succeeded in making crystals of an Elixir of volatile salt *(elixir salis volatilis)* by digesting salt of tartar with essential oil in a sealed vessel kept at moderate heat for three or four months.[100] Soon thereafter, Starkey would modify this process in a way that he evidently considered to be a breakthrough. The secret that he arrived at can be compared in importance only to that of the Philosophical Mercury, for just as Starkey's Philalethes tracts center on allusive revelations of the Mercury's preparation, the two major works published under Starkey's own name center on the secret of volatilizing alkalies. Starkey never openly reveals the crucial part of the secret there, and only by turning to the notebooks can we discover it. Thus, the private notebooks explain the allusive public text and also illustrate the otherwise hidden day-to-day experimental background to Starkey's publications. This is a unique opportunity for the study of seventeenth-century chymistry.

On yet a higher level, the volatile alkalies project became central to a far greater enterprise—the complete reformation of medicine and pharmacy through the development of a single, universal chymical method of preparing medicines. By this one uniform chymical method, Starkey believed that any substance could be prepared into a medicine that was safe, pleasant, and efficacious. While the initial ingredients of this grand design come from several parts of Helmontian chymical theory, Starkey fuses them together with his own ideas in an original way and expands them in the light of his

---

100. BP, vol. 8, fol. 141v, no. 20:

To make Elixir satis [*for* salis] volatilis, Recipe Essentiall oyle 2 Parts, pure Salt of tartar one part (Stirke sometimes told me he tooke 3 parts of oyl & two of Salt) & let them circulate with a Bottome heat 3 or 4 months. The salt will be like sugar-candy & somewhat tincted by the oyle, & will sticke to the sides of the Glasse (which must be large & strong, & exquisitely stopt with Helmonts Lute ex cerâ & colophoniâ) at the Bottom of which notwithstanding som liquor will remaine. Stirkius.

laboratory experiences to create a theoretical and practical system of chymical medicine well beyond that envisioned by the Belgian chymist. Thus this project showcases not only Starkey's methods, originality, and tenacity, but also his desire to reduce his numerous experiments into *generalized principles of chymistry*. We have already shown how theoretical principles guided Starkey's practical laboratory activities; the volatile alkalies project now illustrates how the results of those laboratory activities could be developed back into new generalized systems.

At the beginning of his search for the secret of volatilizing alkalies, however, it is not evident that Starkey had this grand design in mind. His desire to acquire volatilized alkalies, like so many of his studies, began with his reading of Van Helmont. The greatest arcanum that Van Helmont claimed was that of the liquor alkahest, a substance occupying a central position in his chymical system. As mentioned in the previous chapter, the fundamental material for Van Helmont is water. The wide diversity of substances that we see daily arises out of the determination of that water into various forms by the power of *semina*. Simply stated, the alkahest is the way backward. Any compound body treated with the alkahest is first resolved into its ingredients, then upon further treatment these ingredients themselves are reduced into insipid water. Thus the alkahest first resolves bodies, and then reduces those constituents back into their primordial water by mortifying the activity of the seeds. Moreover the alkahest performs these feats without being weakened or, in Helmontian language, exantlated. It is thus called the "immortal solvent," for after it has completed the dissolution and analysis of a compound body into its constituents, it can be separated from the dissolved substances in the same quantity and quality as it was first employed. The alkahest's action is based upon the exceedingly minute size and homogeneity of its corpuscles. On the one hand this means that they can insinuate themselves between the corpuscles of all other bodies and divide them one from another, thus causing dissolution and analysis. On the other, it means that the particles are so small and simple that other particles cannot unite with them.[101]

Thus the alkahest would be a powerful agent for chymistry. It could be used for the analysis of any substance, and thus had potential for the preparation of medicinal and other products from them. In the seventeenth cen-

---

101. On the alkahest, see Bernard Joly, "L'alkahest, dissolvant universal, ou quand la théorie rend pensible une pratique impossible," *Revue d'histoire des sciences* 49 (1996): 308–30; Newman, *Gehennical Fire*, 141–51; Ladislao Reti, "Van Helmont, Boyle, and the Alkahest," in *Some Aspects of Seventeenth-Century Medicine and Science* (Berkeley and Los Angeles: University of California Press, 1969); Paulo Alves Porto, "'Summus Atque Felicissimus Salium': The Medical Relevance of the *Liquor Alkahest*," *Bulletin of the History of Medicine* 76 (2002): 1–29.

tury, the quest for the alkahest was a chymical cause célèbre almost as wide-spread as that for the Philosophers' Stone. Many tracts on this marvelous solvent appeared in the course of the century. Robert Boyle himself was, presumably at least partly through Starkey's influence, fascinated by it and the immense power it promised for the determination of the constitution of mixed bodies and for the advancement of matter theory. Indeed, Boyle apparently wrote (or, at least, began to write) a tract on the alkahest, but never published it and it is now lost.[102]

Of course, the preparation of so great an arcanum was not, and could not be, openly revealed, and so Starkey, like many other chymists, spent much of his life searching for the secret of its preparation in the scattered, terse, and often enigmatic utterances of Van Helmont on the subject.[103] But in this case, Van Helmont seems to have had a measure of pity for the Sons of Art who tried to follow him to the arcana. For he acknowledged that the search for the alkahest was extremely difficult, and so he offered his readers an alternative, namely, the volatilization of fixed alkalies—particularly salt of tartar—to produce a solvent with powers akin to, although still inferior to, those of the alkahest: "If you cannot attain this arcanum of fire [i.e., the alkahest], learn then to make salt of tartar volatile and complete your dissolutions by means of it.[104]

Volatilized salt of tartar is, in Starkey's expression, a *succedaneum* to the alkahest, that is, it is not equivalent thereto, but is capable of performing some of the same feats. Alkalies were known to be not only corrosive but also abstersive (cleansing)—we need think only of the uses of alkalies (like lye and washing soda) in cleaning and in soap production. Thus if alkalies were volatilized, that would imply that their corpuscles had been reduced in size (volatility and particle size had been related in alchemy since the Middle Ages), approaching corpuscles of the alkahest in size. Van Helmont also commended volatilized alkali for breaking up bladder and kidney stones. The practical problem is that alkalies—for example the most

102. Principe, *Aspiring Adept*, 63, 183–84; on Boyle and the alkahest more generally, see Reti, "Van Helmont, Boyle, and the Alkahest."

103. On Starkey and the alkahest, see below, as well as his posthumously published *Liquor Alchahest* (London, 1675), and the items related to the alkahest published in *Notebooks and Correspondence*. It is curious, however, that in spite of the great attention Starkey obviously paid to the preparation of the alkahest, there are no extended sets of experimental studies recorded in the extant notebooks. References to his quest for the alkahest are limited to scattered and brief references regarding the distillation of urine (the starting material Starkey settled upon), some analysis of Van Helmont and comments on the solvent's nature, and the reports on his progress sent in letters to Boyle in early 1652.

104. Van Helmont, *De febribus,* in *Opuscula,* chap. 15, no. 26, p. 58.

common form, salt of tartar (potassium carbonate)—are steadfastly non-volatile. Salt of tartar can withstand hours of brightly glowing red heat without evaporating in the least. The volatile salt of tartar, however, was supposed to vaporize at a fairly gentle heat, below the temperature of incandescence.

Starkey's search for the means of volatilizing salt of tartar began at the very start of his chymical studies. In a set of recollections penned in 1656, Starkey notes that

> towards the end of the month of March in the year 1646 I began the practice of medicine in Boston, New England, and from that time to this very day in the year 1656 I have dedicated my labor with a steadfast mind to the volatilizing of alkalies, especially salt of tartar.[105]

Around the same time, Starkey wrote an extremely valuable reprise of all his erroneous methods, and from this "final project report" to himself we learn that he first attempted to carry out the process using (in turn) spirit of wine, spirit of vinegar, and sal ammoniac.[106] These attempts can actually be located in his surviving records. A piece of a disbound notebook that might conceivably date from his New England years and survives now among John Locke's papers records his treatment of salt of tartar with vinegar and the distillation of the product to produce a "spirit of admirable penetration" that is described elsewhere in the document as a *spiritus tartari volatilis*.[107] The same document records the treatment of salt of tartar with spirit of wine, the process that Starkey recollects as his first attempt to volatilize salt of tartar. This initially gave him the *Balsamus Samech,* a pharmaceutical preparation commended by both Paracelsus and Van Helmont, which he then mixed with potter's clay and strongly distilled, to produce a "most useful menstruum" *[menstruum perutile].*

Although Starkey succeeded in thus preparing two menstrua, he was apparently not satisfied—presumably on the basis of their properties—that either was the true spirit of volatilized salt of tartar described by Van Helmont. Starkey recalls that after using vinegar and spirit of wine, he tried effecting the volatilization with sal ammoniac, again without success, in

---

105. *Notebooks and Correspondence,* document 12; RSMS 179, fol. 19v.

106. *Notebooks and Correspondence,* document 11; Sloane 3711, fols. 3v–5v.

107. *Notebooks and Correspondence,* document 2; Bodleian Library, Locke MS C29, fols. 115–118v, on fol. 116v. We will return to this important document in a later chapter because of the insight it gives onto Starkey's early relationship with Boyle.

spite of the production of "an immensely urinaceous stench" that left behind "a salt . . . not at all distinct in taste from marine salt, but quite diuretic."[108]

We can reconstruct Starkey's thought processes quite confidently from
his initial choices of substances to render salt of tartar volatile, for they are
all volatile things themselves. Therefore, it is likely that Starkey thought he
might be able to induce volatility in the salt of tartar by allying it with
volatile materials. This is analogous to the common practice—extending
well into the Middle Ages—of producing sublimates from fixed materials
by mixing them with sal ammoniac, a material that readily sublimes. The
idea is that the volatile material "carries up" the more fixed component. To
cite but one contemporary example, this well-known methodology is described in Robert Boyle's essay on volatility.[109]

Starkey's further progress can be followed in his spring 1651 letter to
Boyle, where he recounts a "disaster" in his laboratory. He tells of having
made a "pleasant medicine" from the Sulphur of antimony mixed with a
soap prepared from oil of beeswax and salt of tartar and scented with am-
bergris. After he added some sulphur to the mixture, he "intended to boil it
softly" and gave instructions to his operator, probably Francis Webbe, with
whom he was living at the time, to maintain the fire.[110] Alas, the fire "was
made a degree or two too hot, which sent away most of my Confections in
the forme of a vapour which never returned to make report of their virtue
& left me inconsiderable foetid faex almost of no pondus [weight]." But
this misfortune, though costly in materials, was intellectually valuable to
Starkey, for he at once made use of its result, thus indicating the value even
of laboratory accidents. He noted that the residue was "almost of no pondus," in spite of the fact that the weight of the saponary "medicine" from
which it was produced had been at least one and a half pounds, about half
of which would have been salt of tartar. What had become of the salt of tar-

108. *Notebooks and Correspondence,* document 11; Sloane 3711, fol. 4v. Starkey's observations here are quite keen. His treatment of potassium carbonate with ammonium chloride
produces the highly volatile ammonium carbonate (later used as "smelling salts"), the source
of the "urinaceous stench" (in the seventeenth century the term "urinaceous" referred to the
smell of the compound we now call ammonia). The residue would have been potassium chloride, currently used as a "light salt" for those on low-sodium diets and still employed medically, as Starkey himself observed, as a diuretic. Clearly Starkey must have carried out medical
experiments in addition to the chemical ones described in the noteboooks; we have an account
of his trying out a powerful vomitive on his brother Samuel, see Starkey to Boyle, 3 January
1652, in *Notebooks and Correspondence,* document 6, and in Boyle, *Correspondence,* 1:109.

109. Robert Boyle, "Mechanical Origine of Volatility," in *Mechanical Origine of Qualities,*
in *Works,* 8:432.

110. On Webbe, see above at nn. 24–26.

tar? Since the residue was "almost of no pondus," the logical conclusion (using the principle of mass balance to locate missing mass) was that the salt must have been volatilized and driven away by the overly strong fire. "This set me upon a more Eager search after Tartarus Volatilis in this operation then in any other way." Thus, on the basis of observations made upon this experimental "disaster," Starkey turned to a new way of attempting to volatilize salt of tartar, now using the kind of oils he had employed to make the soap that flew away in the fire.

Starkey immediately set about distilling salt of tartar with olive oil, and while he recounts some failed experiments to Boyle, he also manages to distill a soap made of salt of tartar and olive oil to obtain a "white liquor . . . nobly answering the Expectations from Tartarus Volatilis." At the time of the letter, Starkey was still busy at work on the operation and resolved that "the secret of Tartar I shal seeke totally in Oyles mixed with it in forme of a Sapo [soap]." In fact, he describes in detail to Boyle the operations he is about to carry out, and we can recognize from our previous study that this is in the form of a conjectural process. Although none of the notebooks contains trials of this method, in his 1656 review of his quest for volatile alkalies, Starkey clearly recalls this stage of his labors: "I would mix olive oil with strong lixivium of salt of tartar, boiling them in an iron pot until a soap was produced, which I then distilled; but the process did not respond to my expectation." Part of the problem was the "feces of the oil," which kept "infecting" the salt; indeed, an expressed oil like olive oil partly decomposes during distillation, often becoming rank in odor and leaving behind charred products. This consideration makes the rationale behind Starkey's next decision clear—use an oil that has already been distilled and will thus have no feces; in other words, substitute an essential oil for an expressed one.[111]

The substitution of an essential oil for an expressed one must have occurred in 1651, for a fragment of a notebook internally dated to December 1651 contains a confident conjectural process for accomplishing the volatilization of salt of tartar using oil of terebinth (spirit of turpentine). This method involves slowly distilling a lixivium of salt of tartar with oil of terebinth to produce a black precipitate that Starkey calls a "collostrum" and then distilling a spirit out of the collostrum. Since this collostrum is produced slowly during the boiling, Starkey takes the unusual step of distilling the mixture with the beak of the retort inclined *upward*, thus pro-

111. Expressed oils like olive, corn, or sunflower oil are fats (triacylglycerides) produced by pressing vegetable matters to squeeze out the oil. Essential oils like oil of cloves, oil of turpentine, and so forth are terpenes prepared by the steam distillation of vegetable materials.

longing the distillation and thereby maximizing the amount of collostrum produced. This collostrum was then to be powerfully heated in a retort, whereupon spirit of volatile alkali was expected to distill over.[112]

We have no further indications of Starkey's work on the volatile alkalies project until Boyle's mention of an Elixir of volatile salt in his *Philosophicall Diary* of early January 1655. There is no record of Starkey's work on the project between 1652 and 1655, and part of his seemingly slow progress on it is undoubtedly due to his financial and personal disasters of 1653–54. The process documented in Boyle's 1655 work-diary is, however, repudiated by Starkey in his subsequent review of his attempts.[113] Indeed, Starkey's laboratory notebooks from early March of the same year record a critical set of ruminations. An entry there dealing with the volatilization of alkalies takes Starkey's project in a new direction. The process Starkey now writes down to try not only eventually brings him the success he long sought, but also entirely changes the scope of the project. This entry in the notebook bears the rather grand title of the "Arcanum of Alkalies." This new approach involves mixing powdered plant material with the alkali salt extracted from the ashes of the same plant and the essential oil distilled from the plant. The resultant paste is then exposed to the air for two months without heating. At the end of this time, Starkey hopes, the entire mass should be "changed into a volatile salt which is called Elixir."[114]

How did Starkey come up with this new process, particularly its most crucial feature—simple exposure to the air rather than the application of heat? Fortunately, his notebooks give some indication of the background to the process. It appears that Starkey's new process arose from a careful reconsideration of the theory of alkali composition together with his linkage of two widely separated and obscurely described processes in Van Helmont.

The first of the processes incorporated into Starkey's new "arcanum of alkalies" is identifiable when he mentions that the ingredients should be mixed "free from all water" and that the resultant salt is "called Elixir." At the conclusion of one of the many tracts in the *Ortus medicinae*, Van Helmont writes that "when oil of cinnamon, etc. is mixed with its alkali salt, without any water at all, the whole is converted into volatile salt by an artificial and hidden circulation of three months." [115]

Two of Starkey's three ingredients are mentioned here, as is the direction to exclude any water, and while the name "Elixir" does not appear in

112. *Notebooks and Correspondence,* document 5; Sloane 2682, fol. 89r.
113. *Notebooks and Correspondence,* document 11, Sloane 3711, fol. 5.
114. *Notebooks and Correspondence,* document 12, RSMS 179, fols. 63v–60v.
115. Van Helmont, *Ortus,* "Tria prima chymicorum," no. 84, p. 412.

this quotation, the resultant salt is given that name in the synopsis provided at the beginning of Van Helmont's tract.[116]

This terse and incomplete receipt in Van Helmont, however, mentions nothing about the air, but only alludes to an unspecified "artificial and hidden circulation." "Circulation," in chymical practice, generally means what we today call "reflux," that is, heating a substance in a sealed vessel so that the evaporating vapors are recondensed into a liquid in the cooler parts of the vessel and run back down onto the heated substance. But that simple meaning is apparently not Van Helmont's meaning here.

Starkey must have tried to follow this process and consequently to identify the "artificial and hidden circulation." Indeed, in his "arcanum of alkalies" entry he refers to experiments he had previously carried out, and removes any doubt regarding the identification of the cinnamon experiment in Van Helmont as one of his sources:

> I have learned that this mutation of oil into true salt can be performed by no art as successfully as it is done in the open air.
> And this is the "hidden and artificial circulation" which is done by the fire of nature.[117]

It is worth pointing out that if, as seems likely, Starkey carried out the process using oil of cinnamon as mentioned by Van Helmont, he would in fact have observed the change of the oil entirely into salt. As it happens, the essential oil of cinnamon is composed largely of cinnamaldehyde, a liquid having the characteristic odor of cinnamon. Upon exposure to atmospheric oxygen, this aldehyde autoxidizes readily into cinnamic acid, a crystalline solid. The resultant acid can then combine with the alkali carbonates in the "salt of cinnamon" to provide a water-soluble cinnamate salt, which, as Starkey accurately describes in his notebook, "notably differs from fixed sal alkali in taste, and does not liquify in air (as alkali does)."[118] Thus Starkey would have in fact seen the striking transformation of a distilled oil into a crystalline salt by the action of the air—apparently a clear confirmation of his supposition about the identity of the "hidden circulation."

But how did Starkey arrive at this use of the air as the correct interpretation of the hidden circulation? And how did he connect the cinnamon experiment with his project of volatilizing alkalies? It is possible that the discovery was accidental—we know that Starkey worked at improving the

---

116. Ibid., p. 398: "84. Oleum essentiale aromatis, sive crasis ejusdem quomodo fiat Elixir ejus, centuploque potentius."

117. *Notebooks and Correspondence*, document 12; RSMS 179, fol. 60v.

118. *Notebooks and Correspondence*, document 12; RSMS 179, fol. 62v.

cost-efficiency of the isolation of essential oils from spices, including cinnamon. He might then have observed the autoxidation of some oil he had left out in the air into crystalline cinnamic acid. However, Starkey's notebook entry cites another locus in the massive *Ortus medicinae;* indeed, one that we have encountered previously.

> If the air (let him who can, grasp an arcanum) first of all volatilizes the Sulphur of a concrete with complete separation of its Salt, this Salt (which otherwise would be fixed by the fire into an alkali in the coal) will be made entirely volatile and will ascend sometimes in a liquid form and sometimes in the form of a sublimate.[119]

The reader will recall that this passage occurs in the tract "Blas humanum," where Van Helmont introduces his theory on the formation of fixed alkalies from volatile Salt and Sulphur. Starkey apparently linked these two passages, believing that the "arcanum" of the one was the "artificial and hidden circulation" of the other. This linkage was not a stab in the dark, however, for there are textual and theoretical reasons for it. In the first case, the "Blas humanum" reference also notes that the volatile salt produced "sometimes in a liquid form, and sometimes in the form of a sublimate" will "have the whole *crasis* of the concrete." This means that all the essential properties and virtues (the *crasis*) of the original body are preserved in the volatilized salt. The cinnamon process in "Tria prima chymicorum" states that its volatile salt product "expresses the essence of its simple," which is much the same thing as its *crasis*. Thus an astute reader like Starkey might conclude that the two products are identical and that the two sections are therefore related. Additionally, the cinnamon process clearly states that the oil *together with the alkali salt* is rendered volatile, thus linking that process with Starkey's quest for the way to volatilize alkalies.

The theoretical bases of Starkey's linkage of seemingly independent processes reveal more about his practices. Starkey apparently considered once again the nature and theory of alkalies. Recall that according to Helmontian theory, alkalies are produced when a combustible material is heated and a part of its volatile Salt "seizes upon" its neighboring Sulphur so that the two become fixed together as an alkali. The Salt and Sulphur, moreover, remain unchanged in species; they change only in terms of their aggregation, being "imprisoned" together in a fixed alkali rather than "free" in a volatile state.[120] When Starkey eventually perfected his arcanum of alkalies and published a (partial) account of it in *Natures Explication and*

---

119. Van Helmont, *Ortus,* "Blas humanum," no. 45, p. 188.
120. Ibid., nos. 42–43, p. 188.

Helmont's Vindication, he summarized this theory of alkalies and its implications in his own words: "Alkalies are easily volatized, since their generation proceeds, not from seminal beginnings, but is a spontaneous Larva,
which part of the Salt or Sulphur of the Concrete assumes, the better to
withstand *Vulcan's* fury."[121] That is, the transformation into an alkali is not
a fundamental change of species (which would require the action of *semina*) but only a superficial change—the imposition of a *larva* or mask. Fire,
which initiates the production of the alkali, has no *semina;* it therefore cannot by itself effect fundamental changes of substance. Thus, since this is
only a superficial change, it must be reversible. As Starkey records in his
notebook, "although it [the original volatile material] is fixed [it] is not
changed in species; therefore, the alkali is volatilized just as what was previously volatile was fixed."[122] The practical difficulty lies in finding the means
for removing the *larva*. This is what Starkey now locates in exposure to air.
Indeed, the "Arcanum of Alkalies" notebook entry alludes to yet another
Helmontian statement when Starkey writes "the Sulphur of a vegetable
concrete is volatile in every way before its cremation, and by the *fracedo* of
the air, the whole is rendered volatile in the fire, as is exemplified in rotten
woods."[123]

This is a reference to Van Helmont's observation that rotten wood provides no alkali upon burning because all the volatile material needed to
form the alkali evaporated during the rotting process "through the ferment
of rotting."[124] Starkey posits that this "ferment of rotting" exists in the air,
for he refers to the *fracedo* of the air, using a Helmontian word that means
"a power to cause rotting or fermentation," being related to the Latin word
fracidus, meaning overripe or rotten. Starkey's linkage of the two seemingly unrelated Helmontian passages is thus undergirded by the theory that
the rotting power of the air could dissociate the Salt and Sulphur mutually
imprisoned in the alkali, or at least break the dual particles into smaller corpuscles. This after all, is how rotting works in Helmontian theory; it breaks
things down into smaller—and therefore more volatile—particles.

As noted above, Starkey presumably tried this process using cinnamon as
the simple; if he did so, he would have witnessed the conversion of the oil
and alkali into a single salt. But he was not satified with a single result employing cinnamon. Subsequent notebook entries reveal him intent upon a
more universal method, for he immediately sets about trying the process
with the oils and salts of other medicinal simples, such as the salt and oil of

121. George Starkey, *Natures Explication and Helmont's Vindication* (London, 1657), 301.
122. *Notebooks and Correspondence,* document 12; RSMS 179, fol. 60v.
123. *Notebooks and Correspondence,* document 12; RSMS 179, fol. 62v.
124. Van Helmont, *Ortus,* "Blas humanum," no. 41, p. 188: "per fermentum putridinis."

mint, of wormwood, and so forth. He then generalizes this process further to include essential oils and salts *not* prepared from the same source, focusing upon the alkali and essential oil easiest to prepare and cheapest to acquire in quantity—salt of tartar and oil of terebinth. Additionally, he sets down new experiments to be tried in an effort to optimize and streamline the process, such as the direction to "Examine whether this happens in salt with oil, without the addition of a simple."[125] After nearly three months of experimentation, on 29 May 1655 Starkey confidently records that "at last practical experience has revealed to me the whole and entire secret of volatilizing alkalies."[126] Starkey then sets down the improved method of moistening salt of tartar with oil of terebinth, exposing the paste (without powdered plant material) to the air until the oil disappears, adding more oil, and continuing this process for several weeks. Even after this, Starkey apparently continued experimenting, for three months later he concludes that "by examining and by experimenting I know and have learned that plain oil with plain salt becomes the Elixir, soluble in water without oiliness within four months."[127] By March 1656 Starkey has developed his process yet further into a protocol consisting of exposing the paste of salt and oil to air until they appear to unite, dissolving the resultant material in weak spirit of wine, filtering the solution, and then distilling off the solvent in a retort to obtain a dry salt. This final salt—the "Elixir of volatile salt"—is the salt of tartar now made volatile and fully united with the essential oil, which has itself been converted by the hidden circulation into a salt. Starkey is clear that this long-sought success is the result of "having been taught by experiment, the best teacher of all."[128]

Undoubtedly this success, after ten years of work, was extremely satisfying to Starkey. But there was now much more to the project than there had been when he began it as a youthful neophyte in New England. For when Starkey thought to unite the "artificial and hidden circulation" of cinnamon with the "arcanum of the air," he did much more than solve his existing problem of how to volatilize fixed alkalies—he inaugurated what he saw to be a new program for medical chymistry. For even though the secret of volatilized alkali itself had potent medical applications, the linkage of this to the conversion of essential oils into salts made it considerably more far-reaching—namely, the development of a single, uniform method for pre-

125. *Notebooks and Correspondence*, document 12; RSMS 179, fol. 60v.

126. *Notebooks and Correspondence*, document 12; RSMS 179, fol. 56v.

127. *Notebooks and Correspondence*, document 12; RSMS 179, fol. 55v. The "Elixir" Starkey mentions here is not the Philosophers' Stone, but rather the "Elixir of volatile salt," the final product from the union of the salt of tartar and the oil of terebinth.

128. *Notebooks and Correspondence*, document 12; RSMS 179, fols. 44v–40v.

paring any substance into an extremely potent and safe medicine. This significant elaboration of the project on volatile alkalies occurred because in Starkey's final method, not only was the alkali made volatile, but the oil was converted into a salt as well; the significance of this bears explanation. Essential oils are the Sulphurs of plants, and Sulphurs are key to Helmontian pharmacy. Van Helmont asserted that the active part of any substance is its Sulphur. Sulphur is the "life and death" of things, it contains the "vital fire," and is the material basis for all chymical transformations. "Sulphur indeed is life and death, or the dwelling-place of life, in things. For the ferments, *fracedines,* odors, and tastes of the *semina* of the specific needed for any transmutation exist in its Sulphur."[129]

But knowing this is not enough, for these Sulphurs must generally be separated by chymical operations from the mixed bodies that contain them. For vegetables this is not so difficult, but we have already sampled Starkey's arduous attempts to isolate the more powerful mineral Sulphurs of antimony, of its regulus, and of copper. Unfortunately, even when separated these Sulphurs often remain contaminated with "foreign and poisonous" powers that can render them toxic and offensive. A small dose of Sulphur of antimony, for example, induces violent vomiting. When these offending principles are removed, however, the Sulphurs should constitute all the medicines man could ever want for any disease.[130] Thus Starkey heeded the words of Van Helmont in locating a general goal for himself: "I urge beginners to learn how to despoil Sulphurs of their foreign and virulent power . . . for there are certain Sulphurs that, once they are corrected and perfected, the entire army of diseases obeys."[131] This goal is clearly enunciated in Starkey's publications, where he not only paraphrases the pertinent passages from Van Helmont, but also claims in his own words that "the height of medicine . . . is performable by the glorified, spirituated, and perfected *Sulphurs,* which by their eminent purity and perfection, and by their fermentall irradiation, at once mortifie whatever is malignant in the body."[132]

129. Van Helmont, *Ortus,* "Progymnasma meteori," no. 14, p. 70.

130. For these ideas see Van Helmont, *Ortus,* "Progymnasma meteori," nos. 14–17, p. 70, and "In verbis, herbis, et lapidibus," pp. 575–84.

131. Van Helmont, *Ortus,* "In verbis, herbis, et lapidibus," p. 577; "Hortor itaque Tyrones, addiscant sulfura spoliare vi peregrina ac virulenta; sub cujus nimirum custodia abditur ignis vitalis, Archeum in scopos desideratos placidissime deducens. Sunt videlicet sulfura quaedam, quibus correctis atque perfectis tota morborum cohors auscultat: utpote quorum pluralitas in unitatem Archei, tanquam in pugnantem pugnum contrahitur."

132. Starkey, *Natures Explication,* 294–95; see also Starkey, *Pyrotechny Asserted,* 88, 92. Note Starkey's use of the term "fermentall irradiation," the Helmontian action at a distance of *semina* on properly disposed matter, as described in chapter 2 in the context of mercury's ability—without losing any of its own substance—to convert oil of vitriol into alum and common water into a vermifuge.

But even when the Sulphurs are "corrected" there remains a further problem. Sulphurs are by definition oily, unctuous, and generally water insoluble. Therefore they cannot be readily assimilated by the digestion. Here lies the remarkable power of the "artificial and hidden circulation"— it can transform a Sulphur into a Salt. Now a Salt, because it readily dissolves in water unlike a Sulphur, is more able to penetrate all the narrow vessels of the body and act more powerfully. As Starkey eventually wrote when describing his Elixir of volatile salt: "whatever reacheth to the *Balsame* of *Life*, must be Salt . . . nor can any thing be admitted beyond the limits of the first digestion, but it must be of this [saline] Nature."[133] Ingested without union and "glorification," an alkali is neutralized in the stomach and destroyed while a Sulphur is destructively digested or expelled, but if the two are turned into a single Salt, that mild, neutral Salt will pass through the digestion "retaining its virtue" and enter all the parts of the body and work wonders medically.[134] In the Salt prepared by the "secret digestion," Starkey has not only volatilized the alkali, making it more subtle and powerful, but has also converted the Sulphur into Salt, making it capable of penetration into the human body; the two have been united and "coglorified" by this method,

> For between the Oyls essential and Salts Alcalizate, there is a fermentall appetite, whereby they close each with other radically and in the Centrall profundity each of other, which give not a Sapo, nor a Collostrum, (which are the triviall products of erring operators) but a reall Salt.[135]

Here Starkey silently repudiates his earlier operations that yielded a "sapo" or a "collostrum," relegating those to the category of "triviall products of erring operators." The true product is the "reall Crystallizing Salt . . . retaining the whole Crasis or vertue not in the least diminished . . . to the performing of really wonderfull Cures."[136] Now if the alkali salt of tartar and oil of terebinth can be thus converted into a potent, safe, penetrating Salt by the "artificial and hidden circulation" in the air, then what can one say of other alkalies, and especially other Sulphurs, especially those drawn from the mineral realm—the most powerful of all? This investigation opens up the wide horizon of Starkey's grand

133. Starkey, *Pyrotechny*, 96.
134. Ibid., 97–98.
135. Starkey, *Natures Explication*, 325.
136. Ibid., 325–26.

design and explains a feature of his notebook that might puzzle the casual reader. For a total of sixteen months from the time he first discovered the "hidden circulation" in the air, Starkey experimented with numerous variations on the process. During this time he employed a range of alkali salts isolated from different sources, various essential oils, including oil of terebinth impregnated with common sulphur or with Sulphur of antimony, both with and without the addition of various powdered plant materials. A casual reader of the notebooks might well believe all these variations to be either haphazard or no more than the kind of exhaustive combinatorics we have seen previously in Starkey's work. But these numerous trials are actually coherent parts of this grander scheme for chymical medicine that Starkey now had in mind and which he laid out more fully in his published works *Natures Explication and Helmont's Vindication* (1657) and *Pyrotechny Asserted and Illustrated* (1658).

Armed with this background, the myriad variant processes and attempts at generalized preparative protocols recorded in the notebooks now make sense, and we can appreciate the grand design as Starkey lays it out in *Pyrotechny* and *Natures Explication*. Basing himself on the principle he discovered in converting salt of tartar and oil of terebinth into an Elixir of volatile salt, Starkey constructs a coherent system of interrelated protocols applicable to the perfect glorification of every Sulphur.

Suppose one wished to isolate and "glorify" the Sulphur of a vegetable, say, the herb wormwood *(Artemisia)*—Starkey gives several options. First, one could distill off the essential oil, calcine and lixiviate the residue to provide a salt, mix the two together, and expose the pasty mixture to the air. After several weeks, the two would unite and be converted into a "volatile essentiall Salt" of wormwood, containing the "whole crasis or virtue" of the herb in a saline form easily assimilable to the human body. Alternatively (and more conveniently and cheaply), one could mix powdered wormwood with salt of tartar and oil of terebinth, expose the mixture to the air, and generate an Elixir of volatile salt that during its own formation would simultaneously produce and unite with the essential Salt of the wormwood. But third, using the best method (according to Starkey), one could prepare the Elixir of volatile salt from readily available and cheap salt of tartar and oil of terebinth and then dissolve that Elixir in dilute spirit of wine to provide a powerful extracting solvent. The chymist could then digest the wormwood in that solvent and, after a competent period of digestion, evaporate the filtered extract, leaving behind deeply colored crystals that when treated with "dephlegmed" spirit of wine (i.e., concentrated ethanol) give up the extracted and glorified Sulphur of the wormwood to the solvent. This final extract could be evaporated to remove the spirit of wine, leaving the final glorified essence of wormwood pure.[137] The Elixir is able to convert the crude herb into an essential Salt like itself because it contains a "fermental virtue" able to convert other materials into its own nature.

Note here the similarity of this Elixir of volatile salt to the otherwise un-related Philosophical Mercury. By the power of the fermental virtue that common mercury acquires from regulus of antimony, the acuated Philo-sophical Mercury dissolves and digests the heterogeneous metals, converts them into its own Mercurial nature, and unites with them. In the same way, the Elixir of volatile salt has this fermental virtue because it has either reac-quired its own lost *semina,* when the calcined salt is reunited with its own essential oil, or else it acquires new *semina* from a foreign essential oil, such as that of terebinth. In either case, the resulting salt will be capable of con-verting other substances into its own nature. Remember that, according to Helmontian theory, the Salt while in the plant originally had its own *semen,* but this was weakened in the cremation whereby it became an alkali; the "hidden circulation" of the air and the union of the alkali with the Sulphur of the oil restores these *semina.* As Starkey expresses it, "the Alcaly . . . re-covers what it lost by burning, that is a seminal, vital, essential Balsom, and so becomes not only volatile, but fermental and exceeding sociable to our Nature."[138] Thus the Elixir, once prepared, is the means of preparing other substances into its own nature.

Mineral and metalline Sulphurs can be handled by the same general method, which Starkey calls "elixeration." Again, there are several methods possible, but all of them rest on the same foundation. First, the Elixir can be volatilized by mixing it with potters' clay and distilling it in a retort. A "Spirit of volatile tartar" distills over, and this is (finally) the long-sought solvent "succedaneous" to the alkahest. The mineral body can be treated with this solvent in order to isolate its Sulphur.[139] Alternatively, the Sul-phur could be introduced earlier in the process; the Elixir could be mixed with the mineral or calx of the metal and the two submitted to repeated dis-tillations, thus extracting, "Salifying," and volatilizing the Sulphur simulta-neously. The resultant volatile Salt is then treated with dephlegmed spirit of wine (exactly as in the case with wormwood mentioned above) to dissolve out the "metalline tincture from the Salt," thus producing both a residue of the volatile tartar used to make the separation and, when the spirit of wine

---

137. The Elixir used at the beginning, though soluble in ordinary spirit of wine, is insolu-ble in "dephlegmed" spirit of wine and so remains behind alone. Here we see clearly how this Elixir is succedaneous to the alkahest, for it forms (with ordinary spirits of wine) a solvent for isolating the virtues of a mixed substance, and after its use it is recoverable in its original form.

138. Starkey, *Pyrotechny,* 138.

139. Ibid., 85, 90–92.

is evaporated, a "fragrant and very sweet" residue "of wonderful virtue, little inferiour to any glorified Sulphur, by any Alchahestical operation."[140]

A third related protocol calls for the metalline Sulphur to be united with the oil of terebinth prior to its "artificial and hidden circulation" with salt of tartar. The experimental work on this method is recorded in notebook entries dated to the first week of March 1656, where Starkey carries out frequent distillations of oil of terebinth with antimonial substances or with flowers of sulphur in a special still designed specifically for this process. These codistillations produce a red oil, stinking of sulphur, which is elixerated by a method that we have already encountered—mixture with salt of tartar and exposure to the "hidden circulation" of the air. Thus both Sulphurs—vegetable and mineral together—would be converted into Salt and made medicinal.[141] This method, however, does not appear in *Pyrotechny*. Why not? Its absence is explained by a notebook entry dated 7 March 1656—that is, immediately after the series of experiments on codistilling oil of terebinth with mineral Sulphurs—which shows that it was soon superseded by a method Starkey found to be superior. In a section entitled "On Volatilizing Metallic Sulphurs and Preparing them into a Saline Nature," Starkey makes what he terms "A Philosophical Examination *[Disquisitio Philosophica]* of the Process." There Starkey recommends the isolation of mineral Sulphurs by fusing the metal or mineral with alkali rather than by distilling them with oil, and he (as usual) enumerates his "very firm reasons, confirmed on an experimental basis" for this change.[142] One of these "very firm reasons" was provided by a quantitative analytical study.

> I have learned from experience both in the sulphur of antimony and in vulgar sulphur, that a rather small portion ascends, which the weight of the abstracted oil compared with [the weight of the oil] put in shows, even though the color, taste, and odor of the volatile oil of sulphur confirms some degree of marriage, concerning which consult the manual of experiments; I do not think that I have been lazy in this affair.[143]

Clearly, Starkey weighed the oil of terebinth both before and after repeated distillations from mineral Sulphurs, and compared the weights. He then concluded from the small weight difference he found that only a small amount of mineral Sulphur had actually been united with the oil by distillation, despite the significant qualitative changes in the oil. This is yet another

---

140. Ibid., 85–86.
141. For example, *Notebooks and Correspondence*, document 12; RSMS 179, fol. 37v.
142. *Notebooks and Correspondence*, document 11; Sloane 3711, fols. 5v–7.
143. *Notebooks and Correspondence*, document 11; Sloane 3711, fol. 7; the "manual of experiments" cited here is probably RSMS 179.

example of Starkey's use of quantitative mass balance to assess the success of a procedure.[144]

Starkey's improved method for operating upon mineral Sulphurs is proposed conjecturally on 7 March 1656 and described practically in *Pyrotechny Asserted* in 1658. This is the "way I rather choose," and involves fusing minerals or metal calces with the alkali salt of tartar.[145] Remember that this is the method Starkey had employed as early as 1653 to isolate the Sulphur of antimony by extracting it into a saline scoria (note that the success of this operation is guaranteed by the Helmontian theory that salts "seize upon" Sulphurs in the fire). Here, however, Starkey not only borrows from his earlier experience but also generalizes the method to apply to the Sulphurs of all minerals and base metals (gold and silver excluded). The fused mixture of salt of tartar and the extracted mineral or metalline Sulphur is then mixed with oil of terebinth and "circulated" by the air into the saline Elixir in the usual manner. Thus the mineral or metalline Sulphur is converted, along with the vegetable Sulphur (i.e., oil of terebinth), into a "Saline Nature," thereby completing its "glorification" and conversion into a potent and safe medicine.[146]

After ten years of work and thousands of experiments, Starkey has drawn upon his results and theories to develop a single method of preparing all Sulphurs into medicines and elaborated this method into several *convergent protocols*. The difference among these individual protocols rests only on the point in the process where the crude Sulphurs are introduced—beginning, middle, or end. The principle is always the same; Sulphurs are converted into Salts either by the "hidden circulation" of the air or by the fermental activity of the already circulated Elixir of volatile salt. Starkey's sense of the universality of this process, and the importance he attaches to the development of *generalized philosophical principles of chymical medicine,* can be underscored by briefly recounting the usurpation of an early form of the Elixir by an "unlearned alchymist."[147]

---

144. In addition to this quantitative evidence, Starkey also cites the practical issue that because alkalies are nonvolatile, "they will tolerate an ignition one hundred times greater than oils," so the necessary isolation of the crude Sulphur can employ a far greater heat, better able to "open up" the body of the heterogeneous minerals and metals. Finally, Starkey adds theoretical evidence to his argument, noting that since an alkali is itself "true Sulphur fixed with Salt" according to Helmontian theory, it is "far more agreeable for preparing Sulphurs" (Sloane 3711, fol. 6v).

145. Starkey, *Pyrotechny,* 86; 82–83.

146. This completed Saline material can be further "exalted" by mixing it with potter's earth and distilling it into a volatile spirit, whereupon its whole essence is subtilized and thus made even more active.

147. A more detailed account of this controversy is contained in Newman, *Gehennical Fire,* 191–96.

MATTHEW AND THE UNPHILOSOPHICAL PILL

In the late 1650s a medical practitioner named Richard Matthew began selling a substance he referred to as his "Pill." This pharmaceutical was widely
publicized as a universal medicine for all diseases in Matthew's 1660 publication *The Unlearned Alchymist his Antidote.* In the same year, Starkey
protested that he was the "true Author thereof."[148] Nonetheless, after Matthew's death in 1661, his widow Anne continued to profit off the sale of this
medicament, and published an extended edition of the *Unlearned Alchymist* in 1663 in which she denied Starkey's claims. This work was immediately
followed by an *Appendix to the Unlearned Alchimist,* written by George
Kendall, an associate of both Matthew and Starkey, who asserted Starkey's
priority by the use of several witnesses and by publishing both Matthew's
receipt and Starkey's, so that one "shall not stand in need of a judge endued
with the wisdom of Solomon to decide who is the true Father of the
child."[149] From this publication, we can see that "Matthew's" Pill consisted of opium, hellebore, and licorice mixed up into pills with a previously
digested mixture of oil of terebinth and salt of tartar. Thus it is clearly an application of an early form of Starkey's Elixir of volatile salt as outlined
above. Unquestionably, Starkey was distressed by this priority dispute and
by the considerable profit from the results of his own experimental labors
reaped by Matthew and his widow. But Starkey's comments reveal more
about the relative value he accorded to a mere receipt, however profitable,
and the development of a generalized pharmacological system based on
theoretical principles.

The *Appendix* states that Matthew got the recipe for the pill from Starkey in 1655.[150] Some further information has recently come to light in the
diaries of John Ward, who, writing in the midst of the dispute in 1662,
records that he heard that "Mathews had his pill of Starky . . . for 5
pound."[151] In a letter published with the *Appendix,* Starkey claims that
"the secret was known and used by me in the year 1651"; this date must refer only to the use of salt of tartar to "correct" opium.[152] The year 1655,
however, marks Starkey's discovery of the "hidden circulation" in the air
and the further elaboration of the initial "saponary cream" into a crystalliz-

148. Starkey, *The Admirable Efficacy and almost Incredible Virtue of true Oyl, which is made
of Sulphur-Vive, set on Fire* (London, 1660), 13. The sole surviving copy of this first edition is at
the Library of Congress; see Ronald S. Wilkinson, "Bibliographical Puzzles Concerning George
Starkey," *Ambix* 20 (1973): 242–44.

149. George Kendall, *An Appendix to the Unlearned Alchimist* (London, [1663]), 34.

150. Ibid., 1.

151. John Ward Diaries, Folger Shakespeare Library, Washington, D.C., MS V.a.291, fol.
69v.

152. Kendall, *Appendix,* 44.

able salt. Indeed, Starkey remarks that since 1655 he has "far exceeded" his earlier method given to Matthew, referring presumably not only to the production of a true Salt from the "saponary cream," but also to the reduction of these experiments into a universal method, both of which occurred in 1655–56. Neither advancement is included in Matthew's receipt as published by Kendall; thus it appears that Starkey sold not a grand secret but rather only a single process that he had already superseded. More importantly, Starkey argues that Matthew truly was "unlearned" because he was completely ignorant of the real nature of the receipt. "Mr. *Matthews* being no schollar" claimed it to be a universal medicine, but this was no more than "a profitable, but disingenuous trick." Rather, Starkey asserts, "the secret being rather a Mystery of preparation, then a bare receipt, was looked on by me as a store-house or Common place from which various compositions might flow."[153] What Starkey means here is that it is the universal method for preparing medicines partially exemplified in the "Corrector" that is important, not the single specific receipt. As we emphasized at the start of this chapter, Starkey was not a collector of medical miscellanies— the real aim of his laboratory practice was the preparation of higher arcana and the discovery of chymical principles that could be embodied in broad preparative methodologies. Matthew, a self-styled empiric, was ignorant of this wider purpose.

CONCLUSIONS

In this chapter we have explored aspects of Starkey's methodology as revealed in his private laboratory notebooks. We see that his record keeping was surprisingly orderly, methodical, and formalized. Starkey developed and deployed a consistent methodology of data collection and interpretation from both textual and observational sources and used it to direct the course of his laboratory activities. The interplay between theory and practice is clearly apparent throughout, particularly in his habit of drawing conjectural processes from theoretical principles and past experience and then submitting them to the judgment of practical trials in the fire. We have also seen Starkey's further development of the quantitative methods found in earlier writers, above all those of his hero and chief preceptor, Van Helmont. Such methods function for Starkey as probes for the success of a given procedure and as sources of analytical information more precise than that accessible from unaided observation. In addition, we have encountered Starkey's desire to abstract generalized principles and a universalized method from his experimental results to be used in a comprehensive reform

153. Starkey, *George Starkey's Pill Vindicated* (London, [1663]), 1.

of pharmacology. Clearly, Starkey's laboratory investigation was rational and methodical, even in processes relating to transmutation and the production of the Philosophers' Stone, which he elsewhere expressed more publicly in the riotous allegorical symbolism of the Philalethes treatises.

The view of Starkey that the notebooks reveal is at odds with some popular views of seventeenth-century chymical (especially "alchemical") practice. We see clearly that Starkey was far from being an "empiric" (as he has sometimes been called); much less was he the nonrational or "mystical" archetype of the "alchemist" purveyed in varying degrees by so much of the secondary literature on alchemy. Those who wish to perpetuate such images of alchemical workers must explain how the rigorous methodological character of these laboratory notebooks—written by that famous champion of chrysopeia, Eirenaeus Philalethes—squares with their own interpretation of alchemy and its practitioners, particularly in their separation of it from "chemistry."[154]

The present chapter reveals that Starkey's laboratory work shares surprising similarities with that carried out by later generations of chemical practitioners. We do not mean to suggest, however, that Starkey's methods, goals, and attitudes are somehow detachable from his own training as a chymist or from the seventeenth-century context in which he worked. Readers will already have noticed, for example, Starkey's constant reliance upon authoritative texts for the initiation of his experimental projects, an impetus quite different from that in much of later chemical practice. This feature is related to the important topic of Starkey's use of authority—the authority of texts, the authority of observations, and even the authority of the divine revelations to which he occasionally claimed access. Additionally, there is the related issue of secrecy and concealment so pervasive in the chymistry of Starkey's day, especially as regards chrysopoeia and the *arcana maiora:* this is in full evidence in Starkey's work and must be examined further. Finally, there is the perennial nagging question of how and why Starkey, the painstaking laboratory worker, claimed success under the guise of Philalethes in transmutational alchemy even though his notebooks (not to mention the limits of physical possibility) indicate that had not achieved this desideratum. The study of these questions serves to root Starkey more solidly in his seventeenth-century environment; accordingly, these issues constitute the core of the following chapter.

154. This, we fear, may somewhere be done by diagnosing Starkey with multiple personality disorder, but we leave the refutation of such a hypothesis to the reader's good sense of humor.

FOUR

Scholasticism, Metallurgy, and Secrecy in the Laboratory

THE STYLE AND ORIGIN OF STARKEY'S NOTEBOOKS

The previous chapter detailed aspects of the investigative methodology that guided Starkey's practical experimentation. The present chapter focuses on several aspects of Starkey's experimental procedure in the light of his seventeenth-century context, including the training that he received before arriving in midcentury London. We need to account for the remarkable fact that Starkey established a working laboratory in London in less than four months after his arrival there and that he was already turning out products of sufficient novelty to impress such collectors of arcane curiosities as Samuel Hartlib, Benjamin Worsley, Robert Boyle, and the strange Dr. Farrar, who offered him five thousand pounds for his metallurgical secrets. Where did this twenty-two-year-old colonial from the Somers Islands (i.e., Bermuda) acquire such expertise? We now know that New England and Harvard College were surprisingly congenial places to learn about medical and chrysopoetic chymistry, but Starkey's devotion to the economic streamlining of the production of chymical substances, which we mentioned in the previous chapter, suggests an almost industrial attention to manufacturing efficiency. We begin, therefore, with a consideration of Starkey's connection to the metallurgical industry of New England. We shall then turn again to the formalized structure of Starkey's laboratory notebooks; intriguingly, these notebooks—in spite of Starkey's perennial invective against university learning—clearly reflect his education at Harvard College and illustrate the surprising degree to which Scholastic argumentation and practical experimental technique could be integrated. Finally, we turn to the interconnected issues of concealment and textual authority to show how the notebooks shed important new light on the issue of secrecy in transmutational alchemy. Despite Starkey's explicit reliance on textual authorities, we will see that his Scholastic method, combined with the results of his laboratory practice and his determination to make his processes better and more cost-efficient, led Starkey to a continual engagement with and correction of even his most valued textual sources. Even his appeals to divine authority must be seen in this light, for the divine revelations to which

Starkey occasionally claimed access turn out to have been themselves tried in the fire and thus subject to correction by experiment.

SOURCES OF STARKEY'S INDUSTRIAL CHYMISTRY

While still in New England, Starkey came into contact with several men whose chymical and technological expertise must surely have influenced the budding chymist. Starkey's notebooks reveal that he first began experimental chymisty in 1644, under the guidance of Richard Palgrave, a physician of Charlestown, Massachusetts, about whom little is known.[1] In addition to Palgrave, Starkey was able to draw on four figures closely associated with the nascent iron industry in Braintree and Lynn. The first of these was John Winthrop Jr.; Starkey's association with him is sufficiently well known to require only a little comment. Starkey's 2 August 1648 letter to Winthrop, cited early in chapter 1, is composed in such a way as to indicate that the two men had been in established contact for some time.[2] Winthrop, who had attended Trinity College, Dublin, was involved in every aspect of seventeenth-century chymistry: iatrochemistry, chrysopoeia, and perhaps above all the mining and refining of ores. Only about a decade after Boston's founding, Winthrop, the son of the then-governor, traveled to England and began gathering investors for his mining operation. A "Company of Undertakers" was formed, and by 1644 work was begun on a furnace for smelting bog-iron in Braintree. The site was eventually deemed unsuitable for the smelting of ores, and a new ironworks was erected in a part of Lynn that is now included in the town of Saugus. By 1648, the new mill was in full operation, and the Braintree works was reduced to an auxiliary status. Winthrop, meanwhile, had fallen afoul of the Company of Undertakers, presumably because of money lost in the Braintree venture, and was replaced by Richard Leader in 1645.[3]

It was Richard Leader then who actually oversaw the construction of the ironworks at Lynn. Leader had been a minor merchant, probably of Kentish birth, involved in trade with Ireland. Emmanuel Downing, Winthrop's uncle, described him as having "skill in mynes, and tryall of metalls," though it is not clear where Leader acquired this expertise.[4] At any rate, the

1. Newman, *Gehennical Fire*, 48–50, 53, 249–50.

2. *Notebooks and Correspondence*, document 1; John Winthrop Jr., *Winthrop Papers* (Boston: Massachusetts Historical Society, 1943–92), 5:241–42.

3. A valuable reassessment of the Hammersmith ironworks may be found in Stephen Innes, *Creating the Commonwealth: The Economic Culture of Puritan New England* (New York: Norton, 1995). The classic treatment of the subject is still E. N. Hartley, *Ironworks on the Saugus* (Norman: University of Oklahoma Press, 1957), 54–58, 102, 128, 107.

4. Winthrop, *Papers*, 5:6.

new ironworks, dubbed "Hammersmith" by its founders, were impressive. Hammersmith was an example of "high tech" by seventeenth-century standards. In simplified terms, it consisted of a smelting furnace for producing pig iron, bloomery hearths for refining this pig iron into wrought iron (in several stages), and an advanced rolling and slitting mill for producing rods and flat stock that could then be cut into nails, bolts, and other implements.[5] The men who devised this operation knew what they were doing; not only did they emulate the best of English and Belgian ironworking technology, they made local adaptations such as the discovery and use of an igneous rock called gabbro, found at Nahant, for a flux. This was an important breakthrough, since the Massachusetts Bay area was notoriously poor in traditional ironmaking fluxes such as limestone.[6] Hammersmith was a technological success and produced iron for some twenty years, though as a result of litigation, high wages, and shortage of hard currency, it failed to turn a notable profit and was eventually abandoned. Leader, meanwhile, who had also encountered difficulties with the Company of Undertakers, developed an interest in sawmills, and by August 1650 he had severed his connection with the company.[7] It is at this point that we learn of Leader's friendship with Starkey, for during his trip to England of the same year, Leader met with Samuel Hartlib and gave him an extremely favorable report of the young chymist, claiming that Starkey was possessed "of a most rare and incomparable universal Witt" and had cured desperate cases of disease.[8]

Much more data exist for Starkey's relationship with a third member of the ironworks establishment, Robert Child. Child, a university man like Winthrop, was Starkey's close friend in the Massachusetts Bay Colony, as his numerous surviving letters to Samuel Hartlib and the latter's entries in his *Ephemerides* testify.[9] Child received an A.B. and A.M. from Cambridge (1631–32, 1635), and an M.D. from Padua (1638).[10] He was deeply interested in all aspects of chymistry and supplied Winthrop with books on the subject over a period of years.[11] Like Starkey, Child was a Presbyterian, and

5. Hartley, *Ironworks,* 117–38, 167–68, 165–84.

6. Ibid., 149. Robert Child made a special point of this in his "Large Letter concerning the Defects and Remedies of English Husbandry," published in *Samuel Hartlib his Legacie* (1651). See George Lyman Kittredge, "Dr. Robert Child the Remonstrant," *Colonial Society of Massachusetts, Transactions* (1919): 105.

7. Hartley, *Ironworks,* 262, 134.

8. Samuel Hartlib, *Ephemerides* 1650, HP 28/1/57A.

9. Newman, *Gehennical Fire,* 79–80.

10. Kittredge, "Dr. Robert Child," 4.

11. Winthrop, *Papers,* 4:333–38. See William J. Wilson, "Robert Child's Chemical Book List of 1641," *Journal of Chemical Education* 20 (1943): 123–29.

his support of the Presbyterian cause in the Bay Colony resulted in fines, imprisonment, and eventually his departure from New England in 1647.[12] Before his emigration, however, Child had invested the very substantial sum of £450 in the company and had taken an active part in the daily operations at Braintree and Lynn.[13] In a letter of 15 March 1647 to Winthrop, Child refers to an ore sample that the former had sent him, saying that he has "not as yet tried it with the loodstone." (This refers to an assay for ferruginous ores by subjecting them in finely ground, and possibly roasted, form to a magnet to see if the magnet picked up any iron particles.)[14] Child then suggests that if Winthrop could supply him with a ton or two of the ore, he would like to "try it at our furnace." He also reports that "we have cast some tuns of pots this winter," referring to the casting of iron vessels in clay molds.[15] This and other documentary evidence shows that Child was a skilled metallurgist, and that he was involved in a hands-on capacity at the ironworks.[16] Indeed, he seems to have had some responsibility for making personnel decisions, for a letter to Winthrop from one of the skilled workers, William White, states that he "was promised 5s a day by doctor Child for myselfe and my sonn and 2 Cows and house Rent fre and land for me and all my Chilldren: alsoe Covenants for the same."[17] This William White is himself of importance for he too had a relationship with Starkey.

Although the least formally educated of the Hammersmith figures about whom we shall speak, William White was possibly the one who knew most about metallurgy. It may have been White who discovered the fluxing capability of the Nahant gabbro, for his surviving letter to Winthrop states that he "told mr. doctor Child more of the nehaunt mine then I can now spick of." He had acquired skill in siderurgy at the iron mines of Derbyshire and was working at Lynn and Braintree until he had a falling-out with Leader during or before 1648. Although Winthrop tried to convince White to stay in New England, he was persuaded by the would-be adept William Barkeley to ply his trade in Bermuda. Alas, Barkeley too misled White, for his

12. For a new treatment of this episode, see Margaret Newell, "Robert Child and the Entrepreneurial Vision: Economy and Ideology in Early New England," *New England Quarterly* 68 (1995): 223–56.

13. Hartley, *Ironworks*, 77–78.

14. Ibid., 166.

15. Winthrop, *Papers*, 5:140–41.

16. See Child's letter of 1 March 1645 to Winthrop, in which he invokes his own assay of a sample of "black lead" (graphite) sent to him by Winthrop, in order to dissuade the latter from investing heavily in the black lead mine at Tantiusques (Winthrop, *Papers*, 5:10–12). See also George H. Haynes, "The Tale of Tantiusques," in *American Antiquarian Society, Proceedings*, n.s., 14 (1901): 471–97.

17. Winthrop, *Papers*, 5:239.

time in Bermuda was not spent in smelting and refining of ores, but in the soldering of stills, which the members of the Pembroke tribe had worn out in their zeal for distilling fermented figs. There was some recompense in the natural bounty of Bermuda, however, for in a letter from White, dated 8 May 1649, and probably written to Child, White consoles himself with the fact that he has a "hundered turkeys" and "greate makerells as bigg as prit-tye piggs." But White also reveals that Starkey had taken him in just before his departure for Bermuda. After complaining about Leader's slandering of his abilities, White reports that

> mr Leader & his wife did disparidge me telling people that I say somethinge but performe Just nothinge: hade nott mr sturke [Starkey] began to pracktice phisicke & had such practice that he tooke me a great house & gave me 5s a daye 12 weeks fore my passage & there I shewd such works there that gentle & symple saide that I had beene wronged dyvers ways.[18]

Aside from giving testimony to Starkey's flourishing medical practice in the Boston area, this passage immediately raises the question: for what services was Starkey allocating White lodging and the very generous sum of five shillings per day? It is unlikely to be mere coincidence that Starkey paid him the same daily rate as had been promised to him by Child as a worker at Hammersmith. The ironworker's claim that the impressive work that he "shewd" at Starkey's house redeemed his reputation as a metalworker implies that White's service to Starkey was in this area. In all likelihood, Starkey, well off financially from his medical practice, was paying White to teach him the secrets of metallurgy. This is more than just speculation, for we know from another source that White billed himself as an inventor and purveyor of secrets. A "Cattalog of secretts good for a Common welth or plantation" exists among the Hartlib papers; it is attributed to "Mr. White" and appears to be written in the same hand and with the same punctuation as William White's signed 1649 letter. It contains a list of fifteen numbered inventions, most of them relating to metallurgy. Among these one finds high-efficiency ovens, good for "saving much fire & also time," melting pots that are cheaper than those currently available, a jug metal to replace glass in the preservation of mineral acids, a horizontal windmill that "will alsoe doe many things with little tendance," improved horse mills and hand mills, portable ovens and stills, a new way of "making or buildinge of salt-work to save much fire & time," improved waterworks, automated siege defenses, hand grenades, a new type of plow, novel techniques of calcination, an improved technique for cupellation, and a new type of bellows. White con-

18. William White to Child [?], 8 May 1649; HP 15/8/6A–7B, on 6A.

cludes his list by assuring the reader that these are "no tricke but all prof-
ittable things."[19] If Starkey was the beneficiary of this wizard of industrial
efficiency, there is no need to wonder at the former's interest in economic
rationalism in the laboratory or at the attention to furnace construction and
operation that is evident in most of Starkey's works from the notebooks to
the Philalethes treatises.

The picture that we receive of all these men combines an interest in mineralogy and metallurgy with chymistry. Even White, the only one who had
unquestionable experience in the English iron industry before his arrival in
the Bay Colony, was involved in such chymical endeavors as the distillation
of perfumes and strong waters, and while in Bermuda probably helped
Barkeley in his pursuit of "the greate worke."[20] This integration of traditional chrysopoetic concerns and the production of chymical products with
the practical aspects of metallurgy should come as no surprise, given the
perennially close association of these fields presented in chapter 2. At the
same time, however, the intense focus on labor-saving devices and fuel efficiency that we see in White's "secretts" is not very prominent in the chymical literature before the middle of the seventeenth century. Even Van Helmont, in his constant quest to found a new natural philosophy, pays little attention to the outlay of labor and expenses involved in running a chymical
laboratory. The fact that this is not the case with Starkey may well derive
from his association with the clever technological minds of New England,
who combined a Daedalean skill in metals with the cold accountancy demanded by their ever-watchful masters, the Company of Undertakers.

THE STRUCTURE OF STARKEY'S LABORATORY NOTEBOOKS

While Starkey's practical experimentation shows signs of the "industrial"
training he received, the general form of the notebooks in which this experimentation was recorded draws upon the traditions of learned culture as
well. Indeed, one of the most intriguing aspects of Starkey's notebooks is
the coexistence and coadaptation of two intellectual traditions often considered incompatible—the experimentalism of the "New Philosophy" and
the formal Scholasticism of "the Schools." There is an extended tradition of
drawing a strong polarity between these two methods of inquiry, perhaps
most of all in the seventeenth century itself. The rhetoric of Francis Bacon
regarding the sterility of "the Schools" is not only well known, but was also
repeated mantralike by the *novatores* throughout the remainder of the cen-

19. HP 63/11A–B.
20. HP 15/8/6B; immediately after mentioning that he has built better furnaces in
Bermuda than he had in England, White refers to Barkeley's attempts at transmutation.

tury; Van Helmont himself joined the anti-Scholastic chorus. Modern scholarship, however, has begun to blur the lines that were so forcefully drawn in the seventeenth century and perhaps taken too literally in earlier historiography. Recent studies have clearly indicated the continuing vitality of university culture as well as the ongoing contributions of the late Scholastics.[21] Yet Starkey's synthesis gives us a striking example of how the two systems were actually both drawn upon to create a coherent investigative methodology. The fact that this investigation took place in a practical laboratory setting, moreover, points to a largely overlooked contribution of Scholasticism to the methodology of experiment.

Already in the early decades of the twentieth century, Ernst Cassirer— followed by J. H. Randall and A. C. Crombie—argued for the importance of medieval and early modern Scholastic theories of method to the development of experimental science. Scholars such as Robert Grosseteste and Jacopus Zabarella are thus credited with reformulating the discussion of syllogistic method in Aristotle's *Posterior Analytics* so that it could become a tool of experimental research.[22] Yet even if one may grant a similarity between these early blueprints of a "scientific method" and the methodological statements of later scientists (particularly Galileo), the fact remains that the discussion initiated by Crombie and Randall focuses almost exclusively on the pronouncements of bookish scholars who may never have entered a "laboratory," and certainly performed few if any experiments.[23]

The case is obviously very different with Starkey. Here we have a university-trained scholar who buried himself among his furnaces for weeks on end in the quest to discover ever more powerful products of chymistry. As

21. We think, above all, of the pioneering work of Charles B. Schmitt, such as his *Aristotle and the Renaissance* (Cambridge: Harvard University Press, 1983), and the articles collected in his *Aristotelian Tradition and Renaissance Universities* (London: Variorum Reprints, 1984). For a more focused example of the newer appreciation of university learning, see Mordechai Feingold, *The Mathematicians' Apprenticeship: Science, Universities, and Society in England, 1560–1640* (Cambridge: Cambridge University Press, 1984).

22. John Herman Randall, Jr., *The School of Padua and the Emergence of Modern Science* (Padua: Antenore, 1961), 15–68 (first printed in the *Journal of the History of Ideas* 1 [1940]: 177–206); A. C. Crombie, *Robert Grosseteste and the Origins of Experimental Science (1100–1700)* (Oxford: Clarendon, 1953); Brian Lawn, *The Rise and Decline of the Scholastic "Quaestio Disputata," with Special Emphasis on Its Use in the Teaching of Medicine and Science* (Leiden: Brill, 1993), 25–27, 79–82. For Randall's dependence on Cassirer, see Charles B. Schmitt, "Experience and Experiment: A Comparison of Zabarella's View with Galileo's in *De Motu*," in Schmitt, *Studies in Renaissance Philosophy and Science* (London: Variorum Reprints, 1981; reprinted from *Studies in the Renaissance* 16 [1969]: 80–138).

23. For important correctives, see Schmitt, "Experience and Experiment," and B. S. Eastwood, "Medieval Empiricism: The Case of Robert Grosseteste's Optics," *Speculum* 43 (1968): 306–21.

we will show, this pursuit was informed from beginning to end by Scholastic techniques imbibed by Starkey while at Harvard College. Moreover, Starkey's notebooks, as private records, provide an unvarnished behind-the-scenes glimpse of his methods, thereby allowing a more accurate winnowing of private practice from public rhetoric in regard to method. Indeed, Starkey's public comments do not always accord well with his private practice: this study of Starkey may therefore be admonitory for interpretations of other seventeenth-century figures.

Although Starkey occasionally uses his university education as a means of distinguishing himself from mere empirics, he makes very strong declamations against the learning of "the Schools" in *Natures Explication* and *Pyrotechny Asserted*.[24] In these works Starkey, like Descartes, Van Helmont, and so many other seventeenth-century intellectual figures, rails against the uselessness of university training.[25] While Starkey's greatest ire is directed against the traditional Galenist teaching and practice of medicine, there is also considerable criticism given to standard university curricula, partially based on his experiences at Harvard in the 1640s. In the preface to *Natures Explication*, Starkey sarcastically recapitulates the stages of the standard curriculum. At the university, one learns "to dispute according to the Rules of Aristotle . . . Thus at the end of four years upon performing of publick declamations, disputations, and the like, the initiatory title of Bachelor of Arts is bestowed." Starkey notes that as "for the vulgar Logick and Philosophy, I was altogether educated in it, though never satisfied with it." This disaffection took place even though "at length Aristotle's Logick I exchanged for that of Ramus, and found my self as empty as before." At last he concludes that "the foundations of the common Philosophy were totally rotten" and turned instead to chymical writings. He sums up with a wholesale rejection of the Scholastic methods of the university: "my skill in Logick and Philosophy was not worth contemning, yea nothing was in mine eyes more vile."[26]

Along with these strong statements, fit for any of the "New Philosophers" of the seventeenth century, Starkey also tells the story of how he was driven to experimental philosophy by the Scholastic culture of disputation.

24. George Starkey, *Pyrotechny Asserted* (London, 1658), 161; Starkey, *Natures Explication and Helmont's Vindication* (London, 1657), esp. 16–22.

25. On Starkey's English context, see Allen Debus, *Science and Education in the Seventeenth Century: The Webster-Ward Debate* (London: MacDonald, 1970); for university training see also Peter Dear, *Mersenne and the Learning of the Schools* (Ithaca: Cornell University Press, 1988), and Dennis Des Chene, *Physiologia: Natural Philosophy in Late Aristotelian and Cartesian Thought* (Ithaca: Cornell University Press, 1996).

26. Starkey, *Natures Explication,* a4, 19–20, 35, 37; see also Starkey, *Pyrotechny,* 77–78.

He recounts that he once took part in a disputation on whether gold could be made potable, that is, truly prepared into the powerfully medicinal *aurum potabile*. Starkey's role was as the affirmer of the proposition (the *respondens*) and as such he gave "unanswerable" arguments forceful enough that he was encouraged to try to make the substance himself, thinking that "the Logical heads of invention, especially according to Ramus, would not fail to unfold to me this whole mysterie" of its secret preparation.[27] But he found that this Scholastic method completely failed to achieve the desired results in the laboratory, and so he became an "indefatigable prosecutor of experiments" in order to discover the truth about the world. He even wrote a boldly entitled *Organum novum philosophiae* against the "fallacious shew" of Scholastic methods (whether Aristotelian or Ramist).[28]

These published statements on the necessity (and superiority) of experimental trials correspond quite well with the depiction of Starkey's methodology given in the previous chapter. As he remarked in his Helmontian treatise *Pyrotechny*, "This I know, that the subtilties which are oft in *speculative Theorie*, prove dotages in *practice*, this my own experience hath to me put out of question."[29] There can be no doubt that Starkey really was an "indefatigatible prosecutor of experiments" whose commitment to an experimental philosophy would set him among the ranks of "the moderns." Thus it is perhaps surprising that upon returning to the private notebooks to consider their format, we find that their style retains not only the forms but also the language of the Scholastic tradition, reflecting with remarkable clarity the "ancient" dialectical traditions of the schools to which he had been exposed at Harvard. Indeed, the testimony of the notebooks reveals the level of rhetorical exaggeration present in his published comments on the worthlessness of his Scholastic training.

SCHOLASTIC METHODS IN STARKEY'S NOTEBOOKS

The central method of Scholastic inquiry (and of the university curriculum) was built around the *quaestio disputata*. This form was developed in the High Middle Ages and continued to be employed in some locales as late as the end of the eighteenth century. While the question format provides the structure for innumerable Scholastic treatises on everything from theology and law to medicine and natural philosophy, the written format is based ultimately on the oral tradition of the disputation, public and private, central to the activities of the medieval university. In a standard version, the *dispu-*

---

27. Starkey, *Natures Explication*, 36.
28. Ibid., 36–37; this work, cited several times in *Natures Explication*, has not survived.
29. Starkey, *Pyrotechny*, 26.

tatio ordinaria begins with a question asked by the master. This question is generally in a form accommodated to a binary yes-no response: "Whether [*Utrum*] . . . " The question is then disputed by two students: if the *opponens* comes first, he answers the question with a negative assertion supported by a succession of arguments *(argumenta quod non)*. An affirmative response then follows from the *respondens*, refuting the negative points of the *opponens* and supported by his own arguments *(argumenta quod sic)*. After this dialectical exercise has finished, the master gives a summary of the argument and may provide a *vera conclusio* or *vera solutio* to the problem. Other formalized divisions existed in more complex forms of the Scholastic dialectic, such as the *dubitatio* and so on. The key aspect of the Scholastic method is its formalized structure of dialectic used to reach a conclusion.[30] It is such disputations that Starkey denounces in print as useless exercises, and it was in just such a disputation on the topic of potable gold that he participated as the *respondens*.

Turning to the notebooks, we see that despite Starkey's public rejection of the schools, he deployed this formalized structure of argument explicitly in his own private laboratory practice. His notebooks are full of written *disputationes* that he held with himself over *quaestiones* of chymical practice. The formal marginalia, examined in the previous chapter, that give structure to the notebooks draw their origins from the Scholastic disputation. One clear example occurs in Starkey's attempts to discover Suchten's arcanum of antimony. There Starkey presents two numbered lists of arguments, first the negative ones (as those held by the *opponens*) and then the affirmative ones (as in the case of the *respondens*).[31] Similarly, in pursuing the Suchtenian metals, Starkey advances his study by means of formalized questions and conclusions. Again, in another notebook, Starkey questions whether it is right to prepare salt of tartar by mixing saltpeter with the crude tartar in order to calcine it rather than calcining the crude tartar alone. He follows this with a paragraph labeled "Against [*contra*] saltpeter" using citations from Van Helmont and then a paragraph labeled "In favor of [*pro*] saltpeter." The *pro* meets the objection of the *con* by using Starkey's own laboratory experience as evidence. Thereafter he writes a "Conclusion for practice [*Conclusio ad praxin*]" based upon the results of his brief disputa-

---

30. On the Scholastic form and its development see Lawn, *The Rise and Decline of the Scholastic "Quaestio Disputata"*; Bernardo Bazan et al., *Les questions disputées et les questions quodlibetiques dans les facultés de théologie, de droit et de médecine* (Turnhout: Brepols, 1985); Gordon Leff, *Paris and Oxford Universities in the Thirteenth and Fourteenth Centuries* (Huntington, N.Y.: Krieger, 1975), 116–84; and A. G. Little and F. Pelster, *Oxford Theology and Theologians* (Oxford: Oxford University Press, 1935), 29–56.

31. *Notebooks and Correspondence*, document 11; Sloane MS 3750, fols. 6r–v.

tion, deciding in favor of the use of saltpeter but noting that this is only "until I see that the tartar calcined alone produces nobler effects, which I will at some point test."[32]

Thus Starkey unites this formalized type of argumentation with his experimental philosophy, and the two together make up an overall methodology wherein the Scholastic method is propaedeutic to the experimental. Starkey begins with a question about the preparation of a particular product, submits the text and/or his ideas to the formalized logic of the disputation with its *dubia* and its *argumenta quod sic* and *quod non* and finally arrives at the *conclusio*. But this is not the end. As we have seen, this conclusion is only conjectural—it must be tried in the fire, and here the experimental methodology begins. For Starkey it is the fire—that is to say the practical experimentation in the chymical laboratory centered around the furnace—that is the wise master or *praeses* who delivers the *vera solutio* at the end of the disputation.

Starkey's complete investigative methodology is then like a diptych composed of Scholastic and experimental wings. This structure itself is embodied in the layout of the notebooks. We have already noted how Starkey divided his records into sections: some parts, like the "Antimoniologia" of Sloane 3750, are devoted entirely to formalized, "Scholastic" analyses of texts and desiderata, while others, like the subsequent "Deuterai phrontides," contain series of experiments, often dated sequentially, that represent the practical trials resulting from the previous conclusions. Indeed, a letter appended to the end of *Pyrotechny Asserted* witnesses Starkey's conscious division of the two halves of his method as well as their complementarity. When this letter was written in early 1658, Starkey was under "confinement," which may have been some sort of a house arrest, owing to unspecified legal proceedings by William Currer, another chymical practitioner. In the letter, Starkey explains that "the true ground of my patient acceptance of ten months confinement" was that it liberated him from the troublesome expenditure of his time in his medical practice, thus freeing him to do experiments. "In this time . . . I have made it my business to reduce those *Theoricall Contemplations,* and *Conclusions,* (which reading and collaterall Experiments had suggested unto me) unto *practise.*"[33] This seems a clear reference to the trial of Starkey's "conjectural processes," which have been derived theoretically from reading and "collaterall Experiments." He apparently had accumulated a fair collection of them and they were awaiting the test of the fire, and the "freedom" of his confinement allowed him to carry out these tests.

---

32. *Notebooks and Correspondence,* document 11; Sloane MS 3750, fols. 3v–4r.
33. Starkey, *Pyrotechny,* 168–69.

These two wings of Starkey's methodology themselves recall the Scholastic distinction of the sciences into a *theorica* and a *practica*. This distinction is particularly clear in Scholastic medicine and is also very well represented in alchemical writings. Starkey's own chrysopoetic *Marrow of Alchemy* was printed separately in two installments, the first in 1654 "Illustrating the Theory" and the second, in 1655, "Elucidating the Practique of the Art."[34] What is novel about Starkey's use of the two is the degree to which he unites them. While a Scholastic treatise of medicine might well bear the divisions into theoretical and practical sections, there is frequently little interconnection between the two.[35] This is strikingly different from Starkey's methodology, the very importance and uniqueness of which is based upon the way in which he intimately marries theory and practice— basing his practice upon theory and in return amending his theory with practice.

Starkey's use of Scholastic techniques as a private tool for learning accords well with what we know of his Harvard education.[36] A student's experience at seventeenth-century Harvard was thoroughly saturated with disputation and logical analysis from matriculation to commencement, a period that occupied three years in the 1640s. We can reconstruct with considerable accuracy a typical day in the life of a student at Harvard around the time of Starkey's residency. Not surprisingly, the student's day began with Scripture. In addition to demonstrating his ability to read the Old and New Testaments in their original languages and to translate them into Latin, the student was expected to "resolve them *Logically*," an exercise that was rotated among the advanced undergraduates and resident bachelors studying for their A.M.[37] This exercise took place daily, before and after the curricular studies proper, hence before 8:00 A.M. and after 5:00 P.M. An example of such analysis is found in the works of William Ames, a Puritan author particularly favored at Harvard. After reproducing a short passage from the Bible, Ames would analyze it by expanding the meaning

34. George Starkey [Eirenaeus Philoponus Philalethes, pseud.] *The Marrow of Alchemy* (London, 1654–55). Of course, as it is a published work dealing with the Philosophers' Stone, the "Practique" is by no means the clear workable experimentals recorded in the private notebooks.

35. See Heinrich Schipperges, "Die arabische Medizin als Praxis und Theorie," *Sudhoffs Archiv* 43 (1959): 317–28, and Richard Toellner, "Medicina Theoretica-Medicina Practica: Das Problem des Verhältnisses von Theorie und Praxis in der Medizin des 17. und 18. Jahrhunderts," *Studia Leibnitiana*, supp. 22 (1982): 69–73.

36. On Starkey at Harvard, see Newman, *Gehennical Fire*, 18–50.

37. *New Englands First Fruits*, 16, quoted in Samuel Eliot Morison, *Harvard College in the Seventeenth Century* (Cambridge: Harvard University Press, 1936), 1:436. See also the College Laws for 1642–46, published in "The College Book, I," *Publications of the Colonial Society of Massachusetts* 15 (1925), 24–31.

of the passages and dividing this interpretation into major and minor premises, questions, responses, *documenta* (dogmas), reasons, and numbered points.[38] Students were also expected to subject the weekly sermons to this sort of a treatment. A number of seventeenth-century notebooks survive containing these elaborate analyses of sermons—after mentioning the biblical passage that formed the subject of the sermon, the student would supply marginal tags to each division, such as "D." for doctrine (or dogma), "Q." for question, "A." for answer, "O." for objection, "R." for reason, and "U." for use. These tags are often accompanied by numbered subdivisions; similar tags are also found in the records of the debates that were held viva voce.[39] Such marginal tags distributing a sermon or scriptural writing into its dialectical divisions are strongly reminiscent of the style of Starkey's laboratory notebooks, where marginalia are used profusely to articulate not only the sections in analyses of texts and logical arguments, but also stages in practical experimental processes.

After practice at such scriptural analysis, students would convene in their respective classes. President Henry Dunster would lecture to the three classes in order, beginning with the lowerclassmen at 8:00 and proceeding to the other classes at 9:00 and 10:00. His lectures probably consisted of reading from a text or epitome and occasionally explicating it. After Dunster had finished with a given class, the students would work with a tutor or among themselves. The students were expected to memorize the lectures and were responsible for delivering an oral recitation of the lecture to the Tutor.[40] This again provided an opportunity for systematizing and analyzing an oral presentation, as in the case of the sermon synopses. On Monday and Tuesday afternoons, these studies were followed by public disputations on the subjects studied in the morning; during Starkey's tenure these were moderated by Dunster.[41] One such Harvardian disputation—carried out on 3 April 1646, during Starkey's own residence at Harvard—survives in the notebooks of Jonathan Mitchell, a Harvard A.B. of 1647. Mitchell's disputation concerns a metaphysical issue, namely whether a cause remains present in its effect, and the form of the argument is classically Scholastic,

38. Morison, *Harvard College,* 1:89, 268–72.

39. Notebook of Samuel Stoddard, c. 1662, Harvard University Archives, HUD 660mfp. For an analysis of this notebook, see Norman Fiering, "Solomon Stoddard's Library at Harvard in 1664," *Harvard Library Bulletin* 20 (1972): 257. See also Houghton Library MS. Am. 804, which contains similar synopses of sermons made by the Harvard student John Chickering, mostly in 1651 (for Chickering's authorship, see the inscription on the back flyleaf). Unlike Stoddard, Chickering expands many of the marginal tags.

40. Thomas Jay Siegel, "Governance and Curriculum at Harvard College in the 18th Century" (Ph.D. diss., Harvard University, 1990), 224–25.

41. Morison, *Harvard College,* 1:142.

being divided into *Quaestio, Negatio* (possibly *Negatur*), and *Oppositum* and *Responsio* (or perhaps *Opponens* and *Respondens*).[42]

In addition to arguments beginning with a formal *quaestio,* Harvard students were also called upon to defend specific *theses.* A subsequent folio of the Mitchell notebook contains numbered rhetorical theses drawn from the work of Peter Ramus, the famous Calvinist dialectician, who, as we have already seen, was mentioned dismissively by Starkey. These notes are in fact reminiscent of Starkey's numbered journal entries, where he frequently summarizes the main points of an extended text (as in one notebook where Starkey summarizes into twenty-two points the first book of Suchten's *Secrets of Antimony*) or of the main observations from a practical process or experiment.[43] This practice was widely employed at Harvard both for extracting pithy nuggets from written works and for laying down the primary tenets of a discipline. In either case, the theses were characteristically arranged in a numbered sequence. The method was sometimes referred to as "epitomizing" a work or a field, and candidates for the master's degree were supposed to prepare a "Synopsis, or Compendium" of some art, also called a "System," based on these methods.[44] Starkey's *Organum novum philosophiae* may very well have been a system of this sort, though its loss makes it impossible to say anything certain.

Logical analysis of Scripture, public classroom debate, and the organization of knowledge into theses by no means exhausted the opportunities for employing Scholastic methods at early Harvard. The *quaestio disputata* reemerged in a three-week period of "sitting solstices" occurring each summer—a variation on the Medieval *Quodlibeta*—in which "senior sophisters" (our seniors) were required publicly to answer any question put to them by anyone who chose to participate.[45] In addition, graduating sophisters were required to defend theses at commencement, and broadsheets announcing the theses to be defended were printed in advance and distributed. The thesis sheet for 1646 survives—with Starkey's name on it—but unfortunately we do not know which theses he defended.[46]

In addition to all these Scholastic venues, Starkey was also exposed to the use of dichotomy charts for dividing a topic or argument into its com-

---

42. Ibid., 1:143–44.

43. *Notebooks and Correspondence,* document 11; Sloane 3750, fols. 33r–34v; cf. fols. 4r–v.

44. Morison, *Harvard College,* 1:149, 155–57. A good example of such a "system" may be seen in the *Physicae compendium* of Jonathan Mitchell, found in a notebook written by Michael Wigglesworth. See Newman, *Gehennical Fire,* 25–28.

45. Siegel, "Governance and Curriculum," 227; Morison, *Harvard College,* 1:67–68, 206–7, 458.

46. Morison, *Harvard College,* 2:585–87.

ponents. Such charts were widely associated with Ramus at Harvard, though he, of course, was not the first to use such divisions.[47] A famous letter from Leonard Hoar, a Harvard A.B. of 1650 who went on to become president of the college from 1672 to 1675, expressly advises his nephew Josiah Flynt to study the note-taking "method of the incomparable P. Ramus," as a means of navigating his freshman year.[48] The dichotomy charts consisted of classificatory outlines or synopses organized primarily around bifurcations indicated by swung brackets. Starkey's own use of dichotomy charts appears in Royal Society Manuscript 179, in the section of that notebook dealing with the *magnum opus*, that is, the preparation of the Philosophers' Stone.[49] There Starkey constructs a diagram based on swung brackets to explicate the preparation of the Stone, or as he calls it here, the "Greater Bezoar."[50] The chart lays out the various materials and methods that are required in order to produce the Sophic Mercury and to convert it, with an addition of gold, into the Philosophers' Stone (figure 6). Starkey was unable to expand his chart in the usual fashion due to the small format of this notebook; had he been able to do so, it would look more like the diagram made by another Harvard student, John Holyoke, in the early 1660s, shown in figure 7.[51] Like Starkey, Holyoke divides the initial topic, here *substantia*, into its major divisions *(creata, increata,* and *partim increata, partim creata)* and then proceeds to dichotomize these further. Thus here in RSMS 179 we see Starkey in the act of organizing and dichotomizing a very difficult experimental subject—the making of the Philosophers' Stone—in which he was then currently engaged, using the Ramist techniques he learned at Harvard.

While other opportunities for logical analysis and formal argument also existed at seventeenth-century Harvard, we have presented sufficient material to convey the intensely disputational character of the curriculum during Starkey's time. Harvard College inculcated Starkey with the gamut of Scholastic techniques for acquiring and organizing knowledge that were available in the mid–seventeenth century, ranging from the analytical abridgement of sermons to the logical expansion and solution of philosophical questions. These techniques were intended to be universal in scope—

47. On Ramus in English universities, see Mordechai Feingold, "English Ramism: A Reinterpretation," in *The Influence of Petrus Ramus,* ed. Mordechai Feingold et al. (Basel: Schwabe, 2001), 127–76.

48. Morison, *Harvard College,* 2:639–44.

49. *Notebooks and Correspondence,* document 12; RSMS 179, fols. 4r–5r.

50. The term "Greater Bezoar" is one trope for the Philosophers' Stone, which, like the quasi-legendary bezoar stone, was supposed to be a universal antidote.

51. Harvard University Archives, HUC 8662 300 (vt), 39r.

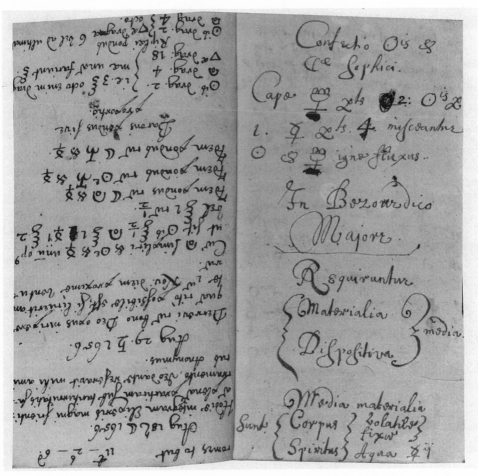

Figure 6. Two sheets from one of Starkey's laboratory notebooks (the second sheet appears on p. 172) showing his Ramist dichotomy chart outlining the preparation of the Philosophers' Stone (here called "the Greater Bezoar"). RSMS 179, fols. 3r and 4r. Reproduced by courtesy of the President and Fellows of the Royal Society.

a properly educated college man could apply them to any area of reasoned discourse. But unlike his colonial peers, Starkey actually carried his Scholastic education into the laboratory and used it there. At the same time he augmented the standard divisions of argument with new categories more appropriate to his practical experimentation, such as his "Observations," "Conjectural Conclusions," "Notables," and "Hands-On Trials." A few of the terms for these categories can be found in the printed literature of early modern chymistry. Most notably, Starkey's "Chimical Evangelist," Joan Baptista Van Helmont, employs the expressions *demonstratio mechanica*

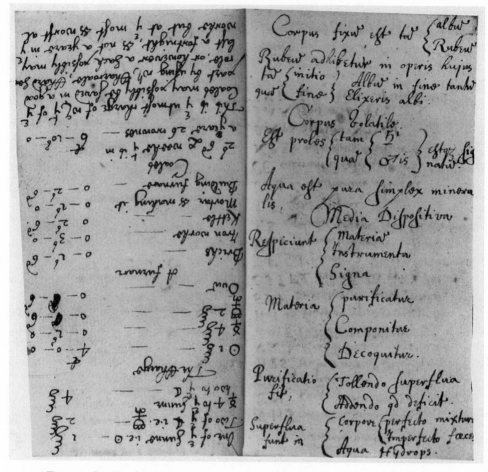

Figure 6. Continued.

and *ratio mechanica* for arguments based on empirical evidence.[52] Andreas Libavius's *Alchemia* of 1597, on the other hand, is famous for its incorporation of dichotomy charts.[53] Another early modern chymist, Angelus Sala, provided lists of "Observations" in his pharmacological studies, while also giving numbered "Problems" *(porismata)* as well as "Objections," "Responses," and "Doubts."[54]

52. C. de Waard, *Correspondence du P. Marin Mersenne* (Paris: Presses Universitaires de France, 1946), 3:111, 144.

53. Owen Hannaway, *The Chemists and the Word: The Didactic Origins of Chemistry* (Baltimore: Johns Hopkins University Press, 1975). Van Helmont himself used a Ramist dichotomy chart in the *Ortus*, "Morborum phalanx," 566.

54. Angelus Sala, *Opera medico-chymica* (Rouen, 1650), 175–76, 183–84, and passim for

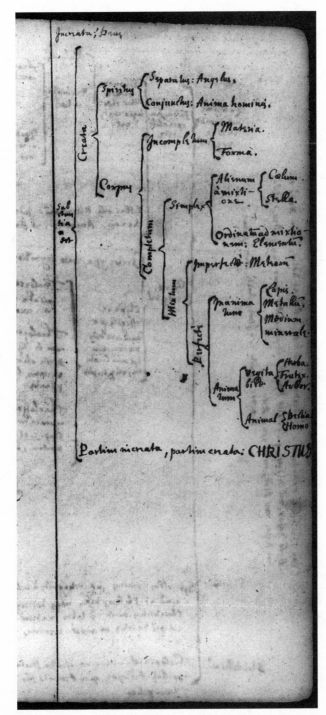

Figure 7. A Ramist dichotomy chart in a notebook of the Harvard student John Holyoke (early 1660s); compare with Starkey's Ramist chart in figure 6. Harvard University Archives 8662 300 (vt).

But these works of Van Helmont, Libavius, and Sala were printed texts destined for the mass market; they were not laboratory notebooks. Their use of Scholastic methodology was intended as a pedagogical technique— a method of conveying existing knowledge rather than a means of discovering new knowledge. This essential difference is revealed in the fact that Starkey's "Conjectural Processes" and "Conclusions" are at times crossed out or explicitly refuted when his experimentation has led him to a dead end. Here we witness the living process of discovery and rejection of a preconceived idea that Starkey's predecessors—and indeed Starkey himself— would never have considered publishing. Yet in these notebooks Starkey presents us with a Scholasticism transformed into part of an integrated experimental methodology. This use of Scholastic methods as tools of discovery surely reflects Starkey's early immersion in the oral and scribal culture of Harvard College, despite his public rejection of university learning.

STARKEY AND TEXTUAL AUTHORITY

Let us examine more closely one aspect of the Scholastic side of Starkey's methodology of investigation, specifically his deployment of texts and authorities in the management of his laboratory activities. One criticism often leveled at "ancient" learning in the seventeenth century was that of excessive adherence to textual authority rather than to the testimony of the senses and new experiential knowledge. This criticism is encapsulated in anecdotes of varying levels of veracity about Aristotelians refusing to amend or discard Aristotle's views on the heavens even when confronted with Galileo's telescopic observations and the like. Indeed, the commitment of the New Philosophers to rejecting classical and textual authorities in favor of experiment and observation forms a ground upon which the Royal Society of London chose its motto "Nullius in verba," an abbreviated version of the Horatian verse "Inclined to swear to the words of no master" (a verse that, incidentally, Starkey himself quotes in *Natures Explication,* possibly paraphrasing Van Helmont.)[55]

It is clear from his laboratory notebooks that Starkey's experimental projects always begin with the text of an authoritative figure. In iatrochemical topics this is most often Van Helmont and sometimes Alexander von Suchten; in chrysopoetic topics Suchten reappears, but is joined with venerable adepti such as Bernard Trevisan, George Ripley, and Artephius. Al-

---

observationes; 212 for *porismata;* 288–95 for *objectiones* and *responsiones;* 345 for *dubia.* As noted in chapter 3, some of Starkey's manuscript titles resemble those of Sala's printed works.

55. Starkey, *Natures Explication,* 33; see Van Helmont, *Ortus,* "Natura contrarium nescia," no. 15, p. 168: "Ego vero sub libertate Philosophica, nemini addictus magistro, sentio."

though some of Starkey's projects surpassed the expectations or intentions of his sources, most notably his grand design for medicine using the Elixir of volatile salt, in general we do not find Starkey beginning an entirely original project or striking out initially toward a new, hypothetical product or goal. Nor do we see him collecting observations in the form of a "natural history" as advocated by the Baconian program and as Boyle would famously do. Like the laboratory work of a modern industrial or pharmaceutical chemist, Starkey's endeavor is highly goal oriented; it is devoted to the manufacture of specific products. Nor is he unlike modern researchers in taking his initial direction from an authoritative text. If a chemist working at Eli Lilly or some other pharmaceutical firm wants to produce a drug having properties similar to those of AZT but without violating the patents held by Glaxo-Wellcome, he will not start from scratch but will first turn to the existing chemical and pharmaceutical literature. In like fashion, Starkey is more interested in producing substances described by others than in creating entirely new products. Yet there are extensive parts of the notebooks that concentrate entirely on the interpretation of textual sources, an activity that would be quite unexpected in more modern laboratory records. These preliminary observations imply that a considerable part of Starkey's work was closely linked to texts—making him in this case more like an "ancient" than a "modern." We must now explore the nature of Starkey's use of textual authority in order to understand it accurately.

Starkey uses authoritative texts at the beginning of a laboratory project primarily as a source of "matters of fact," that is, evidence that a given substance or process exists. Not all texts or authors are of equal importance or validity to Starkey, but when he has decided upon the worth of a particular author, he puts great faith in the veracity of his accounts. Thus, Van Helmont's *Ortus*, or Suchten's *Secrets of Antimony*, or Ripley's *Compound of Alchemy* are like accounts of natural history from distant lands, describing creatures and phenomena seen by their authors but inaccessible to Europeans at home. Starkey trusts that these authors have actually seen and described genuine products; their descriptions then form the basis of his belief that he too can produce them. Although the *method* of actually producing such desiderata in the laboratory may remain uncertain or obscure in the source texts, the fact that these arcana actually *do exist* is not open to question. Indeed, Starkey several times expresses his confidence in the testimony of favored authors in spite of his own continued unsuccessful attempts to reproduce their results. For example, when after months of failed attempts to produce the Suchtenian metals from antimony regulus, Starkey takes stock of his experiences in order to chart future avenues of research, the very first point he records in his notebook reasserts his trust in Suchten despite the experimental failures—"there exists such a work; that is clear from the testimony of Suchten in which I have faith." But Starkey's training in the analysis of texts immediately emerges as he adds that Suchten spoke "hyperbolically" when he claimed that such antimonial metals could be made "in the time that it takes to eat a soft-boiled egg."[56] Hence Suchten's text reassures Starkey in his search for the secret—the secret *is* out there, even if one cannot always rely on the literal sense of the text regarding how to obtain it. Judging from his assiduous pursuit of other topics, it is clear that Starkey thought the same about the accounts of Van Helmont's arcana and about the Philosophers' Stone as well.

Starkey's certainty that the various arcana he pursued were matters of fact is not unwarrantable. Although the transmutatory Philosophers' Stone, for example, seems "obviously" fictitious to modern readers, it was by no means so to Starkey or his contemporaries. The Stone was a logical consequence of prevailing chymical theory and was one part of a rational, coherent body of chymical thought. Moreover, eyewitness accounts of transmutation were not uncommon in the seventeenth century in the form of "transmutation histories."[57] Finally, Starkey had himself succeeded in producing various notable laboratory results—such as making gold sprout and grow into a mineral "tree"—that, since these results were described by some of his authoritative authors, assured him that he was on the right track.[58] Starkey gives the rational grounds for the certainty of his belief not only in the published *Marrow of Alchemy* and other works but also in the draft treatise *Diana denudata,* a work on the Philosophers' Stone found in Royal Society MS 179. There Starkey begins the work with the assertion "that there is in nature such a thing which is called the Philosophers stone, besides the testimony of Credible witnesses, & the Evidence of Reason & Experience, the most assured proofe hath confirmed unto us the truth of the same."[59] As for the other arcana, Starkey had no reason to disbelieve them either, as they too were supported by coherent theoretical frameworks and vouched for by respected authorities like Van Helmont. *That* these substances existed was for Starkey patently clear from text and theory; *how to produce them* was the goal of his practical laboratory experimentation.

56. *Notebooks and Correspondence,* document 11; Sloane 3750, fol. 33r.

57. Principe, *Aspiring Adept,* 93–98, and Newman, *Gehennical Fire,* 3–13.

58. On the surprising phenomenon of the "Philosophical Tree," see Principe, "Apparatus and Reproducibility in Alchemy," in *Instruments and Experimentation in the History of Chemistry,* ed. Frederic L. Holmes and Trevor Levere (Cambridge: MIT Press, 2000), 55–74, where its re-creation in a modern laboratory is described.

59. *Notebooks and Correspondence,* document 12; RSMS 179, fol. 6r; compare this with the passage cited from Starkey's *Marrow of Alchemy* in Principe, *Aspiring Adept,* 110–11.

PROBLEMS OF TEXTUAL COMMUNICATION

This level of confidence in textual sources has a crucial implication for Starkey's practical experimental programs: the failure of practical experimentation to reproduce the results described by his authorities does not arise from errors or falsity on the part of the textual authorities, but rather on account of problems in the *transmission of information* from author to reader. These difficulties vary in their origins, but all have the effect of interrupting or distorting the transfer of information. Starkey's ideas on this topic appear clearly when he must confront the texts with contrary evidence from the laboratory. This engagement between the textual and the practical marks, in fact, the "closure of the circle" of Starkey's research methodology, the point at which results from the fire impact upon conjectural processes or interpretations—where practice returns to look theory in the eye. What is key here is that Starkey never—at least in the records we have—seems to have concluded from his practical experience that a matter of fact regarding the existence of a desirable substance in an authoritative textual source is simply *wrong*. If a reputable text declares that a thing exists, Starkey's failure to produce it comes not from falsity of the report but from problems in its transmission. Thus, the next step for Starkey is not the abandonment of the source or its simple negation but rather an attempt to find and to correct the errors in transmission.

Faults in transmission may occur at several junctures and range from simple to complex. The simplest are "mechanical" problems of transmission. For example, when Starkey first writes out his "Arcanum of alkalies" in 1655, he notes that he was led to the correct method by a consideration of a passage from Van Helmont. That passage states that the "hidden circulation" (which Starkey identified as exposure to the air) is to be done "without any water"; however, Starkey notes that "it is clear from experience" in his own laboratory that the process works much better *with* water, and so he writes a note to himself to "examine *[quaere]* whether there is not a typographical error in that place in Helmont." In fact, Starkey goes even further to suggest the proper reading of the text and how it was corrupted: "Examine whether it is not to be read 'by a hidden and artificial circulation without any fire'; that those who published his works incorrectly changed 'fire' to 'water' because they were not able to understand him. For as a matter of fact, this operation is unsuccessful in the fire."[60] Starkey may even have had in mind that the compositors merely misread Helmont's symbol for fire (a triangle with a vertex pointing upward) as the symbol for water (a triangle with a vertex pointing downward). In any case, Starkey's experi-

60. *Notebooks and Correspondence*, document 12; RSMS 179, fol. 60v.

mental results did not accord with Van Helmont's text, and so there must be a mistake somewhere, but not in Van Helmont or in the experiments. Van Helmont was veracious and the fire cannot err; ergo there must be an error in transmission, in this case on the part of unskilled or sloppy compositors. Note that Starkey does not slavishly adhere to the words of the text in the face of contrary experimental data but rather looks to harmonize the two sources of knowledge by locating and explaining the discrepancy.

Starkey uses similar reasoning in his frustrated attempts to transmute antimony regulus into the Suchtenian metals. After several months of unsuccessful experiments, Starkey reevaluates his methods and Suchten's text, laying out his analysis in what we can now recognize as a characteristically Harvardian format of an enumerated synopsis. After reasserting his belief in the reality of the transmutation based upon the testimony of Suchten, the experimental results impel Starkey to return to the words of the authoritative text, and he sets himself a Scholastic question: "Examine [quaere] whether . . . there is an error of a translator who did not well understand the intent of the author."[61] Again, the experience in the fire did not correspond with the text; therefore this discrepancy must have arisen from a fault in the transmission of information, and in this case Starkey suggests that the error may lie with the translator.[62]

While compositors and translators may introduce errors in the flow of information, a more prevalent source of error lies in the interpretations of the text given by the reader, in this case by Starkey himself. But Starkey's difficulties do not arise from an inability to read content accurately; rather they originate in the very nature of the texts he was using. They belong to the culture of concealment and secrecy so pervasive in early modern chymistry. Very simply, many authors had no intention of being either clear or complete in describing their experimental endeavors. This situation is very different from the majority of modern scientific communication, where clarity and forthrightness, even if not always perfectly executed, are nonetheless generally considered desirable features. This difference in the writing of texts translates into a difference of laboratory methodology; simply,

61. *Notebooks and Correspondence*, document 11; Sloane 3750, fol. 33v–34r.
62. In fact, in this case, assuming that Starkey used the English translation, he actually may have known the translator. The English version of Suchten's *Secrets of Antimony*, which was eventually published in 1670 and listed then as having been "translated out of High-Dutch by Dr. C. a Person of great Skill in Chymistry," may have actually been prepared in the Hartlib circle, for a copy of the translation dating from the 1650s survives as HP 16/1/48A–63B. MS. Ferguson 163, 97, identifies the translator as "Dr Childe." It is not at all unlikely that Dr. Robert Child may have been the translator; he knew German and his interest in Suchten is also attested. See Newman, "Prophecy and Alchemy: The Origin of Eirenaeus Philalethes," *Ambix* 37 (1990):113 n. 50.

since the texts are not transparent, Starkey must spend a great deal more time in interpreting (and reinterpreting) them than he would otherwise have to do. This is made quite clear by the far lesser amount of time he expends upon processes from more straightforward and literal authors like Sala and Hartmann than from the more intricate and metaphorical (i.e., secretive) writings of Van Helmont, Suchten, and the chrysopoeians.[63]

CHYMICAL SECRECY AND LEARNED PLAY IN STARKEY'S NOTEBOOKS

The widespread secrecy of early modern chymistry can be puzzling to those more familiar with modern scientific writing, and it is almost a commonplace to view the development of an ethos of scientific openness as a signal event in the development of modern science.[64] Such a viewpoint often relegates secretive tracts—whether on chrysopoeia or prized medicinal arcana—to a "prescientific" age, or rejects their contents as nonscientific or even as largely or purely fictitious. Whereas many today might question the reality of a thing simply because it is secret, for Starkey and many of his contemporaries, secrecy could have exactly the opposite effect—it marked out the items of greatest value.

The view that real knowledge is not to be freely given away runs deep through many sorts of early modern chymical literature, not only those

63. See the "Antimoniologia" in Sloane 3750, treated above in chapter 3. It must also be pointed out that Sala and Hartmann are "lesser" writers than Van Helmont and Suchten, and so their texts are also "less worthy" of the time required for a close reading.

64. The traditional view that early modern science was inimical to a secrecy epitomized by alchemy is also stressed by Betty Jo Teeter Dobbs, "From the Secrecy of Alchemy to the Openness of Chemistry," in *Solomon's House Revisited*, ed. Tore Frängsmyr (Canton, Mass.: Science History, 1990), 75–94, and to a lesser degree by Pamela O. Long, "The Openness of Knowledge: An Ideal and Its Context in 16th-Century Writings on Mining and Metallurgy," *Technology and Culture* 32 (1991): 318–55. In a more recent work, Long has come to the revisionist conclusion that military secrecy was little evident before the early modern period, but that it is largely the product of our own era; see Pamela O. Long and Alex Roland, "Military Secrecy in Antiquity and Early Medieval Europe: A Critical Reassessment," *History and Technology* 11 (1994): 259–90. For some reassessment of secrecy in early modern chemical literature, see William Eamon, *Science and the Secrets of Nature: Books of Secrets in Medieval and Early Modern Culture* (Princeton: Princeton University Press, 1994); Jan Golinski, "Chemistry in the Scientific Revolution: Problems of Language and Communication," in *Reappraisals of the Scientific Revolution*, ed. David Lindberg and Robert Westman (Cambridge: Cambridge University Press, 1990), 367–96; Principe, "Robert Boyle's Alchemical Secrecy: Codes, Ciphers, and Concealments," *Ambix* 39 (1992): 63–74; Stephen Clucas, "The Correspondence of a XVII-Century 'Chymicall Gentleman': Sir Cheney Culpeper and the Chemical Interests of the Hartlib Circle," *Ambix* 40 (1993): 147–70; Newman, "Alchemical Symbolism and Concealment: The Chemical House of Libavius," in *The Architecture of Science*, ed. Peter Galison and Emily Thompson (Cambridge: MIT Press, 1999), 59–77; Newman, *Gehennical Fire*, 54–78.

dealing with transmutation. Van Helmont, for example, is fond of quoting the maxim that "God sells the arts for sweat," meaning that not even God Himself dispenses knowledge gratis but only as the reward of assiduous labor.[65] Starkey's own *Pyrotechny* contains the same motto.[66] If God sets such an example, who is Van Helmont or Starkey to do otherwise? Thus Van Helmont, like many other chymical writers, teaches not by bare receipt or prescription, but rather only by hints. These hints then must be expanded into a full understanding of the matter by sufficiently talented readers and experimenters (presumably also favored by divine blessing). We have seen a simple example of this in Van Helmont's terse directions for making the cinnabar of antimony. Starkey needed to interpret (and reinterpret) Van Helmont's utterances in the light of his own laboratory experience in order to formulate "conjectural processes" from them. He could not simply follow some receipt set down stepwise in his textual source. The philosophical, or epistemological, lesson here is that careful reading of books is absolutely essential but cannot suffice for the acquisition of real knowledge. When Van Helmont advises readers to learn how to make salt of tartar volatile, he admonishes them that "for these things it is not enough to wear out books, but over and above this you must buy coals and vessels, and spend night after night awake."[67] Indeed, Starkey, as a good student of Van Helmont, heeded his preceptor's advice—he did in fact "wear out books" and "spend night after night awake" in experimental trials. His notebooks bear testimony to both activities, and his printed works echo the Belgian philosopher's advice to his own readers.

In addition to his master's explicit injunctions to secrecy, Starkey had a multitude of other concerns that could have led him to avoid the free disbursement of his own knowledge. Among these were trade secrecy, titillation of the market, and apprehensions about economic unsettlement if the Philosophers' Stone should become common knowledge—topics that have been dealt with elsewhere.[68] But the notebooks reveal yet another side to the chymical practice of concealment. The reasons for secrecy that we have mentioned so far relate primarily to the "supply-side" of chymical concealment. That is to say, they function as a partial explanation of why

65. Van Helmont, *Ortus,* "Puerilis humoristarum vindicta," no. 5, p. 524; *Ortus,* "Supplementorum paradoxum," no. 55, p. 704.

66. Starkey, *Pyrotechny,* "Epistle to the reader" (written by a "Friend of the Author").

67. Van Helmont, *De febribus,* in *Opuscula,* chap. 15, no. 26, p. 58.

68. On Starkey in particular, see Newman, *Gehennical Fire,* 62–78. For more general treatments of secrecy, see Eamon, *Science and the Secrets of Nature,* and Pamela O. Long, *Openness, Secrecy, Authorship: Technical Arts and the Culture of Knowledge from Antiquity to the Renaissance* (Baltimore: Johns Hopkins University Press, 2001).

authors wrote in such a manner but do not address the reason why readers invested so much time in trying to understand their texts. One obvious answer to the latter question lies in the reader's expectation of a compensatory reward, namely the eventual acquisition of the hidden and potentially profitable secrets. But there is another explanation that should be considered of equal importance; namely the early modern taste for riddles, allegories, and intellectual exercises. Many people today still enjoy their newspaper crosswords, word jumbles, and other such diversions. Mind-bending toys like the Rubik's Cube can still gain great popularity, as do murder mysteries and similar genres of literature where the reader is expected to gather clues to solve a mystery or to anticipate the end. The reward in these diversions is neither material nor public, but intellectual and private. In a similar way, the early modern intellectual took great delight in deciphering the meaning of an emblem or motto, or in "reading" the iconography of a pageant, play, or painting. Books of emblems, whether in the collections of Alciati or in the chymical illustrations of Michael Maier, were extremely inviting targets of study and indicate the familiarity that early modern readers had in reading words, signs, emblems, and things simultaneously on a multitude of levels and embedded in a network of correspondences.[69]

All the same, this activity was not necessarily mere entertainment or diversion, but often a key part of the serious intellectual quest for knowledge. The early modern mind took very seriously Heraclitus's assertion that "nature loves to hide." Nature itself was (in the popular Neoplatonic conception) a letter or book written to man by the hand of God, and this writing was full of secrets expressed in allusion and riddle that had to be deciphered and interpreted by the skilled exegete. The quest for knowledge of the world was also frequently visualized as a *venatio*, or hunt, requiring all the stratagems and talents of the hunter in search of a hidden quarry.[70]

The intellectual delight in layered meaning and riddle clearly influenced the writing of texts as well as the willingness to read them. Some chymical authors, especially chrysopoeians, mindful of the reasons outlined above for maintaining secrecy, undoubtedly took as much delight in embroidering the contents of their works into elegant allusive and enigmatic forms as did a contemporaneous painter in incorporating symbolism and allusion into his portraits, or a poet in creating level upon level of wordplay and metaphor. From its beginnings in the Egypt of late Antiquity, practitioners

---

69. See, for example, the classic study of Mario Praz, *Studies in Seventeenth-Century Imagery* (Rome: Storia e Letteratura, 1975).
70. See Eamon, *Science and the Secrets of Nature*.

of alchemy had expressed themselves by means of extended conceits. In cryptic symbols and their later offspring, chymical processes or theories were veiled beneath fanciful accounts of the bizarre transactions between kings and queens, dragons and toads, eagles and hermaphrodites. In addition, the ancient rhetorical device of juxtaposing opposites was employed by the earliest alchemists and carried on by their heirs.[71] The same love of striking and incongruous imagery is at the heart of seventeenth-century "wit" as defined by Samuel Johnson in his life of Abraham Cowley: "Wit, abstracted from its effects upon the hearer, may be more rigorously and philosophically considered as a kind of *discordia concors;* a combination of dissimilar images, or discovery of occult [i.e., hidden] resemblances in things apparently unlike."[72]

This principle of *discordia concors*—harmonious discord—forms the guiding principle behind much of the imagery found in metaphysical poetry, from John Donne to Henry Vaughan.[73] Small wonder, then, that many of the metaphysical poets were attracted to chymistry—Donne's famous poem "Love's Alchemy" is one of the masterpieces of English literature, as is George Herbert's "The Elixir."[74] Indeed, Edward Taylor, a graduate of Harvard College less than a generation after Starkey (A.B. 1671), was an accomplished poet of the metaphysical school who made extensive use of alchemical imagery. Taylor was an active reader of chymical literature, and in true Harvardian fashion, he compiled a synopsis of John Webster's 1671 *Metallographia.* In addition to being pastor of Westfield in the Connecticut Valley, Taylor was a medical practitioner. His *Dispensatory* or collection of medical recipes has survived, and it contains numerous chymical medicaments.[75] Hence Taylor clearly regarded chymistry as an area of high utility, especially in its application to medicine. At the same

71. See the learned commentary of Michelle Mertens in her edition of Zosimos of Panopolis, *Les alchimistes grecs, Zosime de Panopolis, mémoires authentiques* (Paris: Les Belles Lettres, 1995).

72. Samuel Johnson, *The Lives of the Most Eminent English Poets* (Charlestown: Etheridge, 1810), 1:14.

73. Melissa C. Wanamaker, *Discordia Concors: The Wit of Metaphysical Poetry* (Port Washington: Kennikat, 1975).

74. John Donne, *The Complete Poems,* ed. Alexander B. Grosart (printed for private circulation, 1873), 2:199–200; Alistair Fowler, ed., *The New Oxford Book of Seventeenth-Century Verse* (Oxford: Oxford University Press, 1991), 333. See also Stanton J. Linden, *Darke Hieroglyphicks: Alchemy in English Literature from Chaucer to the Restoration* (Lexington: University Press of Kentucky, 1996).

75. Karen Joyce Gordon-Grube, "The Alchemical 'Golden Tree' and Associated Imagery in the Poems of Edward Taylor" (Ph.D. diss., Free University of Berlin, 1990), 1:1–15.

time, however, he employed the imagery of stills, alembics, elixirs, and even the Helmontian alkahest throughout his poetry. The Deity Himself became an alchemist in Taylor's view: "God Chymist is, doth Sharon's Rose distill. / Oh! Choice Rose Water! Swim my soul / herein."[76] And as if to demonstrate the very pleasure that early modern intellectuals took in learned play, Taylor composed a number of his poems in the form of acrostics, sometimes even supplying them with dedications in the form of anagrams.[77]

Starkey himself composed poetry, and if he was not a Vaughan or Marvell, neither did he sink to the rank of a poetaster. Starkey's *Marrow of Alchemy* (1654–55) and his Royalist poem *Britains Triumph* (1660), were both written in six-line stanzas of iambic pentameter and share the same rhyme scheme (ababcc).[78] Although Starkey's decision to compose the *Marrow* in verse may have been stimulated more by conscious imitation of the rhyme royal treatises of his favored authority George Ripley (and other English alchemical poetry in Elias Ashmole's *Theatrum Chemicum Britannicum*) than by a real commitment to poetry, Starkey nonetheless devoted great effort here and elsewhere to the versification of his ideas—that is, the expression of chymical theories and practices in an elegant and contrived format. Moreover, some of the same delight in clever mystification that Taylor displayed is to be found in Starkey's elaborate concealment of his own identity—not so that it will never be uncovered, but so that it will be uncovered only by the clever. Rather than more perfectly concealing his identity by leaving the prefatory epistles to the *Marrow* completely anonymous, he instead signs them with the mottoes *Egregius Christo* ("special to Christ") and *Vir gregis custos* ("A man, guardian of the flock"), which are anagrammatized versions of his own name (spelled "Georgius Stirch" and

---

76. Edward Taylor, "Meditation 4, I am the Rose of Sharon," as quoted in Patricia A. Watson, *The Angelical Conjunction* (Knoxville: University of Tennessee Press, 1991), 103. See also Cheryl Z. Oreovicz, "Edward Taylor and the Alchemy of Grace," *Seventeenth-Century News* 34 (1976): 33–36.

77. Edward Taylor, *Edward Taylor's Minor Poetry*, ed. Thomas M. Davis and Virginia L. Davis (Boston: Twayne, 1981), 19–35. See also Jeffrey Walker, "Anagrams and Acrostics: Puritan Poetic Wit," in *Puritan Poets and Poetics*, ed. Peter White (University Park: Pennsylvania State University Press, 1985), 247–57.

78. For a study of Starkey as a poet, see Cheryl Z. Oreovicz, "Eirenaeus Philoponos Philalethes: *The Marrow of Alchemy*" (Ph.D. diss., Pennsylvania State University, 1972), iv–liv. See also Oreovicz, "Investigating 'The *America* of Nature': Alchemy in Early American Poetry," in *Puritan Poets and Poetics*, ed. Peter White (University Park: Pennsylvania State University Press, 1985), 99–110. On poetry in alchemy in general, see Robert M. Schuler, *Alchemical Poetry, 1575–1700* (New York: Garland, 1995).

"Georgius Stircvs," respectively). In the *Vade-mecum philosophicum,* one Agricola Rhomaeus is called the teacher of the great Philalethes himself; but the clever reader can translate the Latin first name ("farmer") into Greek, and the Greek last name ("strong") into Scottish dialect to reveal Philalethes' master as none other than "Georgos Stark."[79]

While some alchemical riddles and *figurae* no doubt have purely imaginative origins, many others do in fact conceal experimental knowledge and results. In the case of Philalethes, clear evidence of the laboratory operations veiled beneath allegorical language is provided by the juxtaposition of Starkey's private writings with his Philalethes publications. Indeed, the lengthy "vision" in Philalethes' *Ripley Reviv'd* has been comprehensively "decoded" into chymical terms.[80] The central secret concealed allusively in the Philalethes treatises—the making of the Philosophical Mercury—can also be fully decoded into the laboratory practices recorded in Starkey's notebooks. Consider, for example, the crucial section of the *Introitus apertus,* where the experimental practice is veiled with the "classically alchemical" technique of an extended conceit, using bizarre *Decknamen* (cover names) detailing the curious activities of the "Fiery Dragon, which hides the Magical Chalybs in his own belly," a hermaphrodite, a mad dog, the doves of Diana, and a chameleon.[81] In parallel with these extravagant published images, Starkey reveals the very same process in precise, easily comprehensible, and quantitative language in a private letter to Boyle where he states that he has purposefully "not been horrid in Metaphors but would be understood *ad literam* [literally]." This example clearly indicates that the allusive text was intended to conceal a real plaintext recoverable by select readers. If any readers of this book are still skeptical that Starkey's plaintext letter itself contains a real experimental basis, we need only examine the notebooks that record Starkey's work on this process and improvements to it (as noted in chapter 3). Those who still resist the notion of a replicable laboratory practice hidden in chrysopoetic texts may view figure 8, which shows the "Philosophical Tree" grown in a flask according to the directions

79. A fragmentary version of the *Vade-mecum philosophicum sive breve manuductorium ad campum sophiae* is found printed in Starkey [Philalethes, pseud.], *Enarratio methodica* (London, 1678), 189–222. The title page identifies the author as Agricola Rhomaeus, and the student-interlocutor reveals himself to be Eirenaeus Philoponus Philalethes on 191. For other editions, see Newman, *Gehennical Fire,* 268, no. 19. For a more complete manuscript, see *Gehennical Fire,* 272, no. MS13.

80. See Newman, *Gehennical Fire,* 118–33; Starkey [Philalethes, pseud.], *Ripley Reviv'd* (London, 1678), 98–135.

81. Starkey [Philalethes, pseud.], *Introitus apertus ad occlusum regis palatium,* in *Musaeum hermeticum* (Frankfurt, 1678; reprint, Graz: Akademische Druck, 1970) 657–59. Cf. *Notebooks and Correspondence,* document 3.

Figure 8. The "Philosophical Tree" prepared in a modern laboratory from Sophic Mercury and gold according to Starkey's directions. Note that prior to heating, the amorphous starting materials occupy less than a quarter of the height of the spherical part of the flask, while the fully "grown" tree rises to fill nearly all of it.

found in Starkey's private notes and using his Sophic Mercury and gold, exactly as Starkey "and" Philalethes, plus a host of other transmutational authors, described it.[82]

The expectation on the part of both writers and readers that the most important chymical knowledge would be hidden from plain view is then the major source for the errors in transmission of information that Starkey encountered. While end products are generally clearly described—Helmont's spirit of volatile tartar is a colorless liquid that dissolves silver and mercury, the Philosophers' Stone is a dense, red solid that is as fusible as wax and that when thrown upon molten lead transmutes it into gold in a matter of minutes, and so forth—the means of attaining them is not. Thus readers like Starkey are assured of the matter of fact of the existence of the desired products but left with only hints and allusions on how to prepare them. Of course, when teaching is done by hints, allusions, or the fullblown allegorical and emblematic style of some chrysopoetic works, a considerable degree of textual analysis is required to reconstitute the author's meaning, and the chances of misinterpretation rise accordingly. For Starkey, experimental practice was the test of his interpretations of texts, not of the claims of the texts themselves.

METHODS OF TEXTUAL CONCEALMENT AND ANALYSIS IN STARKEY'S NOTEBOOKS

There are various ways in which the full experimental details of a process could be concealed in chymical writings. Different authors and different subdisciplines of chymistry have their own favored methods, as identified and categorized by subsequent historians of alchemy. Besides the use of extravagant allegorical conceits and *Decknamen* so popular in chrysopoetic works (although not limited to them), there are also the methods of dispersion, syncope, and parathesis, used broadly across a range of chymical genres. The technique of dispersion involves scattering the pieces of a single item widely through a text or set of texts. Syncope abbreviates a process, generally omitting one or more steps or ingredients. Parathesis involves the needless multiplication of ingredients or processes—often these substances or steps are mere synonyms for one another.[83] Starkey copes with all four of these techniques in his notebooks.

Dispersion was expected of many chymical writers, and readers recognized their obligation to hunt down, identify, and rejoin these dispersed

---

82. For a further description of this process and the ramifications of it for debunking several unsatisfactory interpretations of alchemy, see Principe, "Apparatus and Reproducibility."

83. These terms are defined and justified in Newman, *Gehennical Fire,* 133–34.

fragments.[84] In Starkey's case, the notebooks show clear evidence of his use of this principle. For example, in the case of his successful project to volatilize alkalies, the solution came only after he had connected Van Helmont's several admonitions to volatilize alkalies with the process given elsewhere on converting oil of cinnamon into a salt by an "artificial and hidden circulation" and then connecting that secret circulation with Van Helmont's comment, found in yet another place, about the activity of the air. After all of these passages were gathered together and assumed to be related, then experimental trials confirmed the correctness of the reconstruction by the successful re-creation of the product described by Van Helmont. It is right to point out, however, that even though the treatment of salt and oil by the air gave Starkey the result he had expected from Van Helmont, there remains the possibility that this was not in fact what Van Helmont had in mind. There is no way to ascertain the correctness of a reader's interpretation except by a comparison of results with those of the writer; this of course can leave the reader in an ultimately irresoluble uncertainty.

Syncope plays a role in several of Starkey's projects. The best example is the attempt to prepare the Helmontian "cinnabar of antimony" detailed in the previous chapter. Helmont's description is terse, and we find Starkey interpreting and reinterpreting each important word (e.g., "cinnabar") to unravel the fuller exposition of the process. Syncope and dispersion come together in the case of Suchten's Philosophical Gold reputedly found in antimony regulus, for Starkey not only assumes that there is an ingredient missing ("syncopated") from the process, but identifies this missing ingredient as the unnamed mystery in the scoria mentioned in passing elsewhere in the text. Starkey uses this method himself in his *Pyrotechny Asserted,* where, although he sings the praises of his Elixir of volatile salt, describes its properties, and notes the need for a "secret circulation," he never mentions anywhere that this is accomplished by the air. There exists only one brief passage that *might* be interpreted as revealing the need for air, but this is recognizable only after one knows what to look for: "Let many Tunnes or never so little quantitie of these fixed Salts, be laid in any Field, and in few months all would be transmuted into a volatile *salt* . . . yet our Philosophers now adaies, have not learned to imitate Nature, in her most ordinary operations."[85]

84. On this issue of "dispersion" see Paul Kraus, *Jabir ibn Hayyan: Contribution à l'histoire des idées scientifiques dans l'Islam* (Cairo: Institut d'Egypte, 1943); for examples in Starkey's published work, see Newman, *Gehennical Fire,* 125–35; for examples in Boyle, see Principe, "Robert Boyle's Alchemical Secrecy," 67–69; for an example in Glauber, see Maurice Crosland, *Historical Studies in the Language of Chemistry* (New York: Dover, 1978), 36–40.

85. Starkey, *Pyrotechny,* 86.

In the case of parathesis, the use of many names for one thing, Starkey is remarkably explicit. His unpublished *Diana denudata* found in RSMS 179 explains this practice.

> Therefore to amuse & to amaze the vulgar erring Artists, wee speake of our ☉ [Sol], our ☾ [Luna], our ☿ [Mercury], our fire, our water, our furnace &c, which is al but one thing . . . It is called our fire, our furnace, our vinegre, &c, with many other infinite appellations which it hath on one account or other.[86]

As we can see, the sages are not content to apply a single *Deckname* to their Sophic Mercury and its ingredients; instead, they multiply these names and employ them in such a way that they seem to be describing a multitude of things when there is only one real subject of discussion. For example, a recipe suggesting that the Sophic Mercury be heated in the fire of a furnace can therefore refer to the hypothetical fire supposedly resident in the Mercury itself and need not really be an injunction to heating at all.

STARKEY'S NOTEBOOKS ON TRANSMUTATIONAL ALCHEMY

We mentioned above how Starkey himself concealed his own experimental processes in the Philalethes treatises, but of course, as Starkey was deeply involved in trying to prepare the Philosophers' Stone—a substance known to the adepti—he had also to decipher their writings to guide his aspirations toward that summum bonum. In published form this activity is seen clearly in *Ripley Reviv'd*, a lengthy (and highly redundant) exposition of the verse treatise *The Compound of Alchemy* by the fifteenth-century English alchemist Sir George Ripley. But the notebooks bear far more dramatic (and far less varnished) witness to Starkey's reading and deciphering of chrysopoetic authors. The first section of the notebook now known as Royal Society Manuscript 179, unlike any other of the surviving notebooks in Starkey's hand, contains three draft treatises dealing with transmutational alchemy. These three works, although all dealing with the same subject—the preparation of the Philosophers' Stone—are actually quite different because their authorial voices very strikingly bridge the gap between the master Philalethes and the aspiring Starkey and thus give us hitherto unavailable insight into the process of both reading and writing chrysopoetic texts.

The first document is dated 1 August 1655 and is entitled *Diana denudata*. The English text is composed of thirty-seven "canons" concerning the initial stages of preparing the Philosophers' Stone. After concluding that the key ingredients for the Stone are gold and mercury, most of the re-

---

86. *Notebooks and Correspondence*, document 12; RSMS 179, fol. 15r.

maining text concerns the need to discover the identity of the appropriate "meane" (mediator) able to join gold and mercury together inseparably so that the two can be concocted into the Stone. The final two canons declare:

> Such a meane the wise Philosophers with all their might have sought & found, & left the record of their search in writing, withall so veyling the maine secret that only an immediate hand of god must direct an Artist who by study shal seeke to atteyne the same.
>
> This meane Substance is the Key of the whole worke, it is the only hidden secret which they in their Bookes have Concealed, concerning which all theyr allegoryes, Metaphors & darke sentences doe treat, learne this, & al the hard, darke sentences of the wise wil appeare plaine & easy to thee.[87]

The penultimate canon records not only the belief that the "wise Philosophers" sought for this mean and found it, but also that their enigmatic works are to be read as a record of that search and discovery. (Note also that Starkey declares that an "immediate hand of god must direct" the reader; we will return to the topic of divine revelation in the next section.) The final canon advises the reader what one thing he must endeavor to discover—assuring him that all the "darke sentences" do in fact contain a decipherable truth. What is noteworthy about this text is that it is written with an air of great authority. The author clearly knows the secret medium and drops hints throughout the text regarding its identity. Indeed, one familiar with Starkey's laboratory practice will have little difficulty in identifying this mean as antimony regulus. The authoritative tone with which this short tract is written suggests that it may have been intended for circulation as one of the Philalethes manuscripts. Certainly the tone is not that of the still unsuccessful Starkey but of the adept Philalethes.

This observation becomes more intriguing if we inspect the next two tracts. The second tract, dated a year after *Diana denudata* on 23 July 1656, is entitled *Clavis totius scientiae*. Unlike the previous tract, which by virtue of its freedom from corrections is most probably the fair copy of an earlier draft, this tract is riddled with so many deletions, changes, and corrections that the manuscript is difficult to read. It also has a more overtly Scholastic style, being divided into orderly sets of *Quaestio, Responsio,* and *Explicatio et Illustratio*. As a *Clavis* or "Key," it deals with the identity of the hidden medium alluded to in *Diana denudata*. Again the style of the *responsiones* is authoritative, but in this case each *explicatio* involves not the speaking of a master in his own words but a quotation from an authoritative author plus an interpretation of the hidden meaning. The significance of the many deletions and substitutions becomes clear only when we move to the final

---

87. *Notebooks and Correspondence,* document 12; RSMS 179, fols. 24r–25r.

treatise, the *Aphorismi hermetici,* written (and dated) the following month, August 1656.

Unlike *Diana denudata* and *Clavis totius scientiae, Aphorismi hermetici* is written not as an authoritative teaching treatise but rather as an exploratory one endeavoring to decipher the enigmatic writings of the adepti. The treatise format—regularly divided into orderly sections with their own headings—remains, yet now Starkey suddenly writes reflexively in the first person and employs a disputational tone striving for the resolution of conflicting positions. Quotations from authorities like Ripley and Bernard Trevisan are lined up against one another to resolve uncertain issues. Even Starkey's trademark conjectural processes (although not explicitly labeled as such) appear here after he has come to a conclusion about the meaning of an obscure text. There are summaries of what is already known and outlines of what is yet to be discovered. Indeed, Starkey is cautious and occasionally even diffident about his interpretations of the sources. After reaching a conclusion about a certain ingredient but before writing out a process employing it, Starkey prefaces the process: "This is my most refinde resolution, confirmed by serious meditation concerning the worke in which I have dealt Candidly, as writing only with intent to informe my selfe, nor designing nor intending to instruct any in the World besides."[88] In the two earlier tracts, Starkey took up the position of teaching others; here he is teaching only himself.

The origin of this "most refinde resolution" gives us important insight into Starkey's methods of treating authoritative texts. The resolution regards the identity of the "Luna of the Philosophers," and its importance is underlined by the fact that Starkey added the date "Monday 18 August 1656" alongside it in the margin to mark exactly when he made it.[89] This "Luna" is the hidden medium alluded to in *Diana denudata* and revealed in *Clavis totius scientiae,* which we have identified as antimony regulus, which in Starkey's method is to be united with both common mercury to prepare the Philosophical Mercury and with common gold to prepare the Philosophical Gold. But now Starkey is in a quandary: which antimony regulus should he use—the stellate martial regulus made with iron, or simple regulus of antimony prepared without the addition of iron?[90] Up until this

88. *Notebooks and Correspondence,* document 12; RSMS 179, fol. 44r.
89. *Notebooks and Correspondence,* document 12; RSMS 179, fol. 40r.
90. We now know that the two are identical chemically. Both reguli are actually pure elemental antimony, only they are prepared from the native sulfide ore by different methods of reduction; the stellate martial regulus is made using iron as the reducing agent while the plain regulus (or regulus *per se*) is prepared using nonmetallic reducing agents such as carbon. In the seventeenth century the martial regulus was believed to retain some part or all of the iron.

point Starkey had always used the stellate regulus in his preparation, but now he has doubts. These doubts may well have arisen from Starkey's experimental results, for although he had been attempting to prepare the Stone with his Philosophical Mercury for five years (since 1651), the digestions with gold—in spite of the encouraging signs of metallic "vegetation"—had so far failed to produce the transmutatory Elixir. Starkey thus perhaps went looking for the source of the problem. Here in the *Aphorismi hermetici* of summer 1656, he fixes on the possibly contaminating presence of iron in the regulus.

In order to resolve his doubts, Starkey turns to the interpretation of the adepts' enigmatic texts, and so here we have the opportunity to see the way in which an experimental chrysopoeian actually goes about deciphering an allegorical text into experimental practice. Starkey begins with a set of four points (the aphorisms of the title) that he considers as certain, including that the "mercuriall Saturnine part of antimony" is the needed key; on this point "Artephus is ful evidence." But now Starkey turns to a "parable" of the early modern author Bernard Trevisan regarding the Philosophers' Stone.[91] This parable, or extended allegorical conceit, begins with Bernard relaxing after a disputation by taking a walk in the open fields. He comes upon a beautifully constructed fountain and meets an old man there who tells him that the fountain's only use is as a bath for the king and that it is tended by a porter who warms it for him. Bernard asks the old man many questions about the king and his odd bathing practices and about the nature of the fountain. Eventually Bernard grows sleepy and accidentally drops a golden book (the prize from his disputation) into the fountain, drains the fountain to retrieve the book, and is thrown into prison for draining the king's fountain. After his release, Bernard returns to the fountain to find it covered in clouds. Bernard concludes by writing that "in this my parable the entire work [of making the Stone] is contained, in practice, days, colors, regimens, methods, managements and connections."[92]

Starkey now sets about interpreting Bernard's parable to answer his question about the kind of regulus to use. First he notes that when the king, whom he easily identifies as gold, comes to bathe, he leaves "behind him al his servants (which are the mettalls)" being accompanied only by a porter. Thus it seems here that iron, one of the lesser metals and thus one of the

91. Although this "Bernard Trevisan" may be modeled on the fourteenth-century Bernard of Trier (for whom see Newman, *Gehennical Fire*, 103–6) it is clearly a matter of a much later pseudepigrapher.

92. Bernard Trevisan, *De secretissimo philosophorum opere chemico*, in BCC, 2:388–99; the parable is on 397–99: "Nam in hac mea parabola totum opus continetur in practica, diebus, coloribus, regiminibus, viis, dispositionibus, & continuationibus . . . "

king's servants, must be "left behind." But Starkey, unsatisfied with a single reason, continues interpreting. This porter is surely the "Luna," or necessary medium between gold and mercury (i.e., the king and his bath), which he has identified as antimony regulus. Now Starkey notes that Bernard affirms "this Porter . . . to be most simple of al things in the world whose office is nothing but day by day to warme the bath (that is by making al fluid) now if it were compounded it could not be sayd to be so simple, which is uncompounded." Here Starkey interprets Bernard's use of the word "simple," which in the context of the parable means that the porter is unsophisticated or naive—"homo valde simplex, imo simplicissimus hominum"—to mean *compositionally* simple, that is, uncompounded, implying that pure "simple" antimony regulus should be joined to the king/gold, not the regulus containing iron. Still not satisfied, Starkey seeks out more verification. He next notes that Bernard asked the old man whether any of the king's servants went into the bath with him, and the "answer is returned not one, & if not one, then not ♂ [iron]."[93]

After providing these arguments from the chemical interpretation of Bernard's allegorical tale, Starkey then supports them further with theoretical considerations. The first of these is drawn from his practical experience of monitoring the weight of the mercury during the treatment with martial regulus. He notes that there is extra "pondus" (or weight) added to the mercury, which the iron adds of its own substance, which "increase it is not fit to admit without a ground." His second theoretical consideration notes that the role of the antimonial component in the Philosophical Mercury is to increase its ability to penetrate and radically dissolve gold, but if the martial regulus is used, the "♁ [antimony] having spent much of its dissolutive virtue on ♂ [iron] receives from it a determination & so being specificated toward a Bodylynes, wants much of its penetrative quality, as A[qua] R[egia] having corroded ☉ [gold] becomes effeat as to action for future."[94] This is an expression not only of the observation that the corrosive abilities of solvents diminish as dissolution occurs, but also of the Helmontian doctrine of exantlation, outlined in chapter 2. Thus, drawing upon the interpretation of Bernard's parable, practical experiments, and theoretical considerations, Starkey resolves now to use regulus per se rather than the martial regulus he had been using since at least 1651.

The effects of this change from martial to simple regulus further elucidate Starkey's textual methodologies. In the first place, this change explains the numerous deletions in the earlier *Clavis totius scientiae*, for on returning to that work, we can see that *every deletion* involves passages or phrases

93. *Notebooks and Correspondence,* document 12; RSMS 179, fol. 42r.
94. *Notebooks and Correspondence,* document 12; RSMS 179, fol. 43r.

stipulating the need for iron. In one case this involves crossing through a paragraph interpreting the *Novum lumen chemicum* of the early seventeenth-century Polish chymical writer Michael Sendivogius:

> Hither glances that most highly refrained enigma of the Philosophers, that the work is to be begun in Aries, for Aries is the House of Mars, and this squares not unaptly with what the author of the *New Light* relates of his Steel, wherefore it may be gathered that iron is the spouse with this our lead.[95]

Here, as in the published *Introitus apertus* of Philalethes, Starkey takes Sendivogius's statement about the "house of Aries" to mean that iron is required, since the astrological house of Aries is ruled by the planet Mars, which in turn is the alchemical planetary symbol for the metal iron. But this text is wholly crossed through, following the reinterpretation recorded in *Aphorismi hermetici*. Iron is no longer a desired ingredient.

Even more illuminating is how Starkey must now change the *Decknamen* he employed in the *Clavis totius scientiae;* his alterations give a behind-the-scenes glimpse of the *encoding* of chymical knowledge into allegory, complementary to the view of *decoding* displayed by Starkey's interpretation of Bernard's parable. Originally, Starkey had hidden the identity of the martial regulus of antimony under the name of *corpus hermaphroditicum*, or hermaphroditic body. The frequent appearance of hermaphrodites in traditional alchemical literature has long titillated those predisposed to a psychological interpretation of alchemy, but here Starkey's notebooks clearly reveal the conscious and rational (rather than sexually suppressed, halluci-natory, or arbitrary) use of the image. In Starkey's set of *Decknamen,* the term "hermaphrodite" is logically and appropriately applied to the martial regulus of antimony simply because that substance is produced from the union of antimony, seen as a female substance, with iron, clearly a mas-culine one. Just as the union of antimony and iron provides the martial reg-ulus, so the union of female and male provides a hermaphrodite. This *Deckname* appears not only here in the notebooks but also in the *Introitus apertus* and other works attributed to Philalethes.[96] Once Starkey rejects the addition of iron, however, that *Deckname* is no longer applicable be-cause there is now no male element present in the regulus. Starkey's cover names are *not* arbitrary; they must be appropriate and decipherable. As a re-sult, he goes back through the whole text of the *Clavis totius scientiae* and

---

95. *Notebooks and Correspondence,* document 12; RSMS 179, fol. 29r; the reference here is to Michael Sendivogius, *Novum lumen chemicum* in *Musaeum hermeticum,* 571. Here Star-key's "our lead" refers to antimony.

96. The hermaphrodite appears as a *Deckname* for the star regulus of antimony in George Starkey [Eirenaeus Philalethes, pseud.], *Introitus apertus ad occlusum regis palatium,* 658. For a treatment of this, see Newman, *Gehennical Fire,* fig. 3B.

systematically deletes every use of the term *corpus hermaphroditicum,* replacing each with the symbol for Venus. Venus, although generally used in alchemical texts for copper, is now a proper *Deckname* for the simple regulus because the regulus without iron would be purely feminine, like the goddess Venus. Looking at it another way, the male "Herm" has been removed from the Hermaphrodite, leaving only Aphrodite, who is, of course, the Greeks' Venus. There may be yet another level of hidden meaning to this new cover name, for the symbol for Venus— ♀ —is merely the inverted version of the symbol for antimony—♂.

Starkey's eventual rejection of iron reappears even in the chymical symbols he uses. On an early page of RSMS 179, Starkey recorded a process for making the "sophic gold" that, along with the Sophic Mercury, is the starting material for the Philosophers' Stone. According to this process, as is repeatedly hinted at throughout *Diana denudata,* antimony regulus is to be fused with gold. For this notebook entry, written in early 1655, Starkey uses the symbol ☿ to signify the regulus; however, at some later point he obliterated the two "wings" at the bottom of the symbol to leave ☿ (see figure 6). This alteration seems arbitrary or at best obscure. But if we consult John Dee's *Monas hieroglyphica,* we find that the symbol Dee called his "hieroglyphic monad" bears a similar feature. Dee asserts that this feature of his monad represents the horns of the Ram, Aries. Now, as we mentioned earlier, the astrological house of Aries is governed by the planet Mars, and the planet Mars in turn represents iron.[97] So Starkey's alteration of his arcane symbol is meaningful and significant; it recapitulates his 1656 decision to omit iron from the preparation of antimony regulus.[98] When Starkey omitted iron from his processes, he returned to this receipt to omit the signifier of iron from his chymical symbol.

Thus we can see clearly that at least some chrysopoeians read allegorical and secretive chrysopoetic texts carefully in order to decode their chymical content and also wrote them just as carefully in order to encode such meaning in a retrievable form. Starkey's interpretations, however, remained ten-

97. John Dee, *Monas hieroglyphica,* in *Theatrum chemicum,* 6 vols. (Strasbourg, 1659–61; reprint, Torino: Bottega d'Erasmo, 1981), 194. On the *monas,* see Nicholas H. Clulee, *John Dee's Natural Philosophy: Between Science and Religion* (London: Routledge, 1988), and Clulee, "*Astronomia Inferior:* Legacies of Johannes Trithemius and John Dee," in *Secrets of Nature: Astrology and Alchemy in Early Modern Europe,* ed. William R. Newman and Anthony Grafton (Cambridge: MIT Press, 2001), 173–233.

98. This change also signals Starkey's familiarity with Dee's text; Starkey cites another work by Dee in the *Diana denudata* (*Notebooks and Correspondence,* document 12; RSMS 179, fol. 7r). Starkey also comments in print on the existence of a Philalethes text entitled "Cabala sapientium, or An Exposition of the Hieroglyphicks of the Magi"; this work is lost, however. See Starkey [Philalethes, pseud.], *Marrow of Alchemy,* pt. 1, fol. A3r.

tative for him, and laboratory practice would again be the judge of their correctness. The interpretation of Sendivogius's secretive text along with Starkey's five-year employment of martial regulus were abandoned when the processes based upon them failed to produce the Philosophers' Stone. Replacement practices and processes were then drawn from new interpretations of Bernard's parable, coupled with experience and theoretical considerations. Again, we find that Starkey's methodology awards final judgment to experimental practice, even in so secretive, allusive, and strongly text-based a pursuit as that of the Philosophers' Stone.

Limits of Textual Authority

The example of the Sophic Mercury has one further contribution to make, namely to help refine further our understanding of Starkey's reliance on texts. We previously noted that Starkey's use of authoritative texts gave him matters of fact—evidence that a given product actually existed because it was seen and described by a creditable author. This action invests texts with a certain level of authority, but we must underline the limitations of that authority in practice. Starkey's use of Suchten, for example, displays a surprising degree of fluidity and development over time, as he subjects Suchten's processes to increasingly rigorous examination. This fact appears clearly in a thumbnail sketch of the history of the all-important Philosophical Mercury.

Starkey's first preparation of this substance was appropriated directly from the process of acuating common mercury with an alloy of silver and stellate martial regulus as described by Alexander von Suchten in his *Secrets of Antimony*. The process for preparing it and the theory behind that preparation, as recounted to Boyle in Starkey's ebullient letter of spring 1651, are taken directly from Suchten. Following Suchten, Starkey initially believed that the putative Sulphur found in iron was superior to the "vegetable Sulphur" found in tartar and that it was therefore necessary to reduce his antimony using iron rather than the more common fluxes containing tartar. Starkey also followed Suchten in insisting that silver was the only mediator capable of making mercury and the antimony regulus amalgamate. Starkey's contribution to the process at this stage was only the assumption that Suchten's mercury was actually that to be used for the preparation of the Philosophers' Stone (a preparation in which Suchten has no apparent interest), plus Starkey's greater application of quantitative methods to monitor the progress and success of the process. But Starkey changed that procedure in 1653, based upon his own practical experience (as we saw in chapter 3), omitting most of the use of silver in order to provide a more convenient, more efficient, quicker, and cheaper production of the Sophic Mercury. In the *Marrow of Alchemy* of 1654 he rejects the use of silver altogether, substituting the cheaper copper instead.[99] This exclusion of silver is directly contrary to Suchten's method and his theoretical basis for the process in which silver is absolutely required for success. Finally, in 1656, Starkey throws out the stellate martial regulus in favor of regulus per se, thus ruling out the use of the iron upon which Suchten had founded the entire success of the process. Hence Starkey, based upon his own experience, improvements, and reconsiderations, successively jettisons virtually all of Suchten's theory and much of his practice based on that theory. His constant quest to streamline the cost of laboratory operations, directed by the analytic method of the *quaestio disputata* and the contributions of cumulative laboratory practice, leads him to strip away all but the essential kernel of Suchten's work. In the end, the only thing Starkey retains from the Prussian iatrochemist is the matter of fact that he acuated common mercury by means of antimony into a solvent capable of radically dissolving metals.

The foregoing examination of Starkey's uses of textual authority indicates that he employed texts in his experimental programs in several ways. Their first use was as evidence simply that a given desirable product had in fact been prepared previously (i.e., it has real existence) and had certain specific qualities and powers. Thus textual authority set the specific goals of Starkey's research. This knowledge—that a given thing was in fact preparable—was reliable and irrefutable for Starkey. Second, texts gave partial accounts of how to prepare these products. These partial accounts could be used in starting assumptions about the methods of preparation to employ. This knowledge remained tentative because the transmission of information from author to reader was incomplete, and generally intentionally so owing to the culture of secrecy embodied in the texts. While the existence of the goal was fixed, therefore, the practical ways of approaching it were quite fluid, being dependent upon experimental results to pass judgment upon the correctness of the interpretations on which the methods were in turn based.

Taken together, these aspects of Starkey's practice differ considerably from those of research bent on the elucidation of nature. Nonetheless, there is a modern practice that Starkey's projects do resemble, namely reverse engineering. Reverse engineering refers to the process, common enough in industrial practice, whereby a finished product is analyzed by one who did not make it, with the view to discovering the method of producing it. This may be a piece of equipment, such as a radio or telephone,

---

99. Starkey [Philalethes], *Marrow*, pt. 2, pp. 16–17; Newman, *Gehennical Fire*, 133–34, 313 n. 72.

or it may be a chemical product like a detergent or pharmaceutical. The important exception for Starkey is that he has no samples of the desired product to analyze—as if one had heard descriptions of, say, an automobile, or seen one drive by, but had never been able to examine one or take it apart. Starkey is thus left with only the knowledge that such a thing does exist, the hints given by his sources to guide his work, and a constantly growing body of experimental results and expertise. Success for Starkey is thus gauged on whether or not he prepares a product whose properties and activities accord with those described in his authoritative texts.

THE PLACE OF DIVINE AUTHORITY IN THE LABORATORY

Although we have examined the interacting roles of textual authority and experimental results in Starkey's methodology and epistemology, there remains one further source of authority—revelations from God, or, as Starkey (quoting Saint James) often says, "The Father of Lights." We saw previously that in *Diana denudata* Starkey asserted that an "immediate hand of God" was necessary for the correct interpretation of the "darke sentences" of the adept philosophers. What, if anything, does this actually mean in practice? While there is a modern tendency to associate religious revelation with visionary or ecstatic experience, the notion of divinely revealed knowledge had a far wider range of meaning in the early modern period, as we shall see here. To Van Helmont and his followers, not to mention many of their contemporaries, "revelation" could indeed mean the experience of a vision imparted in a dream or in an ecstatic state, but it could also refer to knowledge that God allowed or silently assisted the chymist to discover by means of his own sweat and labor. In this latter case, God sometimes "revealed" a particular secret by providing the seemingly accidental circumstances that allowed an experiment to come to fruition. In this case, the revelation by God did not necessarily imply that the chymist was immediately conscious of the divinity's presence when He was dispensing His gift. Hence a chymist could retroactively discover "revelations" that had occurred in the laboratory by attributing his success to divine aid.

Starkey's notebooks and private papers give us a clear perspective upon divinely revealed knowledge and its role in experimental investigations. For Starkey actually *records* the arrival of divine revelations in his notebooks alongside textual analyses and dated experimental trials. Starkey's careful recording of these events is in line with his fastidious record keeping, but it may also owe a debt to seventeenth-century providentialism and the then popular genre of the spiritual autobiography, which sought to identify and record instances of God's providence. Within the Hartlib circle, for example, John Dury proposed a formal "Registring of Illustrious Providences," and Boyle's writings of the 1640s are full of such providentialism.[100] The recording of providential events was intended not only to acknowledge divine goodness but also to help assess and demonstrate the correctness of one's own actions and one's state of grace; this latter may have been of particular importance for the Presbyterian Starkey.

In any event, an examination of how this divine knowledge was deployed will serve to illustrate how this further source of (higher) authority was integrated into the network of Starkey's multifarious sources of knowledge. As we shall see, Starkey's attitude toward divine intervention in the laboratory has some resonances with that of Boyle. This is not merely coincidental, since both Starkey and Boyle received their views not only from the Father of Lights but from their chymical father as well, Van Helmont.

Since we have so much clear and detailed information about Starkey's work on the volatilization of alkalies, that project provides a clear example. In a fragmentary notebook dated to the end of 1651, Starkey writes simply: "God communicated to me the whole secret of volatilizing alkalies."[101] He then follows this summary declaration of divine revelation with the process of forming the collostrum by distilling oil of terebinth with salt of tartar. But as we have seen, Starkey would later reject this method. What kind of a divine revelation can be merely discarded? Very simply, even these divine revelations appear to be subject to trial by fire. When Starkey's laboratory results showed that the distillation of the collostrum did not in fact give him a spirit of volatile tartar, he did not hesitate to reject that process and seek for another one, even though this meant the rejection of what he himself considered a divine revelation. Starkey was no enthusiast, steadfastly clinging to some "inner voice" alone; the trial in the fire had still greater authority.

The position of divine authority and the nature of God's revelations become clearer if we turn to Starkey's summary of this project. After completing the project to his satisfaction in 1656 (after having discovered the secret circulation in the air), Starkey recalls all the various methods he tried and refers his eventual success to God's benevolence. He writes clearly that "the good God finally conceded the whole secret of volatilizing [alkalies] to me."[102] Indeed, Starkey's records of his experimental trials dated 1655–56 are often interspersed with imprecations. In the description of one conjectural process, Starkey writes that it will succeed "God willing," and on 3 March 1656, when he describes the exact quantities of oil, alkali, and other

---

100. John T. Harwood, *The Early Essays and Ethics of Robert Boyle* (Carbondale: Southern Illinois University Press, 1991), xxxvi; on Dury's proposal, see HP 26/8/1.

101. *Notebooks and Correspondence*, document 5; Sloane 2682, fol. 89.

102. *Notebooks and Correspondence*, document 10; Sloane 3711, fol. 5v.

materials he has compounded and his plan to leave them exposed to the open air he concludes the process with "may God bless it."[103]

A clearer sense of the mechanism of God's beneficence to Starkey appears in his "summary of everything which I know and have learned about alkalies, through the experience of what is now nine laborious years" also written in the aftermath of the success of 1656. Starkey recounts that

> I have tried many laborious and costly experiments to this end, yet my unbroken mind could never, from the start, be withdrawn from what had been begun.
>
> I completed many unfruitful trials, yet finally God deigned to direct me into the true Art . . .
>
> The cause of these errors was an ignorance of the nature of the thing sought after, which laborious practical experience took away from me and showed to me the true way.[104]

This passage gives a very interesting sense of how Starkey views his own progress and eventual success. On the one hand, he reiterates his debt to divine help when "God deigned to direct" him to a successful conclusion. Yet on the other hand, he is equally clear that his "ignorance" was removed by "practical experience"—his work in his laboratory—which "showed him the true way."

Much secondary literature, in particular that influenced by Carl Jung and Mircea Eliade, claims that early modern chymistry, especially that grouped as "alchemy," was characterized by a need for divine illumination.[105] This requirement implicitly separates such endeavors from "more modern" chemical ones practiced by people like Boyle. It is true that many writers of chymical, particularly chrysopoetic, tracts emphasize the need for the cooperation of God for success. Starkey himself does the same, claiming that the first obligation of a "true Artist," if he would be at all successful, is to pray and implore God's blessing. In fact, Starkey repeatedly writes that success comes only from the Father of Lights: "It is not the reading of Books, nor is it painful search in the fire, that can do any good; onely the blessing of the Almighty."[106] Starkey's notebooks reaffirm this recognition of divine direction and largesse. But these private notebooks also clarify the actual functioning of this divine assistance for Starkey: it is not that received

103. *Notebooks and Correspondence,* document 12; RSMS 179, fols. 59v, 42v.

104. Ibid., fols. 19v–18v.

105. For examples of the Jungian and Eliadean interpretations of alchemy, see Lawrence M. Principe and William R. Newman, "Some Problems with the Historiography of Alchemy," in *Secrets of Nature: Astrology and Alchemy in Early Modern Europe,* ed. William R. Newman and Anthony Grafton (Cambridge: MIT Press, 2001), 385–434, esp. 401–15.

106. Starkey, *Pyrotechny,* 10–11.

by a quiet contemplative, but rather that gained by a laborious worker whose experimental practice leads him finally to a good idea or a correct solution. In some instances it is perhaps sudden ideas—events to which we still refer as "flashes of inspiration"—that Starkey identifies as divine revelations. These "inspirations" may turn out to be wrong or, perhaps more correctly stated, falsely identified as emanating from God when they are put to the test. The "divine revelation" of 1651 was presumably such an idea that was in due course rejected.

A similar set of circumstances can be seen in the three revelations regarding the alkahest that Starkey recorded in various ways. The first of these was in the form of a dream in January 1652. Starkey recounts how after having fallen asleep at work in his laboratory, he dreamt that a figure—who identified himself as his tutelary Eugenius, or good spirit—appeared to him. Starkey asked him about the alkahest, to which the Eugenius answered using obscure language yet, as Starkey records, "with the response an ineffable light entered my mind, so that I fully understood." Starkey describes this event as a case of instantaneous cognition through the *intellectus*—an act described by Van Helmont as a divine way of understanding, supplemental and superior to the methods of linear, logical argumentation in time. Indeed, Starkey explicitly calls this experience an "intellectual revelation" *[intellectualis revelatio]*.[107] For months thereafter Starkey labored on the alkahest according to this new understanding, finally producing in August a substance that seems to have satisfied him—at least for a while.[108] But Starkey must have been ultimately unsatisfied with this revelation, presumably on the basis of dissatisfaction with its practical product, in spite of what might seem to be from his account its highly authoritative origins and method of delivery. Thus in 1656 he recorded another revelation about the alkahest, with place and date carefully noted.

> At Bristol, 20 March 1656
> God revealed to me the whole secret of the liquor alkahest; let eternal blessing, honor, and glory be to Him![109]

Here there is no trace of this revelation coming in the form of a dream, nor is there such anywhere else in the notebooks. Given this absence, it is possible that Starkey cast a less dramatic sort of "revelation" into the oneiric form described in his letter of January 1652, using Van Helmont's dreams as

---

107. *Notebooks and Correspondence,* document 6; Starkey to Boyle, 26 January 1652; also Boyle, *Correspondence,* 1:121.

108. See below, chapter 5; Starkey, *Pyrotechny,* 34; Starkey, *Liquor Alchahest* (London, 1675).

109. *Notebooks and Correspondence,* document 11; Sloane 3750, fol. 19v.

a model, at least partly as a means of impressing Boyle and Hartlib, to whom the 1652 dream account was told. On the other hand, he may in fact have had the dream he described and believed it to be a divine message in the form of an "intellectual revelation." In either case, what is important is that Starkey seems to have rejected even this in due course based upon practical experience and was obviously delighted and thankful four years later to have received another revelation of the practical process he had not yet attained. The practical details of this 1656 revelation remain unrecorded, and since we have no notebooks describing Starkey's work on the alkahest, we cannot say precisely what he did with it in the laboratory. But this revelation as well seems to have eventually proved unsatisfactory. For in a notebook from late 1657 or 1658 there is a further record of Starkey's attempts to discover the secret of the alkahest, not by anxiously awaiting or praying for a fresh revelation, but by the more mundane means of *ratio*, using Scholastic method and textual analysis.[110] Yet something Starkey describes as God's revelation came once again in 1658, as shown by the records of yet another notebook.

> Monday 20 September 1658
>
> From the year 1647 to this very year and day, I have exerted myself in the search for the liquor alkahest with many studies, vigils, labors, and costs. Today for the first time it has been granted and conceded to my unworthy self by the highest Father of Lights, the best and greatest God, to attain complete knowledge of it and to see its final end. To Him let there be eternal praise both now and forever. Amen.[111]

Here Starkey declares that on this date he received full knowledge of the alkahest from the Father of Lights "for the first time," implicitly rejecting the previous two revelations. And yet, Starkey continued his practical labors toward the alkahest for the rest of his life. Starkey thus seems to view reason and revelation as cooperative powers that assist the chymist in an ultimately pragmatic way. They are not mutually exclusive. On the one hand, God's revelatory assistance is prepared for by laboratory practice and rational analysis, and on the other, revelations are channeled into further experimental practice that eventually evaluates their validity. This concept of God's activity corresponds extremely well with the Helmontian view of the status of knowledge, that "God sells secrets for sweat." Starkey did not imagine that hopeful waiting or even fervent prayer would suffice on its own to discover the secrets of nature; God may cooperate in the laboratory but He does not simply operate.

110. *Notebooks and Correspondence,* document 14; Sloane MS 631, fols. 198v–199r.
111. *Notebooks and Correspondence,* document 15; Harvard University, Houghton Library Autograph File. Also printed in Newman, *Gehennical Fire,* 250.

Let us briefly return to Starkey's chief model, Van Helmont, in order to sample his view of God's action in the laboratory. Van Helmont also claimed to be the recipient of divinely revealed knowledge in the form of dreams. The *Ortus medicinae* is replete with these colorful visions, and they are possibly patterns for Starkey's own dream revelations.[112] But somnial revelations were not the only means by which God could affect the chymist's laboratory practices and success. According to Van Helmont, the situation is far more complex, as can be seen from his *De lithiasi,* one of Starkey's favorite works. In *De lithiasi,* Van Helmont describes his chymical analysis of urine, which he undertook in order to determine the cause of kidney and bladder stones. Van Helmont had serious doubts about the Scholastic theory that urinary calculus was caused by the action of salt on a "slime" hidden in the urine. He ridicules the university doctors, saying that they are so lazy that they do not even throw a handful of salt into a flask of urine to test their theory.[113] He then recounts his own laborious experiments. For a long time, he tried unsuccessfully to "dissect" urine into its parts in order to separate the *Duelech*—the insoluble calculus—from it. Finally, Van Helmont let a quantity of his own urine putrefy for forty days and then distilled half of it into a large and beautiful crystal vessel. After having done this, Van Helmont was called away for two weeks, first on family business, then by the festivities of Pentecost. When he returned to the laboratory, Van Helmont found that the distilled urine had deposited an unseemly stain on his expensive vessel. Infuriated by his inability to remove this stain, even with scouring, Van Helmont impetuously threw the distilled urine away. He soon regained his composure, however, and grew curious as to the nature of the discoloration—was it due to a corrosion of the glass, or had the urine left a deposit on the receiver? Suddenly, Van Helmont realized with excitement that the urine had indeed precipitated a deposit on the glass. At this point in the narrative, Van Helmont interjects his appreciation of the divine largesse:

> Then I was filled with admiration: I praised God, that He had taken care of me, for I realized that what I thought I had done out of my own carelessness had actually happened by the action of divine goodness. For whom He wishes, to be sure, He turns all things to good.[114]

In other words, the two weeks of inactivity—seemingly mere accident—were actually the result of divine intervention. This was just the right

---

112. For a few examples, see Van Helmont, *Ortus,* "Imago mentis," no. 13, pp. 269–70; *Ortus,* "Potestas medicaminum," no. 3, p. 471; *Tumulus pestis,* in *Opuscula,* pp. 5–7.

113. Van Helmont, *De lithiasi,* in *Opuscula,* no. 30, p. 27.

114. Ibid., no. 33, p. 29.

amount of time for the urine to deposit its *Duelech*. But the story does not end here. Van Helmont then decided that he could probably remove the *Duelech* from his receiver at some later date with nitric acid, so he reattached it to his still and distilled off the remainder of his putrefied urine. Then, as he says, "urged on by a new favor from the divine largesse, I saw the individual drops of distilled urine dissolve the dark, adhering *Duelech*," leaving the glass as clean as it had been originally![115] Furthermore, when he added this distilled urine to spirit of wine, he found that it did not produce the normal deposit of *offa alba* (ammonium carbonate) that he had come to expect from a mixture of spirit of wine and spirit of putrefied urine; therefore, the second portion of distilled urine had lost its coagulating power. It seemed to Van Helmont then that he had discovered within urine itself the very agency that could dissolve the stones that form in the bladder and kidneys. This wonderful gift leads Van Helmont to interrupt his narrative once again, and to expostulate:

> Led thus by the divine will *[divino ductus nutu]* (which others might think a chance event), I found part of that which I had long been seeking anxiously with much expenditure. Hence I praised God, that he had given understanding to one who was small and poor. For if He had not commanded that I be called away from the work, nor detained me in festivities until the *Duelech* congealed on the receiver, and if the receiver had not been so clear and precious, indeed, if I had completed the whole operation in one go, then I would have done all in vain. Therefore God has considered the needs of mortals, and has not spurned the prayers of the lowly.[116]

As we can see, while Van Helmont attributes this "lucky accident" to the providence of God, this revelation also required that the chymist buy his secrets with sweat. God did not reveal the secret of *Duelech* and its dissolvent in the form of a narrative communication or even as a cryptic dream. Instead, he altered the external circumstances of the chymist's life just enough to allow Van Helmont's experiment to succeed. The successful outcome required first Van Helmont's initiative, in setting up the experiment. It then called for God's intervention, in distracting the chymist from his procedure. Finally, it necessitated that Van Helmont overcome his childish temper tantrum upon discovering the apparent damage to his expensive glassware, and upon regaining his composure Van Helmont had to exercise his own judgment in interpreting the final outcome. God's input in all this is expressed very well by the term that Van Helmont uses for the divine will—

---

115. This same experiment was repeated and recorded by Starkey in a notebook from c. 1650. See *Notebooks and Correspondence*, document 2; Locke MS C29, fol. 117r.

116. Van Helmont, *De lithiasi*, in *Opuscula*, no. 33, p. 29.

divinus nutus—literally "the divine nod." A nod is a silent symbol of approbation, like a wink or a nudge. It is not primarily by the direct approach of visions and dreams that God reveals knowledge, but rather by such silent hints.

Van Helmont and his American acolyte were not alone in accepting the presence of God in the laboratory. Starkey's experimental collaborator Robert Boyle was also willing to acknowledge divine aid in the cause of a good experiment. But Boyle—undoubtedly fearful of appearing an enthusiast—was publicly leery of any claims to special revelation, as he reveals in *Usefulnesse:*

> And though I dare not affirm, with some of the *Helmontians* and *Paracelsians,* that God discloses to Men the Great Mystery of Chymistry by Good Angels, or by Nocturnal Visions, as he once taught *Jacob,* to make Lambs and Kids come into the World speckled, and ring-streaked; yet perswaded I am, that the favor of God does (much more than most Men are aware of) vouchsafe to promote some Mens Proficiency in the study of Nature, partly by protecting their attempts from those unlucky Accidents which often make Ingenuous and Industrious endeavors miscarry; and partly by making them dear and acceptable to the Possessors of Secrets, by whose Friendly Communication they may often learn that in a few Moments, which cost the Imparters many a Years toyl and study; and partly too, or rather principally, by directing them to those happy and pregnant Hints, which an ordinary skill and industry may improve, as to do such things, and make such discoveries by virtue of them, as both others, and the person himself, whose knowledge is thus encreased, would scarce have imagin'd to be possible.[117]

Thus while explicitly distancing himself from the Helmontians, Boyle at the same time expresses a view akin to Van Helmont's own doctrine that God looks over His laboratory workers and leads them on by means of silent hints. Here what might appear to be a rejection of the Helmontian insertion of God into the laboratory is in reality only a partial disclaimer. Mirroring Van Helmont quite closely, Boyle goes on to say that God not only prevents experimental mishaps, but actually provides "accidental hints," like the Helmontian accident of the stained receiver. It is "as if God design'd to keep Philosophers humble, and (though he allow regular Industry, sufficient encouragement, yet) to remain Himself dispenser of the chief Mysteries of Nature."[118] Here Boyle points to the necessary interplay of sheer hard work—"regular Industry"—and God's voluntary granting of

---

117. Boyle, *Usefulnesse,* in *Works,* 3:276.
118. Ibid., 3:277.

the "chief mysteries." As in the case of Starkey and his Belgian preceptor, these accidents depend on the assiduous labor of the chymist if they are to provide new knowledge. It is interesting to note that the perspectives on God's laboratory participation shared by Van Helmont, Starkey, and Boyle imply similar theological perspectives on God's activity and the nature of revelation despite the differing doctrinal attachments (Catholic, Presbyterian, and Anglican, respectively) of the three.

Returning to Starkey then, we may say that his view of God's activity in his laboratory investigations is to be compared to that of the farmer, who, even though he plants the seed and nurtures the plant, is nonetheless dependent upon God for the final harvest. These are likewise the sentiments behind the Psalm "If God build not the house, in vain do the workers labor." The constant awareness of the activity and presence of God in early modern thought appeared quite naturally in such expressions of the indispensability of divine aid in difficult endeavors. Thus we see that expressions of pre-Enlightenment piety in chymical or other writings do not automatically imply that the *nature* of experimental labors or the expectations of laborers were different from those of more modern laboratory workers, only that they were differently enunciated.

CONCLUSIONS

The present chapter has shown us many aspects of the labors of Starkey, and by extension of seventeenth-century chymical practices in general, that are visible only by examination of his notebooks. First, there is the matter of Starkey's dogged intentness on cost-efficiency and labor-saving methods in the laboratory. This is not something that emerges clearly from his published works, and as we stated earlier it is not a prominent feature in the printed literature of earlier chymistry. Certainly chymistry was a favored subject of economic projectors in the seventeenth century, such as Johann Rudolph Glauber and Johann Joachim Becher.[119] For that matter, alchemy had throughout its history engaged in a polymorphous range of salable chemical technologies, from the manufacture of pigments to the refining of salts. But rare indeed is the sort of evidence provided by Starkey's notebooks, with their interest in determining (and maximizing) efficiency, often using quantitative methods such as mass balance of intake and output. It was the Helmontian program of quantitative *spagyria* in conjunction with the laboratory balance that allowed Starkey to proceed in this direction,

119. See Pamela H. Smith, *The Business of Alchemy: Science and Culture in the Holy Roman Empire* (Princeton: Princeton University Press, 1994), and Smith, "Vital Spirits: Redemption, Artisanship, and the New Philosophy in Early Modern Europe," in *Rethinking the Scientific Revolution,* ed. Margaret J. Osler (Cambridge: Cambridge University Press, 2000), 119–35.

while the American chymist's concern with industrial efficiency is probably the result of his early exposure to the metallurgical industry of New England, no doubt abetted, of course, by the Hartlibian enshrining of "the lucriferous."

As we also showed, the structure of Starkey's notebooks is another feature that is new to the historiography of chymistry, and to some extent to that of experiment more generally. The unusual fusion of formalized Scholastic methodology with the experiences of the chymical laboratory presents an unknown chapter in the history of science and opens further windows on the issue of Scholastic contributions to the development of experiment. As we have argued, this too is partly the result of Starkey's background in the New World, specifically his education at Harvard, though again he added measures of Van Helmont, Sala, and other chymical writers to this product of his colonial experience. We have also demonstrated that Starkey's method of experiment coalesced with formal textual analysis in such a way that he could correct and revise authoritative figures in chymistry such as Alexander von Suchten. This belies the dismissive picture of the "alchemist" stuck forever in a loop of failure, supported by elaborations and lucubrations that were guaranteed to explain away his lack of success.[120] Similarly, Starkey's notebooks have revealed the dynamic process of ciphering and deciphering between laboratory processes and enigmatic chrysopoetic texts—the notoriously secretive writings of chrysopoeia were carefully written and just as carefully read for their practical, although carefully concealed, content. In a similar vein, the present chapter has revealed that Starkey's reliance on God in the laboratory is not to be seen as the product of a visionary fervor, but as the pious attribution of all gifts to God—Starkey rejected neither reason nor the ineffable light of God's grace and favor. Thus Starkey's private notebooks reveal details of a new face for early modern "alchemical" laboratory practice. In the following chapter we examine how Starkey's developed laboratory practice and chymical expertise affected his young associate Robert Boyle, and determine the depth to which Boyle's views were colored by the chymical tinctures of the man beneath the mask of Eirenaeus Philalethes.

---

120. We think of the classic study of Evans-Pritchard, *Witchcraft among the Azande* (Oxford: Clarendon, 1937), 339, 475–78, where magic is treated in such terms.

FIVE

Starkey, Boyle, and Chymistry in the Hartlib Circle

In the previous chapter we presented and analyzed George Starkey's experimental methodology. This material—the details of the actual interplay between theory and practice, between Scholastic logical training and laboratory experimentalism, and between the reading and the writing of secretive texts in the daily work of an influential seventeenth-century chymist—is of great interest, for it reveals a world that has hitherto been closed to the modern spectator. These aspects of Starkey's investigative program shed much new light on the nature of the chymical enterprise and the chymist in the mid–seventeenth century. We have elsewhere noted the influence of Starkey (qua Philalethes) on later seventeenth-century chymical theorists such as Sir Isaac Newton and Wilhelm Homberg. We have also noted the importance of Starkey's legacy of the recipe for a Philosophical Mercury to the chrysopoetic pursuits of Robert Boyle. But now we must return to the topic with which we introduced this study—George Starkey's direct interactions with Boyle in the 1650s.

The juxtaposition of these two chymists during a period crucial for Boyle's own intellectual evolution as an experimental natural philosopher naturally evokes the question of how much influence Starkey's chymical practices had on Boyle. Thus, this chapter delineates the impact of Starkey's developed experimentalism and chymical expertise on the young Boyle by considering Boyle's early chymical training in some detail. Despite some claims that have been made for Boyle's early circle of scientifically minded friends—the "Invisible College" of the 1640s—we will show that Boyle's first exposure to the full range of laboratory chymistry occurred at the hands of Starkey. Moreover, we will show that the Helmontian principles informing so much of Boyle's early thought were conveyed to him first and above all by Starkey. As we argue in the following chapter, some of these notions persisted within the permanent foundations of Boyle's mature chymistry, for it appears that even some of Boyle's most lasting contributions had a substantial Helmontian component. We also examine the wider circle of the young Boyle's chymical contacts by looking at the activities and interests of three of his other chymical associates of the 1650s—Benjamin Worsley, Frederick Clodius, and Sir Kenelm Digby.

STARKEY AND THE DEVELOPMENT
OF BOYLE'S EARLY CHYMISTRY

Our understanding of Boyle's early life has been significantly revised during the last ten years. Until quite recently, Boyle's development as a natural philosopher was taken as somehow inevitable, and consequently his interest in the topics that made him famous later in life were routinely read backward into his earliest years. Boyle's predilection for experiment in general and chymistry in particular was thought to have developed in the mid-1640s, when he was still in his late teens and early twenties. This view found support in works that he wrote during these years but only published much later—such as *Seraphic Love*, written in 1648 but published only in 1659.[1] In that work, which even in its published form bears a dedication dated 1648, Boyle praises authors like Van Helmont and Paracelsus, refers to himself as "a converser with Furnaces," and gives numerous other indications of his familiarity with chymical authors and practices. But recent closer inspection, coupled with the discovery of new manuscript material, shows unambiguously that all these comments relating to Boyle's natural philosophical interests are later interpolations and thus are not representative of his state of knowledge or interests in the 1640s.[2] Indeed, Boyle's first successful initiation of experimental endeavors dates to the summer of 1649. Prior to this time, as Michael Hunter puts it, "Boyle was at best a reader of scientific books with a generalized curiosity about low level technology."[3]

How much of an impact did the young American have on the development of Boyle's most celebrated interests and attributes? Fortunately, we have three sources that bracket the meeting of Boyle and Starkey in early 1651. These sources—Boyle's own surviving manuscript treatises written during the early 1650s, his annual "Memorialls Philosophicall" (collections of natural philosophical and medical information) and the records relating

---

1. The comments published in 1659 in *Seraphic Love* are taken as evidence of Boyle's 1640s knowledge and interest in, among others, Marie Boas Hall, *Robert Boyle and Seventeenth-Century Chemistry* (Cambridge: Cambridge University Press, 1958), 20; J. J. O'Brien, "Samuel Hartlib's Influence on Robert Boyle's Scientific Development," *Annals of Science* 21 (1965): 4; Malcolm R. Oster, "The 'Beaume of Diuinity': Animal Suffering in the Early Thought of Robert Boyle," *British Journal for the History of Science* 22 (1989): 154.

2. Robert Boyle, *Seraphic Love*, in *Works*, 1:78–79, 85, 105; Lawrence M. Principe, "Style and Thought of the Early Boyle: Discovery of the 1648 Manuscript of *Seraphic Love*," *Isis* 85 (1994): 247–60.

3. Michael Hunter, "How Boyle Became a Scientist," *History of Science* 33 (1995): 64.

to him in Samuel Hartlib's *Ephemerides*—can act as barometers of Boyle's chymical interests, opinions, and expertise. By means of these sources we can compare Boyle's interests and knowledge immediately before he met Starkey with his interests and knowledge a year or so after he had taken up with the colonial émigré. These sources are unanimous in their witness to a sudden and significant increase in Boyle's knowledge and sophistication in chymistry immediately after his introduction to Starkey and to the fact that the two men were on close terms throughout the 1650s. Moreover, much of the new knowledge that Boyle acquired (or advertised) at this time is directly traceable to Starkey, and in some cases Boyle cites Starkey by name.

During the 1640s Boyle was a writer of moral and devotional treatises. He was recognized in this capacity even in his connection with the members of the Hartlib circle. For example, Boyle's first published work, "An Invitation to a Free and Generous Communication of Secrets and Receits in Physick," was requested by Hartlib, yet is nonetheless part of the program of moral epistles on which Boyle was engaged in the late 1640s.[4] Far from being Boyle's manifesto of an ideal of public knowledge based upon a new culture of science, the "Invitation" is the work of a Christian moralist drawing heavily upon sacred Scripture and showing little familiarity with the actual technical or social practice of experimental philosophy. Instead, Hartlib seems to have recognized Boyle's program of moralizing and requested an item of this genre that also overlapped with the utopian and utilitarian goals of his circle. But there are two treatises written by Boyle in the early 1650s whose contents show both a continuity with his moral-devotional program and also a turn toward his developing interests in natural philosophy. Given the dates of their composition, these texts allow us to compare the pre- and post-Starkey Boyle. The works in question are "Of the Study of the Booke of Nature" and an "Essay of the Holy Scriptures." The first was written c. 1650, shortly before Boyle met Starkey; the other was begun around the end of 1651—less than a year after the two first met—and largely written in 1652–53 during the period of Boyle's Irish travels. Since these two works belong to the same genre—both are inherently and expressly theological but occasionally deploy notions from natural philosophy—the two treatises provide a good locus for comparison of Boyle's levels of chymical sophistication.[5]

The tract on the "Booke of Nature" was intended "for the first Section of my Treatise of Occasionall Reflections," a work eventually published in

4. Principe, "Virtuous Romance and Romantic Virtuoso: The Shaping of Robert Boyle's Literary Style," *Journal of the History of Ideas* 56 (1995): 377–97, esp. 383–85.

5. On these treatises see Hunter, "How Boyle Became a Scientist," 67–76; they are published in Boyle, *Works*, 13:145–223, with comments on dating, context, etc. at xxxvii–xlii.

1665. While the text advocates the study of the natural world, this is not for its own sake, but rather to render it a promptuary able to redirect the student's gaze toward its beneficent, almighty Creator. Boyle's exhortatory, sermonizing style developed in the 1640s remains intact here. The Bible provides the bulk of Boyle's citations, although there are a few from ancient authorities like Galen and Aristotle, and several from moderns like Bacon and "the subtill Campanella." One point of interest is the apparent affinity Boyle had at this time for the *corpus Hermeticum* (encountered in John Everard's 1650 translation of *The Divine Pymander*) and the notion of the *prisca sapientia,* as recorded, for example, in "the excellent" Pico della Mirandola.[6] This interest seems to have been very short-lived, for Boyle recapitulated it only twice elsewhere: in another tract from this period where he recounts the tale that Aristotle met a Jew more learned than all the Greeks, and in the portions of *Usefulnesse of Experimental Naturall Philosophy* that are dependent upon "The Booke of Nature."[7]

The chymical content of "The Booke of Nature" is concentrated into one section of sufficient brevity that it can be quoted here in its entirety. In a discourse on how men tend to ignore the true values of things and thereby undervalue God's providence in supplying them to mankind, Boyle writes that

> the Juice of the Clouds, really containes that true Mercurius & Sulphur Philosophorum which so many Laborious Chymists have celebrated, wish't, & Dream't of & that's all. What Creature is there more despicable then Sand & (what are the naturall Loaves of it) Flints: & yet have I made of them, lasting & orient Gems; & yet they are the true Metallicke Wombes and Paps, & not unfrequently praegnant with the pretiousest of them; yeelding a Liquor in which all Metalls grow into Lovely Trees compos'd of Roote And Branches, & the usuall Parts constituent of those Plants.[8]

This passage reveals two things. First, it makes a rather vague reference to a contemporary chymical theory claiming that the long-sought Mercury (and Sulphur) of the Philosophers is contained in rainwater—presumably

6. Boyle, "Of the Study of the Booke of Nature," in *Works,* 13:153–54, 156.

7. Boyle, "Essay of the Holy Scriptures," in *Works,* 13:175–223, on 197. On Boyle and the Hermetic tradition, see Principe, "The Alchemies of Robert Boyle and Isaac Newton: Alternate Approaches and Divergent Deployments," in *Rethinking the Scientific Revolution,* ed. Margaret J. Osler (Cambridge: Cambridge University Press, 2000), 201–20, esp. 211–13, and Jan J. Wojcik, "Pursuing Knowledge: Robert Boyle and Isaac Newton," in *Rethinking the Scientific Revolution,* ed. Margaret J. Osler (Cambridge: Cambridge University Press, 2000), 183–200.

8. Boyle, "Booke of Nature," in *Works,* 13:159.

this is what Boyle means by "the Juice of the Clouds." The juvenile Boyle is right to claim that this Mercury is a thing that "Laborious Chymists have celebrated, wish't, & Dream't of," but his location of these valuable substances in the "Juice of the Clouds" places him in a distinctly nonmetallic school of chymistry with strong ties to the natural magic tradition of the sixteenth century, and promoted perhaps most notably by the Polish adept Michael Sendivogius.[9] Another contemporaneous view sought the Philosophical Mercury and Sulphur in metals; indeed, we have already encountered Alexander von Suchten's work in this area in chapter 2 and Starkey's in chapter 3, where the Mercury is prepared from common quicksilver. Boyle's mention of the "Juice of the Clouds," though intended to express not only God's providence in unlooked-for places but also Boyle's own knowledge, actually advertises Boyle's relative unfamiliarity with the traditions of metallic chymistry.

In singing the praises of sand in the next line, Boyle clearly reveals his immediate source. He mentions that he himself has made factitious gems from sand—a rather simple chymical operation—and also that he knows about a "Liquor of Flints" in which "all Metalls" are supposed to "grow into Lovely Trees." This fluid is an aqueous solution of sodium or potassium silicate, produced by fusing powdered flint or sand with an alkali carbonate, and then dissolving the fused, glassy mass in water. When pieces of metallic salts are thrown into this solution, they "grow" into treelike excrescences due to the slow production of insoluble metal silicates as the metal salts dissolve. Boyle also provides the reader with an aside regarding the presence of the "pretiousest" of the metals (i.e., gold) in common sand. All of these comments on sand and flint are derived from one section of the second part of Johann Rudolph Glauber's *Novi furni philosophici*. Glauber not only prepared this liquor of flints, or *liquor silicum*, but also described the treelike growths produced by its action on metal salts.[10] He also gave recipes in the same work for the production of factitious gems from flints and sand like those mentioned by Boyle.[11] Furthermore, in the same work Glauber also called sand "the mother of metals," asserted that alluvial gold was actually produced in sand, and published a process for extracting this gold from its sandy matrix.[12] Thus Glauber is clearly the source not only of

---

9. Newman, *Gehennical Fire*, 87–90, 211–22. Also on Sendivogius's niter theories see Paulo Alves Porto, "Michael Sendivogius on Nitre and the Preparation of the Philosophers' Stone," *Ambix* 48 (2001): 1–16, and below, 238, 248–51.

10. Johann Rudolph Glauber, *New Philosophical Furnaces*, in *Works*, tr. Christopher Packe (London, 1689), 44–48; see also 7, 11.

11. Ibid., 82–83.

12. Glauber, *Furnaces*, in *Works*, 45 (cf. 48), and *The Mineral Work*, in *Works*, 100–147. The first part of the latter text (101–14) gives the process for separating gold from sand using spirit

Boyle's denomination of sand as a "Metallicke Wombe" of gold, but also of all the other comments on sand and flints made in the "Booke of Nature." Even Boyle's subsequent comment that "Paracelsus forgot to lye when he say'd, that oftentimes a Flint is better than a Cow," is lifted directly from the same source, for in describing the process for making the liquor of flints Glauber remarks that "*Paracelsus*... saith in his book (concerning the vexations of Alchymists) that many times a despicable flint cast at a Cow is more worth than the Cow."[13] Glauber's citation of Paracelsus's claim is even set into exactly the same context of man's ignorance of God's bountiful providence as is Boyle's subsequent use of it.[14]

The pertinent parts of Glauber's *Novi furni philosophici* were published in German by 1646–47, well before Boyle wrote "Booke of Nature." But Boyle could not read German, and the Latin edition did not appear until 1651, after the composition of "Booke of Nature." How then did Boyle acquire this Glauberian material? The Hartlib circle maintained a keen interest in the entrepreneurial Amsterdam chymist and his *New Philosophical Furnaces*, and by the mid-1640s several Glauber manuscripts had reached Hartlib, who farmed them out for translation and subsequent circulation.[15] It is quite possible that Boyle obtained a copy of one of these manuscript translations.

It is also likely that Boyle's knowledge of Glauber and his processes came at least partly through Benjamin Worsley. Worsley had been sent to Holland by the Hartlibians in 1648–49 to work with Glauber and learn his secrets. Worsley was ultimately disappointed in this mission both by Glauber's reticence and the fact that he and Glauber had no common language.[16] At any rate, we know that Boyle and Worsley were on close terms both before and after the latter's Dutch trip and that Worsley shared recipes with Boyle. The mediation of Worsley would also explain Boyle's belief in the value of the "Juice of the Clouds," for Worsley aligned himself with those "Sendivogians" who believed that an aerial *sal nitrum* was the key to chymical operations. Such theories included the notion that rainwater and

of salt while the second (114–24) deals with the generation of metals, locating the origin of gold on the surface of the earth in sand. Glauber apparently attempted to sell this extraction process to the Dutch States; see Hermann Boerhaave, *A New System of Chemistry* (London, 1727), 41–42.

13. Glauber, *Furnaces*, 45.

14. Ibid., and Boyle, "Booke of Nature," in *Works*, 13:159. The Paracelsus reference is to the end of his *Coelum philosophorum*, sometimes called the *Book of Vexations*.

15. On Glauber's reception in the Hartlib circle and the translation of his works, see John T. Young, *Faith, Medical Alchemy, and Natural Philosophy: Johann Moriaen, Reformed Intelligencer, and the Hartlib Circle* (Brookfield, Vt.: Ashgate, 1998), 183–216, esp. 198–200.

16. On Worsley's 1648–49 mission to Glauber in the Netherlands, see Young, *Faith, Medical Alchemy*, 217–26.

dew—particularly at certain seasons of the year—were impregnated with this spiritual principle. This view of a philosophical *sal nitrum* contained in rainwater was promoted by the contemporaneous writer Thomas Vaughan (1622–66, alias Eugenius Philalethes), himself relying on the writings of Sendivogius, among others.[17] As we shall show below, Worsley's chymical commitments were predominantly Sendivogian; indeed, he engaged in a protracted argument in 1654 with Frederick Clodius, who denied the importance of the Sendivogian *sal centrale*. Thus Boyle's comment here in the c. 1650 "Booke of Nature" reflects Worsley's early influence. We will return to Worsley's chymistry and his influence on Boyle later in this chapter.

The rather low-level chymical knowledge of "Booke of Nature" is in marked contrast to the contents of the "Essay of the Holy Scriptures," which displays a significantly more developed understanding of chymical theories. Since this work was written during Boyle's time in Ireland, where it was "impossible" to do anything chymical, its contents must reflect the state of Boyle's chymical knowledge before he left for the Emerald Isle in the summer of 1652; that is, within eighteen months of meeting Starkey.[18] Even though the topic of "Holy Scriptures" is expressly theological, chymical (as well as a few other natural philosophical) items surface unexpectedly throughout it. Indeed, a fairly lengthy section filled with chymical comments and allusions interrupts the flow of the theological text. This material suffices to indicate that during those months that separate "Booke of Nature" from "Holy Scriptures" Boyle's understanding of prevailing chymical theories and goals, his practical expertise in preparative laboratory work, and his knowledge of key chymical authors burgeoned.

Boyle's views on the Mercury of the Philosophers, for example, have changed markedly. When speaking of God's ability to "keepe a Naturall Body from the Violence of Outward Agents," citing the young men preserved in the fiery furnace and the manna kept incorruptible in the Temple, Boyle suddenly remarks that " 'twould be hard for all Mankind, (except by the true Mercurie of the Philosophers; wherein it will truely dissolve, & putrefy with it; & one or two more unknown Arts) essentially to destroy or transmute into any other Body one ounce of pure Gold."[19]

Boyle then notes that "Common Mercury" is itself impossible to alter "by reason of its more exquisite Homogenëyty" unless it be treated with

17. See Vaughan, *Aula lucis* (London, 1652); Michael Sendivogius, *Novum lumen chemicum* and *Tractatus de sulphure*, in *Musaeum hermeticum* (Frankfurt, 1678; reprint, Graz: Akademische Druck, 1970), 545–645.

18. Boyle's complaint about his inability to do chymistry in Ireland occurs in a letter of April/May 1654 to Frederick Clodius; see Boyle, *Correspondence*, 1:165–68.

19. Boyle, "Essay of the Holy Scriptures," in *Works*, 13:211.

the "Alkahest, or Elixir." Boyle goes on to demonstrate that he is now quite familiar with a theory of how the alkahest and elixir act upon mercury— "the former but sequestering its impure & adventitious sulphur, & the latter Ripening & exalting it's Internall & solary one."[20] Boyle notes that these effects have been told to him by "Ey witnesses."

Just as the chymical comments in "Booke of Nature" point to Glauber and Worsley as sources, these comments in "Holy Scriptures" point to Starkey and his mentor Van Helmont. Employing the Mercury of the Philosophers to "dissolve & putrify" gold is exactly what Starkey was doing at this time with his Sophic Mercury and telling Boyle about. In the 1651 "Key" letter to Boyle, Starkey describes how his process yields "a mercury that dissolves al mettals ☉ [gold] especially . . . It also makes ☉ [gold] to puffe up, to swel, to putrefy"; Boyle's treatise uses similar language to describe this same action of the Philosophers' Mercury on gold.[21] A few months of interchange with Starkey have evidently converted Boyle from the Sendivogian position adopted in "Booke of Nature" to a Mercurialist one that he would never relinquish.[22]

Similarly, Boyle's explanation of the action of the alkahest and Elixir on mercury is in exact accord with what we know of Starkey's own theory of metallic composition and transmutation. We have already encountered Starkey's belief (gleaned ultimately from Van Helmont) in a twofold Sulphur in metals—one internal and one external—these are the "internall" and "adventitious" Sulphurs mentioned here by Boyle. Moreover, in the *Marrow of Alchemy*, written by Starkey at about the same time as Boyle's "Holy Scriptures," Starkey explicitly teaches that the alkahest acts upon common mercury by removing its external, impure Sulphur as an oil and leaving behind a quicksilver resistant to corrosion.[23] This same view that the alkahest removes the Sulphur "that is external to the Internal nature of Mercury" is also presented in *Sir George Riplye's Epistle to King Edward Unfolded* of Philalethes, which began circulating in the Hartlib circle no later than mid-1653.[24] Starkey was in fact working on the isolation of this external Sulphur from mercury in mid-1653, and recounting his efforts to

20. Ibid.
21. Starkey to Boyle, April/May 1651, in *Notebooks and Correspondence*, document 3; also in Boyle, *Correspondence*, 1:90–103, on 95.
22. On the Mercurialist school and Boyle's subsequent adherence to it, see Principe, *Aspiring Adept*, 153–179.
23. Starkey, *Marrow of Alchemy* (London, 1654–55), pt. 2, pp. 19–22.
24. Starkey [Eirenaeus Philalethes, pseud.], *Sir George Riplye's Epistle to King Edward Unfolded*, in Samuel Hartlib, *Chymical, Medicinal, and Chyrurgical Addresses* (London, 1655), 25–27; Newman, *Gehennical Fire*, 59–60.

Boyle.[25] The Philosophers' Stone, on the other hand, acts upon mercury's *internal* Sulphur according to Starkey, thereby transmuting the liquid metal into gold. Thus it is clearly Starkey's own theory that Boyle recapitulates in "Holy Scriptures." Indeed, the "Ey witness" to transmutation mentioned by Boyle is quite probably Starkey himself, who by this time had become renowned among the Hartlibians for having witnessed great arcana, including transmutation, at the hands of his acquaintance, the mysterious adept Eirenaeus Philalethes. Curiously, Boyle's now apparently dismissed Worsleian-Sendivogian notion about the value of the "Juice of the Clouds" is itself mentioned by Starkey in the *Marrow*, where he refers to those who seek the Mercury of the Philosophers "in simple water / Such as from clouds is caught" as "grosse sots."[26]

But these are not the only signs of Starkey's influence visible in the "Essay of the Holy Scriptures." Boyle also claims to have "seen an Alcali volatile," clearly a reference to Starkey's long-term project of volatilizing alkalies, partly recounted to Boyle by Starkey in his letter of April/May 1651.[27] Boyle also refers here to "a venereall Body" made by "a (secret but an) Easy Sublimation" from vitriol—no doubt a reference to the *ens veneris*, a Helmontian pharmaceutical on whose preparation Starkey and Boyle collaborated in 1651, of which we shall speak more later.

The signs and scope of Starkey's immediate and deep influence on Boyle can be seen likewise in a second source: the dated "collections" that Boyle kept in the 1640s and 1650s. During the 1640s these collections were repositories of literary and rhetorical ornaments and tropes, quotations from romances, and fragments of clever phrasing for use in Boyle's own romance-influenced devotional and ethical writings. The sudden change in 1650 of these collections to accumulations of medical, chymical, and technical recipes was first pointed out by Michael Hunter as an unmistakable indication of Boyle's change of interests.[28] While these annual

25. Hartlib, *Ephemerides* 1653, HP 28/2/69B.

26. Starkey, *Marrow*, pt. 1, 58–59. Indeed, Starkey here ridicules beliefs held by several members of the Hartlib circle—he criticizes those who "Saltpeter do the matter judge," perhaps another reference to Benjamin Worsley, and those who "attempt our hidden stone to finde / In Sunbeams eke to powder dry calcin'd," possibly a reference to Sir Kenelm Digby, who told of such a process in his *Two Treatises* (Paris: Gilles Blaizot, 1644), 51. Starkey's rejection of the Sendivogian *sal nitrum* theory may lie behind the comment given in early 1651 that he was "about to refute Vaughan"; Hartlib, *Ephemerides* 1651, HP 28/2/7B.

27. Boyle, "Essay of the Holy Scriptures," in *Works*, 13:204–5. Starkey to Boyle, April/May 1651, in *Notebooks and Correspondence*, document 3; also in Boyle, *Correspondence*, 1:90–103.

28. Hunter, "How Boyle Became a Scientist." The texts are now in the process of being electronically published at www.bbk.ac.uk/boyle/; in this regard see Michael Hunter and Charles Littleton, "The Work-Diaries of Robert Boyle: A Newly Discovered Source and Its Internet Publication," *Notes and Records of the Royal Society* 55 (2001): 373–390.

records indicate the major shift in Boyle's intellectual evolution from a moralist to a natural philosopher, they also tell us more about Boyle's changing contacts, and in particular about the development and nature of his relationship with Starkey.

The first of the nonliterary collections, begun on 1 January 1650, is brief and mentions only two informants—Benjamin Worsley and "Dr. Boate" (either Gerard or Arnold). Most of the recipes are medical, and they prescribe predominantly herbal and other organic remedies. Dr. Boate recommends nutmeg and alum for ague (from which Boyle had suffered seriously in 1649) and tells of an infallible cure using snails and cobwebs. Worsley commends a poultice that is to be applied to the feet during "malignant fevers," prepared by pounding leaven, onions, garlic, and pigeon dung into a paste with turpentine.[29] The only indication of Boyle's newfound interest in experiment is a mention of the luting he uses for lining furnaces. With these "Memorialls Philosophicall" of 1650 we find Boyle accumulating diverse recipes, particularly medical ones, as was common practice in the seventeenth century. These medical recipes show no particular novelty or method, nor any influence from recent chymical medicine.

But this situation changed very quickly after Boyle met Starkey. The first indication of this is a sheet, possibly intended as all or part of Boyle's collection for 1651; it contains some recipes from Dr. Boate and Johann Moriaen, but most are either explicitly attributed to Starkey or identifiable as his.[30] Unlike the earlier recipes from Dr. Boate and Worsley, these processes are fully chymical and rely on implicit Helmontian principles—the earliest indication of Helmontian notions from Boyle's pen. Moreover, the

29. BP, vol. 28, fols. 310–11.

30. The sheet is in Boyle's early hand and is undated and untitled, but we locate it in 1651 (or at latest, early 1652) for the following reasons. The last recipe—dealing with Sulphurs of antimony and their extraction using oil of terebinth distilled from salt of tartar—is very similar to an experiment described in one of Starkey's notebooks and dated December 1651 (Sloane MS 2682, fol. 94r). Similarly, the *ens veneris fixum* mentioned here is likely to have been developed during 1651 when Starkey was at work on this project with Boyle's assistance. A terminus ad quem is Boyle's departure for Ireland in June 1652, for by the time Boyle returned, Starkey had superseded the process given here for correcting opium. The recipe given here for corrected opium is possibly the laudanum mentioned by Starkey in his Spring 1651 letter to Boyle. Furthermore, there is no set of Memorialls by Boyle dated 1651 (the only year unaccounted for from 1647 to 1657—the set for 1653 is now lost but was extant as late as the 1740s; see below), and this sheet may represent all or part of that year's formal collection. It might be mentioned that its present position in the Boyle Archive is directly preceding the Memorialls for 1652, but the papers have been sufficiently churned over during the intervening years that current proximity must be used as evidence of original affiliation only with extreme care. The Boate mentioned here is probably Arnold, as Gerard had died in 1650, although it is possible that a receipt attributed to Gerard may have still been transmitted to Boyle by a third party.

details of the way in which these items are written suggest something significant about the nature of Starkey's relationship to Boyle at this period. The processes from Starkey (unlike an adjacent one from Boate) are in Latin and contain several idiosyncratic abbreviations and symbols. One entry uses the symbol ♉ for regulus of antimony and the curious half-Latin, half-Greek "*oleo θερ*" for oil of terebinth. Both of these are characteristic of Starkey's own writing; the rounded regulus symbol, in particular, is extremely rare. The use of these symbols (and the Latin text) implies that this material was not received orally, but copied directly from a written source. Now the style does not accord with that of Starkey's letters but is instead strongly reminiscent of Starkey's laboratory notebooks. For example, the concluding paragraph to a process for the isolation of the Sulphur of copper is typical of Starkey's notebook style: "Proceed thus with iron, and with corals, and with zinc, and with regulus martis, & you will admire the result. Also with Ludus."[31] Here it sounds as if Starkey, having had apparent success in the isolation of one kind of Sulphur, now orders himself to try generalizing the procedure to other substances—a habit we encountered frequently in the notebooks and one that is characteristic of Starkey's methodology.

But it appears that Boyle's copying out of Starkey's laboratory notebooks was not merely transcription. Following a process in Latin for a solvent that works *sine repassione*—that is, without being acted upon, thus identifying this as an attempt at the alkahest—Boyle switches suddenly to English:

> NB. After the third Destillation the spirit will be about a fourth of the whole Urine. NB. Equall quantity of onely Good Spirit of Wine & Spirit of Urine. The Coagulum comes over first & the Subsequent Spirit dissolves it. NB. Equall quantity of Vinegar in relation to the Spiritus Compositus: the more Vinegar the fewer Cohobations will serve. Distill it till the Phlegm (made of the Vinegar) begin to ascend.[32]

Each of these sentences presents a clarification of the practical details of the process set out in Latin directly above. Why would English expositions suddenly intervene? A likely explanation is that Boyle was copying Starkey's notebooks under his supervision, with Starkey available to answer questions or to provide expositions based upon his direct experience of the process or upon his more complete understanding of the Helmontian the-

---

31. BP, vol. 25, p. 341: "Sic procede cum ferro, item cum corallis, item cum Zink, item cum Regulo ♂ⁱˢ & successum mirabere. Item cum Ludo."
32. Ibid.

ory upon which they depend.[33] Other processes from Starkey show a kind of simultaneous glossing, where Boyle inserts comments giving finer detail within the flow of the text, but sets them off by enclosing them in square brackets. The impression then is of Boyle as student to the more experienced Starkey, copying out of the master's books and receiving oral elaborations.

A similar interaction is preserved in a document we have recently discovered among John Locke's papers.[34] This is the only known manuscript that preserves Starkey's handwriting and Boyle's together—and the way the two hands interact is very interesting. The document is a fragment of a disbound laboratory notebook by Starkey, and the date of its use by Starkey is not later than early 1651, since the processes recorded in it for attempting to make tartar volatile with spirit of wine and with vinegar were superseded by him before April/May of that year.[35] Starkey's original entries throughout are overlaid with notes in Boyle's hand—all of them requesting further information about the process. Sometimes these jottings take the form of a simple "q" (presumably for *quaere*, "ask") in the margin. In other cases they are more specific *Q. Quant* ("ask the quantity") or *Q. Prop* ("ask the properties"). In some cases Boyle adds details of the process, such as inserting the word *paulatim* ("gradually") to the instruction to mix distilled vinegar with salt of tartar (a sensible insertion considering the vigorous effervescence that occurs when the two are mixed). In one case, after Boyle had glossed Starkey's Latin process for a powerful solvent, he recopied it in English and in an expanded format (incorporating the additions) on another page. Presumably Starkey handed the pages to Boyle, Boyle read them and marked his questions on them, then queried Starkey, and finally recopied and translated the expanded process in which he was most interested, entitling it more grandly than had Starkey, as an "S. T. V.," that is, *spiritus tartarus volatilis.*[36] Once again, this document seems to preserve evidence of Starkey's more or less formal tutorship of Boyle.

If we move on to Boyle's work diaries for subsequent years, the importance and uniqueness of Starkey's contributions continue. The Memorialls for 1652 show a curious bifurcation. Starkey's recipes involve cutting-edge chymical medicine, heavily dependent upon the operations of the chymical

---

33. The note that the "Phlegm" is "made of the Vinegar" suggests an implicit reliance on the theory of exantlation.

34. Bodleian Library, Locke MS C29, fols. 115r–118v.

35. For further information on the dating of this document, and for a full transcription and translation see *Notebooks and Correspondence,* document 1.

36. Bodleian Library, Locke MS C29, fol. 116v (Starkey's original process) and fol. 115r (Boyle's copying).

laboratory and Helmontian principles, while all the other recipes continue to be herbal or organic, employing items like boar's tooth, wood sorrel, plantain water, poppy seed, and so forth. Judging from the recipes that Boyle had collected up to 1652, Starkey seems to be Boyle's first, and only, chymically sophisticated contact at this time. Moreover, the same signs of formalized learning from Starkey persist. The 1652 collection begins in large letters: "Dr. George Stirke" and here again we find twofold entries containing first a process and then a gloss beginning "NB" that provides clarifications or expositions.[37] In the case of a "true correction of opium," the initial text is again in Latin, and begins with a general principle drawn from Van Helmont: "Omnis Opii vis Narcotica in Alcali amittitur" ("all the narcotic power of opium is lost in alcali"). A clause beginning *quare* ("wherefore") then introduces the precise process, written in Latin in the first person: "I take one ounce of pure salt of tartar . . ." The following paragraph then suddenly switches to English and the third person, and provides a gloss. Thus this entry (like several others from Starkey) has the form of principle-consequence-process-explication. Again, this seems likely to represent a combination of Boyle's transcription from Starkey's first-person Latin notebooks and a subsequent exposition given by Starkey, very possibly orally, in English.

---

37. BP, vol. 25, pp. 343–58. The dating of this collection has been problematic because its first page bears two different dates. The heading reads: "Memorialls PHILOSOPHICALL Beginning this First day of the Yeare 1651/52. And by God's Assistance to be constantly continued during my life. A.D. M.DC.LIII/LIIII." There is more than one conceivable resolution for this dating quandary. One possibility is that the record was begun in 1652—as is first stated—but was interrupted by Boyle's departure for Ireland in June 1652. When Boyle returned permanently to England in July 1654, he recommenced using this record, and this second stage of writing is denoted by the second date. Indeed, the second date bears some, but inconclusive, evidence of having been altered by the addition of two further I's at the end of the sequence of Roman numerals—thus changing 1651/52 to 1653/54. Alternatively, the second date may simply be a mistake. It should be noted that the document itself has a clear first section that runs from page 343 to page 346—the last of these pages has the lower third left blank, and the soil and wear on this sheet show that it was the outside leaf of a bundle (folded in sixths) for some extended period. This section, we believe, represents Boyle's Memorialls for 1652—the Starkey processes recorded here fit with what we know of his activities and the point of evolution of his processes for that year. The immediately following leaves represent a second section, without an independent heading, beginning on page 347, where a new series of contemporaneous pagination begins, even though these new page numbers are not written in until the fourth page (page 350). This section, we believe, is material dating from 1654. This dating is reinforced by the sudden appearance of "Mr. Smart" at the start of the second section; he appears also in Hartlib's *Ephemerides* beginning in 1654 (HP 29/4/26A–27B, and passim.) It may be that the second section now follows the first only on account of a post-Boyle ordering of the archive. Nonetheless, regardless of how the contradictory dating of the heading on page 343 is resolved, the dating of the materials—pages 343–46 dating from 1652 and pages 347–58 dating from 1654—still seems likely based on the evidence presented above.

It is important to notice the prominence given to theoretical principles as guides to experimental processes—a feature absent from processes gathered from informants other than Starkey. In this regard, the expositional paragraph following a preparation of vitriol of iron witnesses Starkey teaching Boyle Helmontian theory: "NB. 1. The Crocus remains undissolved; & communicating but an Odor to fix the Sal Vitrioli; remaines fere eodem pondere quo prius [in nearly the same weight as before]." Here Starkey presents Boyle with the Helmontian principle of radial activity, whereby the iron, acting solely by an "odor," can fix the salt of the vitriol without losing any of its own substance. We encountered this theory in chapter 2 in the course of examining Van Helmont's experiments with the action of mercury and iron on oil of vitriol. Thus Starkey's contribution appears not to be limited to giving Boyle bare recipes, as is usual with Boyle's other informants; rather Starkey often presents such recipes within a specific framework of chymical (Helmontian) theory along with detailed supplemental information on the practical manipulations involved.

These entries thus seem to be records of a more or less formalized instruction in chymistry. The method of copying an authoritative text and then providing oral glosses or explanations of it is of course a classic method of formalized university instruction. We know that Starkey himself was instructed by such a method at Harvard College. It is thus very possible that Starkey's recollections of that tutoring system formed the basis of the way he introduced his own chymical methods and Helmontian theories to his friend Boyle. Given how evident Scholastic university methods are in Starkey's own private notebooks (regardless of his published rhetoric declaiming against the "method of the Schools"), it would not be unlikely that Starkey would turn to the same sources for a didactic model in teaching Boyle.

It must be stressed, moreover, that only the entries from Starkey have this format of a process followed by a gloss. The Memorialls of 1652 contain recipes from Dr. Boate, Dr. Coxe, and others besides Starkey, but these occur without the explanatory "NB's" usually found following Starkey's contributions. Similarly in later collections, this format is very rarely found except in connection with Starkey.[38] This argues that the relationship between Starkey and Boyle was unique—while Boyle collected medical or

---

38. The only exceptions to this occur late, in Boyle's collections for 1655; there, recipes from Clodius and from Mr. Smart are followed by an "NB" and observations (BP, vol. 8, fols. 143v, 148r). Perhaps Boyle was imitating the style he had used with Starkey since 1651. In any event these are the only such cases in all of Boyle's collections for the 1650s, and even in this later collection, it is predominantly Starkey's recipes that take this form (e.g., fol. 147r).

chymical recipes from many individuals, no one else seems to have had the kind of a role as interpreter and teacher that Starkey did.

The impression of Starkey as Boyle's teacher is corroborated by Boyle's own recollections of some of his earliest interactions with Starkey. In *Usefulnesse of Experimental Naturall Philosophy* Boyle recalls the circumstances surrounding the production of the *ens veneris*, a pharmaceutical preparation lauded by Van Helmont. Boyle writes that

> an Industrious Chymist (of our Acquaintance) and I, chancing to Read one day together that odde Treatise of *Van Helmont*, which he calls *Butler*, when we had attentively perus'd what he delivers of the Nature as well as scarce credible Vertues of the *Lapis Butleri* he there mentions, we fell into very serious thoughts, what might be the matter of so admirable a Medicine, and the hopefullest manner of preparing that matter.[39]

Boyle then notes how after reading Van Helmont together and comparing "Butler" with "other Passages of the same Author," and after "having freely propos'd to one another our Conjectures," he and this unnamed "Industrious Chymist" settled on a sublimation of sal ammoniac from colcothar (the residue from roasting copper vitriol) as the way to approach this Helmontian arcanum.

There is no question that this "Industrious Chymist" was George Starkey. Not only do Starkey's notebooks witness his work on the *ens veneris*, but he also notes in a later pamphlet how he prepared the substance for "the Honourable *Robert Boyl* Esq; one of the Royal Society, who hath wrote of its excellency, as his extant Treatise [i.e., *Usefulnesse*] thereof can testifie." Starkey also gives the date of this preparation as 1651—only a few months after Starkey and Boyle's first meeting in January of that year.[40] The key point is that this collaborative project arose from a meeting where Boyle and Starkey were studying Van Helmont *together*. Now in 1651, Starkey had been reading Van Helmont carefully for at least three years and vigorously pursuing, in both America and England, the reduction of Van

---

39. Boyle, *Usefulnesse of Experimental Naturall Philosophy*, in *Works*, 3:500; compare also Boyle's other mention of this episode at *Usefulnesse*, 391–92.

40. Starkey, *George Starkey's Pill Vindicated* (London, [1663]), 6. Note that Starkey's treatise gives two different dates for the preparation of the *ens veneris* with Boyle: 1651 and 1652. The earlier date is surely the correct one, for a notebook fragment dated to late 1651 indicates that the *ens veneris* preparation was finished by then (Sloane 2682, fol. 88r), and Hartlib recorded information about Starkey's *ens* in his 1651 *Ephemerides* (HP 28/2/26A). The erroneous date of 1652 (just a few lines from the correct date) in the pamphlet is probably a typesetter's error induced by Starkey's handwriting in the original manuscript, for Starkey habitually writes the numeral "1" in a figure very much resembling a "2"; see, for example, the numerals in his account reckonings at Sloane 3750, fol. 20r, and RSMS 179, fol. 1r.

Helmont's often obscure directions into experimental practice. At that same time, Boyle was a newcomer not only to chymistry and experimental philosophy, but to the realm of learning as well. The kind of close textual analysis that led to the *ens veneris* project was the daily bread of Starkey's Harvard education, but an ability that Boyle was only beginning to develop in 1651.[41] Therefore, during that day in 1651 when Starkey and Boyle were reading Van Helmont together, it is unlikely that Boyle was a superior or even equal participant in the discussion, despite his implication to the contrary. Rather, it is more likely that their discussion of the *ens veneris* occurred during a session where Starkey was in effect tutoring Boyle—the same kind of meeting suggested by Boyle's Memorialls, where the shared text was one of Starkey's notebooks rather than Van Helmont's *Ortus*. Boyle's chief contribution probably consisted of funding the experimentation and encouraging Starkey to carry it out.[42]

THE ORIGINS OF BOYLE'S HELMONTIANISM

The Helmontian content of Boyle's early natural philosophical writings has long been recognized, and Boyle's Helmontianism has already been cited in several studies.[43] Now that we know that references to Van Helmont apparently made by Boyle in the late 1640s (e.g., in *Seraphic Love*) are in fact later interpolations, we must redate Boyle's introduction to the Flemish natural philosopher. We suggest that it was predominantly through Starkey that Boyle acquired a firm grounding in Helmontian theory and practice. Throughout 1651 and 1652, all but one of Boyle's known references to Van Helmont appear in the context of processes and notes from Starkey. This is not to say that Starkey was the first or only devotee of Van Helmont whom Boyle encountered. The late 1640s and early 1650s saw an increasing interest in Van Helmont in England. Sir Cheney Culpeper, Walter Charleton, and Noah Biggs were all enthusiastic about Van Helmont in the late 1640s, before Starkey's arrival in England.[44] Yet among all of Boyle's associates in

---

41. On Boyle's introduction to and development of scholarly textual skills, see Hunter, "How Boyle Became a Scientist."

42. Newman, *Gehennical Fire*, 70–72.

43. For example, Antonio Clericuzio, "Robert Boyle and the English Helmontians," in *Alchemy Revisited*, ed. Z. R. W. M. van Martels (Leiden: Brill, 1990), 192–99; Clericuzio, "From van Helmont to Boyle: A Study of the Transmission of Helmontian Chemical and Medical Theories in Seventeenth-Century England," *British Journal for the History of Science* 26 (1993): 303–34; Charles Webster, "Water as the Ultimate Principle of Nature: The Background to Boyle's *Sceptical Chymist*," *Ambix* 13 (1966): 96–107; Michael T. Walton, "Boyle and Newton on the Transmutation of Water and Air, from the Root of Helmont's Tree," *Ambix* 27 (1980): 11–18.

44. See Clericuzio, "From van Helmont to Boyle."

the first years of the 1650s, when Boyle's interest in natural philosophy was beginning to develop, Starkey was by far the best acquainted with Van Helmont and was the one with whom Boyle was most intimate.[45]

The first appearance of Boyle's nascent Helmontianism in a treatise occurs in the c. 1652 "Essay of the Holy Scriptures." Indeed, Hunter noted the greater emphasis and interest Boyle places on chymistry in general in this essay.[46] It is in particular Boyle's sudden interest in Van Helmont, a character wholly missing from his earlier writings, that is noteworthy for our present purposes. Boyle's first explicit citation of Van Helmont derives from Charleton's translation of three essays from the *Ortus*, which he published in 1650 as *A Ternary of Paradoxes*. The reference Boyle makes here to Van Helmont concerns a theological point regarding man's reason and is taken from the tract "Imago dei."[47] But most of the further explicit or implicit references in "Holy Scriptures" to "the Acute & most Ingenious Van Helmont," are chymical in nature, and such material is not to be found in Charleton. For example, in one place Boyle writes of his ability and equipment to fuse sand and alkali together and then recover the sand from the resultant glass "as confidently as many Chymists have esteemed Glasse indestructible by Art or Nature"—a process found in the *Ortus* and that we discussed in chapter 2.[48] In another place he names Van Helmont's notion of the *vita media*, and then refers to the "admired Butler," the Irish physician to whom Van Helmont devoted the tract that inspired Starkey and Boyle's collaboration on the *ens veneris*. Still elsewhere Boyle praises "our Van Helmont" (along with Kircher, Sennert, Bacon, and Campanella) in preference to "any Galenicke, or Peripateticke Names."[49] He also mentions a "sort of sand" that resists normal chymical operations; this is probably a reference to Van Helmont's inert and indestructible *Quellem*. Finally, here for the first time, Boyle mentions the objections "I think I could bring against the Four Elements in Mixt Body's," the topic to which he de-

45. Besides the copious evidence presented in chapter 3, recall that Starkey had asked John Winthrop Jr. for a copy of Van Helmont in 1648 while Boyle was still writing moral treatises; Hartlib noted in 1650 that Starkey knew "almost all Helmont by heart" while Boyle's experimentalism was in its infancy (Hartlib, *Ephemerides* 1650, HP 28/1/40B), and early during his time in England Starkey actually began a translation of the *Ortus* that is preserved in one of his notebooks (Sloane 3708, fols. 79–102). Hartlib noted Starkey's translation work on Van Helmont in early 1651; *Ephemerides* 1651, HP 26/29/5B.

46. Hunter, "How Boyle Became a Scientist," 77.

47. Walter Charleton, *A Ternary of Paradoxes* (London, 1650); the three tracts Charleton translated are "The Magnetic Cure of Wounds" ("De magnetica vulnerum curatione"), the "Nativity of Tartar in Wine" ("Tartari vini historia"), and "The Image of God in Man" ("Imago dei"); Boyle's allusion to this work occurs in "Holy Scriptures," *Works*, 13:187.

48. Boyle, "Essay of the Holy Scriptures," in *Works*, 13:204.

49. Ibid., 206, 197.

voted a key part of the *Sceptical Chymist,* and which—as has been known for some time—he drew largely from Van Helmont.[50] It is reasonable to conclude that Boyle's new familiarity with Helmontian chymistry shown in "Essay of the Holy Scriptures" has the same source as his newfound notions on chrysopoeia and metallic composition that the same essay displays— George Starkey.

The closeness between Boyle and Starkey in 1651 and Boyle's simultaneous increase in chymical interests are recorded also in a third source—Hartlib's *Ephemerides.* Prior to Starkey and Boyle's meeting in early 1651, the citations of Boyle in Hartlib's records rarely involve chymistry. One reference from 1648 deals with a request for the preparation by Glauber of the ludus—a cure for the stone, from which Boyle thought he was suffering at this time—but the rest have no chymical component.[51] Of the score of entries dealing with Boyle for 1649, only two mention chymistry: one records that he and Worsley obtained a chrysopoetic recipe, and the other mentions Etienne de Clave, "a Dr. in Physick that hath written 2. or 3. treatises very singular in Natural and Chymical Philosophy, one of which Mr. Boyle hath."[52] The 1650 *Ephemerides* associate nothing chymical with Boyle, but the 1651 records are quite different. Here half of the entries dealing with Boyle involve chymical topics; strikingly, almost all of these entries mention Starkey as well. In fact, Hartlib's information on Starkey for 1651 comes exclusively through Boyle. This argues not only for the close relationship the two young men quickly struck up but also for the distinctly chymical component of that relationship.

It is widely assumed in the secondary literature that any impact Starkey had on Boyle waned suddenly in 1654. Almost every citation of Starkey's role in the Hartlib circle has made much of Starkey's "degeneration" in that year, invoking the term used by Hartlib in a much-cited letter to Boyle written on 28 February 1654.[53] The episode referred to is Starkey's bankruptcy, imprisonment for debt, and departure from central London—a sad chapter in Starkey's life whose impact we have seen in his laboratory notebooks. This expression of dissatisfaction from Hartlib has sufficed to argue for an end to Starkey's productive contributions to the Hartlibians and by extension to Boyle. Yet it is important to recognize the possible overinter-

50. Ibid., 190; on the origin of these objections in Van Helmont, see Allen G. Debus, "Fire Analysis and the Elements in the Sixteenth and Seventeenth Centuries," *Annals of Science* 23 (1967): 127–47.

51. Hartlib, *Ephemerides* 1648, HP 31/22/2B.

52. Hartlib, *Ephemerides* 1649, HP 28/1/9A–B. Etienne de Clave is therefore one of the very first chymical authors Boyle read. He is mentioned by name in the "Essay of the Holy Scriptures" (*Works,* 13:205) for his description of a palingenic plant.

53. Hartlib to Boyle, 28 February 1654; Boyle, *Correspondence,* 1:154–63.

pretation of Hartlib's comments. They themselves are possibly exaggerated. No doubt Hartlib's presentation of the situation to the absent Boyle was aggravated by Starkey's unwillingness or inability to turn his attention to the kind of lucrative schemes so much desired by the Hartlib circle, presumably the "ungrateful obstinacy" Hartlib cited later in the same letter.[54] Furthermore, by all accounts Starkey was hard to get along with—Robert Child had previously counseled Hartlib to overlook Starkey's erratic nature, saying he was like "a good vessell with much saile & little ballast."[55] (One might even speculate that part of Starkey's inconsistent behavior was due to heavy metal poisoning, since it is known that chronic poisoning by mercury—one of Starkey favorite subjects of study—induces paranoia and mental instability.)[56] Moreover, Hartlib no longer needed Starkey, since he now had his countryman Frederick Clodius as a son-in-law and as the new chymical light of his life. It is likely that Clodius now not only occupied Hartlib's attentions, but may also have actively discouraged Hartlib's interest in Starkey in order to advance his own standing.[57] Yet most importantly, while Starkey's bankruptcy and confinement to debtor's prison may reflect badly on the quality of his financial sense, it cannot say anything about his chymical or intellectual acuity. Nor can it be blandly assumed that Hartlib's clear dissatisfaction with Starkey (and the disgust at chymistry in general expressed by Worsley in the same letter) was adopted by Boyle himself.

The notion that this unfortunate episode spelled the end of Starkey's productive career or of his constructive influence upon Boyle is easily dispelled. While Hartlib's interactions with Starkey never recovered, Boyle's good relations with Starkey resumed immediately upon his return from Ireland, as Boyle's work diaries attest. Boyle's collection for 1653, presumably made while he was in Ireland, is now no longer extant. But upon his return

54. Ibid., 159. See Newman, *Gehennical Fire*, 78–83, for the likely financial origins of Hartlib's disaffection with Starkey.

55. Child to Hartlib, 2 February 1653, HP 15/5/18A–19B, on 18B.

56. A similar cause has been invoked for Newton's mental instability in the early 1690s; see P. E. Spargo and C. A. Pounds, "Newton's 'Derangement of the Intellect': New Light on an Old Problem," *Notes and Records of the Royal Society* 34 (1970): 11–32; and L. W. Johnson and M. L. Wolbarsht, "Mercury Poisoning: A Probable Cause of Isaac Newton's Physical and Mental Ills," *Notes and Records of the Royal Society* 34 (1970): 1–9. Ironically, we have suggested that Newton's mercury poisoning (if that diagnosis is correct) was brought about by carrying out George Starkey's process for the Philosophical Mercury; see Principe, *Aspiring Adept*, 178–79.

57. This suggestion was made by Betty Jo Teeter Dobbs in *The Foundations of Newton's Alchemy, or The Hunting of the Greene Lyon* (Cambridge: Cambridge University Press, 1975), 74. Clodius exhibited the same behavior toward Johann Moriaen, who had not only worked with him but had actually introduced him to Hartlib; see Young, *Faith, Medical Alchemy*, 54.

in the second half of 1654, he continued making such collections.[58] Here Starkey's processes again appear prominently; these are now joined with others from one "Mr. Smart" and from Frederick Clodius.[59] Likewise, Boyle's "Philosophicall Diary" of 1655 still contains more entries derived from Starkey than from anyone else.[60] Even Clodius's name comes in a distant second, for there are nearly twice as many items derived from Starkey as from Clodius. For Starkey's part, he had recovered sufficiently from his financial debacle to return to experimentation by mid-1654, and his lab work was again in force by 1655 at the latest.[61] This return—or partial return—to solvency may have been assisted by contributions from Boyle upon his return from Ireland in mid-1654, but there is no clear evidence for this, save the close contact that the two resumed as witnessed by Boyle's work diaries.

A potentially more important division between Starkey and Boyle occurred toward the end of 1655. At this time Boyle relocated to Oxford to join the experimental group gathered around John Wilkins, and Starkey went to Bristol to oversee a metallurgical refining project. The nature and extent of the contact between Starkey and Boyle after their respective relocations is impossible to judge directly given the absence of any letters or other such evidence. Yet it seems that their relationship, and Starkey's influence on Boyle, did not come to an end. One testimony of this is that when Johann Moriaen wished to contact Starkey in 1658, he hoped to have a letter delivered to him through Boyle.[62] More significantly, in 1658 Starkey

58. A manuscript described as "Promiscuous thoughts 1653" was catalogued by the Reverend Henry Miles in the early 1740s, when he was collecting and sorting Boyle's papers. In one of his lists, Miles marked this manuscript as "N.W.," i.e., "No Worth," and presumably threw it away; see Michael Hunter and Lawrence Principe, "The Lost Papers of Robert Boyle," *Annals of Science* 60 (2003). On the issue of identifying/dating the 1654 collections, see above, 219 n. 37. On these collections in general, see Hunter and Littleton, "Work-Diaries of Robert Boyle."

59. Mr. Smart first appears in Hartlib's *Ephemerides* in 1654 (HP 29/4/26A–27B and passim). He is presumably Thomas Smart, and the same Smart referred to by Hartlib as "an artist" living "next unto Vaux-hall" (Hartlib to Boyle, after 1 Sept. 1657; Boyle, *Correspondence*, 1:229–32, on 230) and referred to as a "drudging operator" living at Dorchester House by Samuel Collins in a letter to Boyle dated 1 October 1663; ibid., 2:104–8.

60. BP, vol. 8, fols. 140–48v.

61. It is, however, unlikely that Starkey ever fully recovered financially after 1654. The dowry that Israel Stoughton had no doubt settled on his daughter had been spent in experiments that failed to remunerate. The pamphlet wars that Starkey entered into in the 1660s suggest a degree of acquiescence to the marketplace that he was clearly unwilling to make in the early 1650s, again probably the product of need. In *Pyrotechny* (1658) he mentions being under house arrest, and in a letter of 1663 (*Notebooks and Correspondence*, document 15) he complains of wanting his liberty and being "uncertain where to reside."

62. Moriaen to Hartlib, 1 January 1658, HP 31/18/25A.

dedicated *Pyrotechny Asserted*—a book containing the outlines of his grand project for the correction of all substances into medicines—to his "very good Friend" Robert Boyle. While Boyle did not explicitly record (to our knowledge) his opinions of its contents, there are two sources of importance for assessing it. First, another member of the Oxford group, John Ward, kept diaries of his reading, conversations, and other activities at Oxford.[63] These diaries, now at the Folger Shakespeare Library, show that Starkey's works were carefully read in Boyle's circle at Oxford. It is even possible that Boyle, as the dedicatee, introduced and circulated *Pyrotechny* there himself.

BOYLE AND THE "ESSAY ON POISONS"

While these notes indicate how seriously Starkey's work was taken among the club of Oxonian experimentalists of which Boyle was a member, we have more direct evidence for Starkey's continuing influence on Boyle's thought. While at Oxford Boyle engaged in a flurry of writing on natural philosophical topics. The flood of publications that Boyle released in the early 1660s, and upon which so much of his later reputation rests, draws its origins from this period. One of these Oxford works was an "Essay of Turning Poisons into Medicines." The treatise survives only in a partial transcription by Oldenburg made about 1660. The text was probably written in 1656–57.[64] While the treatise cites several printed authorities (such as Johannes Hartmann and Basil Valentine) and several of Boyle's informants (Sir Kenelm Digby and Thomas Coxe), the work is heavily dependent upon George Starkey.

Among the many signs of this reliance on the work of the American chymist is Boyle's mention of the "correction" of arsenic by fusion with

63. We thank Bruce Janacek for alerting us to Starkey references in the Ward diaries. The notes from Starkey appear in Folger Shakespeare Library, V.2.290. On the diaries in general, see Robert G. Frank Jr., "The John Ward Diaries," *Journal of the History of Medicine* 29 (1974): 147–79.

64. On dating and context see Boyle, *Works*, 13:xliii. The essay cannot predate 1655 (pace A. Rupert Hall and M. B. Hall, in *The Correspondence of Henry Oldenburg* [Madison: University of Wisconsin Press, 1965–73], 1:135; and Hall, *Robert Boyle and Seventeenth-Century Chemistry*, 29) owing to its incorporation of a process that Boyle obtained from Sir Kenelm Digby in January 1655 and recorded in his "Philosophical Diary" (BP, vol. 8, fol. 140, no. 5; the process is presented in Boyle, "Poisons," in *Works*, 13:241.)

The complete work apparently survived until the 1740s, when the Rev. Henry Miles inventoried the Boyle papers. On 10 February 1743 Miles listed both a manuscript on "turning poisons into remedies, 72 p[ages]" and "another same, 40 [pages]," see BP, vol. 36, fol. 143 (cf. the derivative list at BP, vol. 36, fol. 146, for a list of several items dealing with poisons including "turning poisons to remedies, 112 pages.") The length cited here would suggest an original text considerably longer than Oldenburg's transcript. On the loss of Boyle's papers after the 1740s see Hunter and Principe, "The Lost Papers of Robert Boyle."

saltpeter.[65] The process that Boyle gives in the essay is directly paraphrased from a Latin fragment of one of Starkey's notebooks that was in Boyle's possession.[66] Although Boyle sets the process in the context of Helmontian medicine, and quotes directly from Van Helmont's "Scabies et ulcera scholarum,"[67] the contribution of Starkey is clear—in spite of the fact that he is nowhere mentioned. Van Helmont does not give quantities or precise operational directions for the preparation: he merely intimates an unspecified "dulcification of it with spirit of wine." Starkey's notebook, on the other hand, gives an exact method for carrying out this "dulcification," and that method is closely recapitulated in Boyle's essay.[68] The process is also followed by a set of enumerated points "for the better understanding of this proces" in the same format that we have seen previously in connection with other processes presented to Boyle by Starkey. These latter points contain further details of practical considerations or explications in terms of Helmontian theory, just like the "NB" explications of Starkeian processes in Boyle's surviving work diaries.

The section of Boyle's tract that deals with animal poisons also shows reliance on Starkey. There Boyle mentions a "viper-wine . . . that we make, by taking out a live-vipers heart, and, before it hath done panting, with a sharp knife dividing it in a glas of wine, which . . . must be immediately drunk up." This very recipe was commended by Hartlib as good for the eyes in his April 1653 *Ephemerides*, where he attributes it to "Stirk."[69]

But Boyle's reliance on Starkey is most dramatic in the section "Of vegetable poisons." There the entire opium "correction" process recorded in the 1652 Memorialls (and studied above) appears nearly verbatim. Exactly as in the case of the Memorialls entry, the presentation of this process here begins with Van Helmont's theoretical principle "omne Narcoticum perit in Alcali," and then continues with the practical consequences of this maxim: "this Hint hath directed some of us to attempt the correction [of opium] with salt of tartar."[70] Boyle's use of the indefinite "some of us" elides the

---

65. Boyle, "Of Turning Poisons into Medicines," in *Works*, 13:247.

66. *Notebooks and Correspondence*, document 1; Bodleian Library, Locke MS C29, fol. 118r.

67. Van Helmont, *Ortus*, "Scabies et ulcera scholarum," pp. 320–28, esp. pp. 324–25.

68. Boyle rightly notes that this operation is only "intimated by Van Helmont" but that "according to our preparation . . . the spirit of wine must be very often abstracted from the white matter (15 or 20 times will be litle enough, and every new distillation requires the affusion of fresh spirit of wine)" (Boyle, *Works*, 13:247). This line paraphrases Starkey's Latin direction: "Cohoba aqua vitae rectificata ad vices 15, renovato spiritu vice singulâ" (Locke MS C29, fol. 118r).

69. Boyle, "Poisons," in *Works*, 13:252; Hartlib, *Ephemerides* 28/2/58A: the "heart and liver of a Viper being taken out hott or fresh out of the Viper put into Wine and druncke."

70. Boyle, "Poisons," in *Works*, 13:254.

fact that this is Starkey's process, indeed, the basis of Starkey's trademark "Vegetable Corrector." Boyle continues then to give the process directly out of his 1652 Memorialls (though translated into English), and here too the process is followed by an "NB" along with several of the same practical points enumerated there. For example, the 1652 collection records Starkey's direction that "the signe of Dissolution's being accomplish't, is the residence of vegetable fibres & sagges at the Bottome; & that residence after Decoction will wholly loose it's tincture." In the "Essay on Poisons," Boyle writes identically that "when it is competently dissolved, [it] usually leaves at the bottom a residence with divers vegetable fibers and saggs, which after digestion will loose its tincture."[71]

The remainder of Boyle's essay (as it survives) derives entirely from Starkey, including the presentation of an alternate opium correction that Boyle had previously recorded in another set of "Collections" where he explicitly attributed it to Starkey.[72] But Boyle also notes that this preparation "useth a litle to suspend urine," and thus "I therefore add to it" a salt that is "the strongest innocent diuretique, I have met with." Notably, the discovery of this diuretic salt (which is identifiable as potassium chloride) and its method of preparation given in "Of Poisons"—which Boyle eventually published in Usefulnesse—are clearly recorded first in one of Starkey's notebooks.[73] Thus even "Boyle's" correction of the corrected opium is really Starkey's.

Just as Starkey's direct influence on Boyle's "Essay on Poisons" is undeniable in terms of practical processes, much of the theoretical foundation of the document likewise devolves from Starkey as well. The idea of using poisons in medicine is a fairly common part of seventeenth-century iatrochemistry, being prominent, for example, in the writings of Basil Valentine (whom Boyle cites in connection with two antimonial processes). Nonetheless, the date of composition of this essay and the particular theories it employs ally it closely with Starkey's Natures Explication and Pyrotechny Asserted. Starkey had completed the former and prepared a draft of the lat-

71. Cf. ibid. and BP, vol. 25, p. 344.

72. Boyle, "Poisons," in Works, 13:255–56, and BP, vol. 25, p. 341. This entry is marked "Laudanum St[arkeii]," which was later extended by "& Mor[iaeni]." The addition of Moriaen's name may mean that Boyle received some indication later that Moriaen was using a similar process, but since Moriaen was in direct contact with Starkey, it is possible that Starkey's process came again to Boyle through the Amsterdam chymist. In any event, the attribution to Starkey of the use of salt of tartar for correcting the toxic side effects of opium is unquestionable.

73. Notebooks and Correspondence, document 10; Sloane 3711, fol. 4v; Boyle, Usefulnesse, 514.

ter by November 1657, around the time Boyle was preparing the "Essay on Poisons."

The main thesis of Starkey's two books is that the poisonous nature of pharmacological simples hides great virtues, and accordingly Starkey's "grand design" for medicine involved the preparation of even toxic materials into medicinally useful products by "glorifying" their Sulphurs to free them of "virulent and alien" toxic properties.[74] *Pyrotechny* assails the Galenists' use of "churlish Vegetals" and points out how the "impression of the venome" must first be blotted out by digestion and alkali. Here Starkey notes that even arsenic "loseth its venom" by means of fusion with the alkali in saltpeter, a process we have seen already in Boyle's "Essay of Poisons."[75] *Natures Explication* likewise spends much time on the issue of correcting poisons—including some of the same substances treated by Boyle—and devotes a whole chapter to various species of poisons.[76] Starkey's study and classification of poisons was a long-term interest, for in a conversation on 20 February 1662 Starkey told John Ward that poisons were "2fold"; that is, "venenum corrosivum such as in vitriol which is lost in much liquor, and Venenum fermentale as Arsenici 2 drachms of which retains its poisonous nature put it in what vehicle you please." Apparently, Starkey divided poisons into those that act mechanically by corroding the body, like oil of vitriol (sulphuric acid), and whose toxic effects are eliminated by dilution, and "chemical" poisons that operate by a "fermental" action on the body, like arsenic, of which a toxic dose remains toxic even if diluted.[77] Given Starkey's interests, expertise, and publications, it seems quite plausible that the intent and theoretical background—besides the more obvious case of the specific processes—of Boyle's essay came from Starkey through direct contact, correspondence, or perhaps book drafts that Starkey may have shared with Boyle.

As a final example, note that in *Natures Explication* Starkey deals briefly with animal venoms and claims that some do not really fall into the class of poisons. Rather, the poisons from "the biting of Serpents, the biting of mad dogs, &c. are remote from this our purpose, such poysons being only in the power of that angry beast that inflicts it . . . So that in vain should we get the teeth of dead Serpents . . . all the virulency being then extinct."[78] The

74. See chapter 3 and Starkey, *Natures Explication*, 281 ff.
75. Starkey, *Pyrotechny Asserted*, 100–102.
76. Starkey, *Natures Explication*, 92–106.
77. John Ward Diaries, Folger Shakespeare Library, V.a.292, fol. 11r. Starkey makes a similar claim about oil of vitriol and other acids in *Natures Explication*, 97–98.
78. Starkey, *Natures Explication*, 93–94.

idea that the poison of vipers (and mad dogs) arises only from the anger of the animal became a topic of controversy in the seventeenth century. While this belief was famously disseminated by Moyses Charas in 1669, and refuted by Francesco Redi, its early popularizer and possible originator was Van Helmont. Van Helmont used this claim in the context of his study of the causes of plague in order to exemplify the power of the imagination to impress real effects upon matter.[79] Starkey's mention of this theory in *Natures Explication* is an early repetition of this Helmontian notion. Thus it is significant that Boyle makes the same claim about vipers and mad dogs in the "Essay on Poisons," and for the same purpose (to exclude vipers from the category of poisonous animals). As it turns out, Boyle's assertion was of sufficient novelty and interest that Henry Oldenburg singled out this passage for special comment in a letter to Boyle.[80]

Boyle's "Essay on Poisons" employs Helmontian notions implicitly and explicitly throughout. Boyle mentions Van Helmont by name over a dozen times in the transcript left by Oldenburg; indeed, the whole essay is a thoroughly Helmontian piece.[81] While we have argued that in the early 1650s Boyle's Helmontianism was a product of Starkey's tutelage, one might be tempted to argue that now in the late 1650s Boyle had bypassed Starkey and gone directly to Van Helmont for his inspiration. But this conjecture, as likely as it might seem, is weakened not only by the fact that so many specific processes that come directly from Starkey retain the same Helmontian context in which Starkey developed them, but also by the fact that the choice of key passages from Van Helmont found in the "Essay" parallels Starkey's own use of Van Helmont. Very specific sections out of Van Helmont's voluminous writings, in particular from the *Tumulus pestis,* the

79. Moyses Charas, *Nouvelles experiences sur la vipere* (Paris, 1669); on the attribution of this notion to Van Helmont, see Paul Ammann as cited in Lynn Thorndike, *A History of Magic and Experimental Science* (New York: Columbia University Press, 1958), 7:235. The original locus in Van Helmont is *Tumulus pestis,* in *Opuscula,* pp. 46–47. On Van Helmont's notions of the imagination's action in sickness and health, see Guido Giglioni, *Immaginazione e malattia: Saggio su Jan Baptiste Van Helmont* (Milan: Francoangeli, 2000).

80. Oldenburg to Boyle, 29 August 1657, in Hall and Hall, *Correspondence of Oldenburg,* 1:133. Boyle writes that "it may be justly doubted, whether [vipers] be to be reckoned among poisonous creatures . . . for, it may be supposed, that the venom of vipers consists chiefly in the rage and fury, wherewith they bite, and not in any part of their body." It may be worth remarking that when Boyle used this section of the "Essay on Poisons" in *Usefulnesse* (in *Works,* 3:324) he added to it a comment about how some poisons fatal to man are not so to dogs and other animals—this is the very same remark made by Starkey immediately following his own claim about the anger of vipers in *Natures Explication* (94).

81. The Helmontian character of this essay has been briefly remarked upon by Clericuzio, "From van Helmont to Boyle," 315.

"Pharmacopolium ac dispensatorium modernum," and the "Blas humanum," which are freely alluded to by Boyle are also among those most important to Starkey. Additionally, the interpretations that Boyle gives to obscure Helmontian passages are identical to Starkey's. We have already seen one example in the way Boyle uses Starkey's practical interpretation of Van Helmont's hints regarding the use of spirit of wine in preparing arsenic. But another example of Starkey's continued mediation between Van Helmont and Boyle shows up again in Boyle's explication of a curious line from Van Helmont's "Blas humanum." In a singularly obscure passage following a discussion of salt of tartar, Van Helmont writes that "Quocirca etiam, est sua inter alkalia, respublica & adulteratores monetae, plurimum circa sal tartari laborant, spreto alkali salis petrae."[82] The English translation published in 1662 renders this directly as "wherefore also there is among *Alcalies,* their own Common-wealth, and the Adulterators of money do labour very much about Salt of *Tartar,* the Alcali of *Salt-peter* being contemned," but this does remarkably little to explain what Van Helmont is trying to say.[83] Starkey, on the other hand, gives his own interpretation of the text in *Natures Explication.* There he explains that "Alcaly of Tartar hath deserved and gotten the name of *Respub. Alcalium;* since whatever vertue is to be found in any Alcaly, may be found in and demonstrated from the Alcali of Tartar."[84] Thus Starkey seems to be reading the phrase "est sua inter alcalia, respublica" not in the more straightforward manner (like the Chandler translation) as "there is among alkalies their own republic" but rather as "[salt of tartar] is among alkalies their republic." Starkey's notebooks show the same interpretation of Van Helmont's obscure utterance.[85] It is therefore significant that Starkey's idiosyncratic interpretation is *exactly* the one given by Boyle in his own exposition of the passage in the "Essay on Poisons." Like Starkey, Boyle reads the passage to mean that the salt of tartar *is* the "respublica Alcalium," and that Van Helmont calls it such because it is "eminently endowed with Alcalizate properties, and thereby enabled to perform, whatever may justly be expected from lixiviate

82. Van Helmont, *Ortus,* "Blas humanum," no. 40, p. 188.
83. Translation by John Chandler, published as *Oriatrike, or Physicke Refined* (London, 1662), 184. It should also be noted that Knorr von Rosenroth's German translation of the *Ortus* gives a clarification of this phrase, and certainly not a reading akin to Starkey's; see Van Helmont, *Aufgang der Artzney-Kunst,* 2 vols. (Sulzbach, 1683), 1:238.
84. Starkey, *Natures Explication,* 300.
85. *Notebooks and Correspondence,* document 10; Sloane 3711, fol. 4r: "teste Van Helmontio tartari sal est alcalium respublica quod prae Alcali Salispetrae ab adulteratoribus monetae desideratur." This entry is dated 25 January 1656.

salts, as such."[86] Starkey's identification of the salt of tartar with the "republic of alkalies" even shows up in a set of Boyle's own laboratory records dating from the late 1650s where Boyle uses "the republican Salt" as a *Deckname* for salt of tartar.[87]

It appears, then, that Boyle adopted not only Starkey's processes but also his predilections for Helmontian theory and his interpretations of Helmontian obscurities. By the time Boyle wrote the "Essay on Poisons," he was a thoroughgoing Helmontian, and he seems to have owed this Helmontianism predominantly to Starkey's tutelage. Starkey's clear importance to the "Essay on Poisons" makes it all the more remarkable that it nowhere mentions the New England chymist explicitly, even though other living persons such as Digby and Coxe are freely cited by name. Only at the end of the section on vegetable poisons, when Boyle mentions "Alcalisate Correctors" does he finally allude to "an Ingenious physitian, who having taken great paines about the improving of poysons, do's now, by remedies prepared of them, get the greatest part of his credit and substance." This "Ingenious physitian" is unquestionably Starkey, who at this time was profiting from his "Corrector" or "Pill."

Boyle never published his "Essay on Poisons." He did, however, transfer the short section on angry vipers to *Usefulnesse*, where he noted that this section was taken from a "Treatise, I am like, for certain reasons, to suppress."[88] Boyle did not specify what these reasons were, but now that we have shown this essay's debt to Starkey and its similarity to Starkey's own publications, the possibility emerges that Boyle's suppression of the essay stemmed from a concern that its lack of originality would have been apparent. This is especially true if, as is very probable, Boyle wrote the essay before Starkey's *Pyrotechny Asserted* had cleared the press in 1658. We know that Starkey had written a work called *The Art of Pyrotechny explained and confirmed* by late 1656, and it is possible that he shared drafts of it with Boyle, its dedicatee.[89] Assuming that this was essentially the same work as *Pyrotechny Asserted*, the period between the composition and publication of *Pyrotechny Asserted* fits well with the best estimates of when Boyle wrote his "Essay on Poisons." Thus the appearance of Starkey's major chymical publications would have precluded the subsequent printing of Boyle's essay.

86. Boyle, "Poisons," in *Works*, 13:254.

87. BP, vol. 25, p. 88; note also Boyle's coining of the analogous term "respublica acidorum" on p. 81.

88. Boyle, *Usefulnesse*, in *Works*, 3:324.

89. Starkey's "Prefatory Epistle" to *Natures Explication and Helmont's Vindication* is dated 20 November 1656, and notes that *The Art of Pyrotechny explained and confirmed* had been written by that time.

BOYLE'S LATER RELATIONS WITH STARKEY

There is some evidence that Boyle's interactions with Starkey may possibly have continued into the 1660s, up until Starkey's premature death in the Great Plague of 1665. This evidence comes from some two hundred pages of transcriptions made from Boyle's papers in 1692, shortly after his death. These transcriptions were made for John Locke by one of his amanuenses during Locke's abortive term as an executor of Boyle's chymical papers. The transcriptions preserve material that has subsequently disappeared from the Boyle archive, and since very nearly none of the original documents transcribed currently exist in the Boyle archive, Locke's transcriptions underscore how extensive the losses from Boyle's original cache of papers really are.[90]

Starkey shows up in these transcriptions several times. There is, for example, a complete transcription of his 1651 letter to Boyle; at present only part of Starkey's original letter survives in the Royal Society's Boyle archive, and so Locke's transcription is our only surviving source for the last third of the original English text.[91] In addition there is a series of processes that includes the full account, with all the operational details, of Starkey's successful volatilization of alkalies using the medium of air—clear evidence that Starkey shared this prized process, perfected only in 1656, with Boyle.[92] There is also a transcription of the opium correction process that appeared in Boyle's 1652 collection and several other items that may be Starkey's.[93] But in addition there are some intriguing references to someone with the *Deckname* "Americus" or "the American." This same person is also concealed under the name "Vesputius." Now Boyle is known to have used codes to hide the identity of ingredients in chrysopoetic processes (which foiled Locke's attempts to organize Boyle's papers); he also used them to hide the identity of various informants. Indeed, a curious note among the Boyle papers describes his use of "fain'd Names" in a "Collection of Receipts, Processes, &c"; this note was intended to accompany a decipherment list that is now lost.[94] Other "fain'd Names" to be found among the

---

90. Bodleian Library, Locke MS C44. On losses to Boyle's papers, see Hunter and Principe, "The Lost Papers of Robert Boyle."

91. Locke MS C44, pp. 142–53; note that most of the original manuscript is unpaginated; the page numbers given are ours. For the letter, see *Notebooks and Correspondence*, document 3; also published in Boyle, *Correspondence*, 1:90–103.

92. Locke MS C44, pp. 32–33.

93. Ibid., p. 195.

94. It is clear from the entries on Locke MS C44, p. 36, that "Americus" and "Vesputius" are the same person. On Boyle's use of coded names, see BP, vol. 36, fol. 16, and Hunter, "The Reluctant Philanthropist: Robert Boyle and the 'Communication of Secrets and Receits in Physick,'" in *Robert Boyle (1627–1691): Scrupulosity and Science* (Woodbridge: Boydell, 2000),

Boyle Papers (and the Locke MS) include "Parnassus," "Marcellus," and "Morgana."

Given Starkey's origins, it seems plausible, though by no means sure, that he could be the person disguised as Americus/Vesputius or the American. The information supplied under this name is chymical and deals with the kind of materials Starkey pursued, and one such entry is juxtaposed with a recipe for Sophic Mercury. In another place, Boyle begins an entry with "the American sayes" and refers to the same sort of operation (a calcination using "Helios," possibly meaning sunlight) linked previously to the name Americus.[95] Yet another entry begins with "the American says" and concludes with the statement "Vade Mecum, the marrow & one piece more are spurious but the Ruby the Version & the preface are genuine."[96] These titles refer to various Philalethes treatises, and it is conceivable that this sentence represents Starkey's verbal repudiation of some Philalethes tracts as "false"; Starkey could be rejecting some of what were secretly his own compositions as his experimentation progressed. The *Marrow*, for example, published in 1654–55, insists upon the use of a copper-antimony alloy for making the Sophic Mercury, a procedure that (as we have seen) Starkey rejected around 1655. Despite the temptation to identify Starkey with Boyle's American, there is nonetheless a reference elsewhere in the Locke MS to "Am." where it is quite clear that this "Am." is Philalethes, not Starkey (since there is reference to a page in Philalethes' *Secrets Reveal'd*).[97] Yet in some other instances Americus does seem fairly clearly to be a living person, as when "Americus says to Parn[assus]," referring by a coded name to another of Boyle informants (indeed, one who brought Boyle specifically chrysopoetic information).[98]

218–19. Boyle's "Collection" referred to here is now lost, but the Locke transcription was in part taken from the lost original. For Boyle's use of codes in transmutational receipts, see Principe, "Robert Boyle's Alchemical Secrecy: Codes, Ciphers, and Concealments," *Ambix* 39 (1992): 63–74.

95. It is of course possible that the obvious sense of "Helios" hides a different meaning, since the sun is often used as a *Deckname* for gold, and sometimes, by virtue of its fiery heat, even for sulphur (be it common sulphur or a hypothetical mineral or metallic Sulphur).

96. Locke MS C44, pp. 69 and 95; "version" is probably a scribal error for "vision," i.e., the Philalethes tract entitled *An Exposition upon Sir George Ripley's Vision*. The citation on p. 69 reads "The American says to Parn[assus]," implying that whoever Parnassus was (all the items attributed to him deal strictly with metallic transmutation), he and Americus were in communication.

97. This usage of "Am[erican]" occurs also in an interesting document transcribed from Boyle's papers (though probably not written by Boyle) at Locke MS C44, pp. 110–14. There is also the case elsewhere of Boyle's reading notes from one "Parrac" (BP, vol. 26, fols. 33–36v, on fol. 35v), where the term "the American" is used for the author of a text rather than for a personal correspondent.

98. Locke MS C44, pp. 69, 95.

A key point to note here is that if "Americus" in any of these cases should in fact refer to Starkey, then the colonial chymist's relationship with Boyle would have endured much longer than has previously been known, for two of the entries are dated to 1664, the last full year of Starkey's short life and fourteen years after Boyle and Starkey first met.[99] At any rate, Starkey's processes continued to be of importance to Boyle late into his life, for they were included in the now-lost "Collection," made probably in the 1680s, that the Locke MS partially preserves.

From all the foregoing material it is clear that Starkey exercised enormous influence on Boyle in the 1650s, and also that their relationship may possibly have even continued into the mid-1660s. But of course Boyle's contacts were not restricted to Starkey. Therefore, we must examine three other chymically inclined figures with whom Boyle had contact in the 1650s. We will show that these three men did not have the degree of influence on Boyle's early chymistry that Starkey did, although they are still of interest for deepening our knowledge of Boyle's early chymistry and his relations with Hartlibians, as well as for better delineating the internal workings of the Hartlib circle. By looking at the interactions among these figures we will be able to construct a more comprehensive—if not exhaustive—depiction of practical chymistry in the Hartlib circle. We begin then with Benjamin Worsley, one of Boyle's earliest contacts with an interest in chymistry.

THE ROLE OF BENJAMIN WORSLEY IN BOYLE'S "CHYMICAL EDUCATION"

Earlier in this chapter we briefly considered Boyle's earliest "scientific" notebook (or work diary), the "Memorialls Philosophicall" beginning 1 January 1650. This document contains the first concrete, dated references to two figures who have been considered key to the initiation of Boyle's chymistry, and both of whom have been associated with the "Invisible College": "Dr. Boate" and "Mr. B. Worsley."[100] The first refers either to Gerard Boate (1604–50) or to Arnold Boate (1606–53). The Boates were two Dutch brothers; both were physicians and were closely associated with Boyle and his family in the 1640s.[101] The recipes attributed to Dr. Boate are purely

99. Ibid., p. 36, "April 2 64"; p. 40, "Feb 1. 64." The fact that these entries are dated strengthens the supposition they in particular record information verbally acquired on those dates, rather than Boyle's gleanings from texts.

100. BP, vol. 28, pp. 309–11.

101. Charles Webster, *The Great Instauration* (London: Duckworth; New York: Holmes and Meier, 1975), 64–65.

medical in character, and there is little evidence that the Boates were involved in chymistry; there is virtually none that they were teaching the discipline to Boyle.[102] The situation is very different with Benjamin Worsley, however, and he must be treated at some length.[103] Charles Webster and R. E. W. Maddison have argued for Boyle's early acquisition of chymical knowledge from Worsley on the basis of two letters apparently written by Boyle in late 1646 and early 1647.[104] The first of these letters requests general information about chymistry from its unnamed recipient. In addition, Boyle laments the delay in receiving some "*Vulcanian* implements" or laboratory apparatus that had been sent to him by wagon. The first concrete notice that we have about information given to Boyle by Worsley is found in the 1650 "Memorialls Philosophicall." But these recipes concern surgery and brewing, and are conspicuously lacking in the technical language and theory of seventeenth-century chymistry.[105] Indeed, the arguments for Worsley's impact on the young Boyle were made against the backdrop of the belief that Boyle's natural philosophical interests were already well developed in the 1640s—a view now criticized.[106] However much Worsley might possibly have been encouraging Boyle toward the study of chymistry, this seems to have had little practical effect beyond the failed attempt on Boyle's part to acquire a furnace and apparatus, the "Vulcanian implements" alluded to in the letter, for his manor at Stalbridge. It is worth noting that when the furnace broke in transit in 1647, Boyle made no apparent effort to repair the loss, but instead continued writing *Amorous Controversies* and moral epistles until mid-1649, when his experimenting did in fact commence.

102. Hartlib's *Ephemerides* for 1649 through 1653 contain a number of references to Gerard and Arnold Boate, but without clear connections to chymistry. The only two entries of interest are a mention by Arnold Boate of an "everlasting mine" of saltpeter in a copy extract dated 29 March 1653 (HP 39/1/6B), and a note by Hartlib in 1650 that an unspecified book by Etienne de Clave "concerning Natural Philosophy" was sent by one brother to the other and that this book is "highly commended by them and Mr Boyle" (HP 28/1/56B).

103. T. L. Leng of the University of Sheffield is currently writing a dissertation on Benjamin Worsley that will cover many aspects of Worsley's life and thought that we must pass over here.

104. Webster, *Great Instauration*, 59–60; R. E. W. Maddison, "Studies in the Life of Robert Boyle, FRS, part 6: The Stalbridge Period, 1645–1655, and the Invisible College," *Notes and Records of the Royal Society* 18 (1963): 111. The letters are found in Boyle, *Correspondence*, 1:42–44 and 47–49.

105. BP, vol. 28, p. 309; Clericuzio, "Robert Boyle and the English Helmontians," 193, suggests that these recipes contain Helmontian concepts because of the reference to ferments and spirits, but the ferments mentioned here are expressly for making beer, and are not related to Van Helmont's notion of a formative ferment.

106. Hunter, "How Boyle Became a Scientist," 83.

It is, however, worthwhile to explore Benjamin Worsley's chymical endeavors in greater detail, for despite the rather disappointing character of the 1650 recipes, we know from other sources that Worsley entertained high hopes indeed for chymistry. Besides clarifying Worsley's own interests, this will allow for an elucidation of Worsley's possible interactions with Boyle and then a comparison of these influences with Starkey's and those from other associates of Boyle. In spite of Worsley's interests in chymistry, we will show that his competence in metallic and pharmaceutical chymistry was limited and that his allegiances belonged primarily to a branch of the discipline that was antithetical to the highly technical operations characterizing Starkey's work and that the young American imparted to Boyle.

Worsley's long-term sympathies lay with those who believed that the first matter of metals, and hence of the Philosophers' Stone, was not to be found in common mercury or any other metal, but in some more "universal" principle. This idea was intimately linked to Worsley's belief that the fundamental principle of life and growth—whether animal, vegetable, or mineral—was to be found in saltpeter. The main proponents of this theory in the late sixteenth and early seventeenth centuries, such as Michael Sendivogius and Clovis Hesteau, Sieur de Nuysement, claimed that the genuine starting point of the Philosophers' Stone was a *sal nitrum* or philosophical niter that had some—but not all—of the properties of normal saltpeter. Partly because this "philosophical" niter was the universal matter from which the metals and the Philosophers' Stone were made, it was called by Sendivogius and others "philosophical mercury."[107] Although this *sal nitrum* existed within saltpeter, one could not necessarily extract it therefrom; indeed, one finds chymists recommending the most varied substances, such as dung, urine, May-dew, and humus as the basic matter of the *lapis philosophorum*. As Boyle himself stated in "Of the Study of the Book of Nature," the philosophical mercury (or *sal nitrum*) is to be found in the "Juice of the Clouds." Despite a lapse from 1648 through 1651, Worsley was a committed adherent to the Sendivogian *sal nitrum* theory. As noted above, Boyle's reference to "Juice of the Clouds" probably reflects the influence of Worsley's notions.

107. Porto, "Michael Sendivogius on Nitre," esp. 7–10; Newman, *Gehennical Fire*, 87–90; W. Hubicki, "Michael Sendivogius' Nitre Theory: Its Origins and Significance for the History of Chemistry," *Actes du Xe Congrès International d'Histoire des Sciences* (Paris: Hermann, 1964), 2:829–33. See also Zbigniew Szydlo, "The Alchemy of Michael Sendivogius: His Central Nitre Theory," *Ambix* 40 (1993): 129–46; Szydlo, "The Influence of the Central Nitre Theory of Michael Sendivogius on the Chemical Philosophy of the 17th Century," *Ambix* 43 (1996): 80–97; Szydlo, *Water Which Does Not Wet Hands* (Warsaw: Polish Academy of Sciences, 1994). Szydlo's scholarship is uneven and has been called into question.

WORSLEY, SALTPETER, AND THE *SAL NITRUM*

Thanks to detailed studies by Charles Webster and John Young, we now have a fairly solid picture of Worsley's early life and affairs.[108] Born around 1618, he attended Trinity College, Dublin, in the early 1640s, but there is no evidence that he obtained a degree. He seems to have met Samuel Hartlib in 1645 and soon became a regular correspondent. As Webster points out, Worsley may have received his first interest in saltpeter from another friend of Hartlib's, the impoverished Kentish knight Sir Cheney Culpeper, who was an avid reader of Sendivogius, Nuysement, and another *sal nitrum* theorist, Blaise de Vigenère.[109] By March 1646, Worsley had submitted a project for a new method of producing saltpeter to a committee of London aldermen; it would soon be discussed by the House of Lords.[110] But what was Worsley's real part in this scheme? His proposal, entitled *Certaine Propositions in the behalfe of the Kingdome concerning Sallt-Peter,* contains no description of a process, but only promises that his method will obviate the need for having saltpeter dug from the grounds around basements and animal yards.[111] In his study of Worsley's proposal, Webster turns to another document for concrete information about the process—the undated *Acte for a new way of making Salt Peeter,* which is identified in the manuscript itself as a "Draught of an Act of Parliament for the making of a corporation of Saltpeeter-makers in England Ireland and Dominion of Wales to Sir William Luckin Mr Joyner and others."[112] Worsley's name appears nowhere on this document, and if we dig deeper in the Hartlib papers, it becomes clear that Worsley was negotiating with the "mistress" of "Joyner" to acquire his process for a minimum of two hundred pounds up front and fifty pounds per annum. Several cautionary letters from Culpeper discuss the terms of the deal, and the legal propositions suggested by "Joyner's Mistresse" survive.[113]

---

108. Charles Webster, "Benjamin Worsley: Engineering for Universal Reform," in *Samuel Hartlib and Universal Reformation,* ed. Mark Greengrass et al. (Cambridge: Cambridge University Press, 1994), 213–15; Young, *Faith, Medical Alchemy,* passim. See also T. C. Barnard, *Cromwellian Ireland: English Government and Reform in Ireland 1649–1660* (Oxford: Oxford University Press, 1975), and G. E. Aylmer, *The State's Servants: The Civil Service of the English Republic 1649–1660* (London: Routledge, 1973), 270–72. Additions to this biography will no doubt be forthcoming from T. L. Leng, see above, note 103.

109. M. J. Braddick and M. Greengrass, *The Letters of Sir Cheney Culpeper (1641–1657),* Camden Miscellany 33 (Cambridge: Cambridge University Press, 1996), 135–37, 332–34, 363–67, and passim.

110. *Journals of the House of Lords,* 8:573–74; Webster, "Benjamin Worsley," 216.

111. HP 71/11/14A–15A.

112. HP 71/11/2A.

113. HP 71/11/13B. For Culpeper's comments, see Culpeper to Hartlib, 17 February 1646, HP 13/127A–128B, and Culpeper to Hartlib, 4 March 1646, HP 13/136B–137A. See also Braddick and Greengrass, *Letters,* 251, where they identify "Joyner" as Francis Joyner.

It thus appears, to put it bluntly, that Worsley recognized that he had *not* discovered a workable process of his own for making saltpeter and was trying to buy Joyner's. But before we can draw further inferences, we must deal with other evidence for Worsley's knowledge of saltpeter in 1646. Webster has argued that around the time of the 1646 proposal, Worsley wrote his *De nitro theses quaedam*, an undated tract found in the Hartlib papers. In Webster's view, this short treatise provided an important element of "scientific reasoning" that gave it an advantage over similar schemes presented by others.[114] If the dating of Worsley's composition of *De nitro* to 1646 were correct, it would indeed have important ramifications for the state of his chymical knowledge in the mid-1640s at the time when his proposals were under parliamentary consideration, for the treatise does reveal an unusual awareness of the chymistry of niter. But in fact, as we will show, *De nitro* was almost certainly composed during or after 1654, after Worsley had made his extended trip to the Netherlands, where he had encountered some of the most committed chymists of the day, including, as we mentioned above, Johann Rudolph Glauber.

Worsley's treatise *De nitro* belongs to a group of documents in which he laid out his plan to provide saltpeter to the army of the Interregnum. The first of these texts, in all likelihood, is the one entitled *Observations about Saltpeter*. Attached to it is a brief notice in German by Johann Moriaen, dated 18 May 1653, and apparently referring to these *Observations* as "Worsl[eys] communication."[115] Here Worsley argues that he has found a "ferment" that will act upon a "matter," converting it into saltpeter cheaply and rapidly. This section is followed by "Animadversions" that announce that the aforementioned ferment is simply rich earth containing unrefined saltpeter, and the matter is "tops of grasse, when ready to bee cut and fullest of seed & of that nitrous Universal spirit." Relying on the chymical theory of *sal nitrum*, Worsley has determined reasonably enough that grass that has gone to seed is a cheap source of the concentrated nitrous spirit. One should therefore take seed-bearing grass, mix it with nitrous earth as a "ferment," and place the mixture in clay-lined pits, along with lime and wood ashes, to be exposed to the heavens.

The next significant reference to Worsley's scheme is found in an extract of a letter from him to Hartlib, dated 16 May 1654, and sent from Dublin. He refers to a proposal that a captain made to the auditor general of the army for producing saltpeter out of excrement such as dung and urine. The

114. Webster, *Great Instauration*, 378–79.

115. HP 39/1/11A–12B; Hartlib's note on the *Observations* reads "Ex lit. Mor. 18 Mai 1653. Mr Worsl. communication von [sal]petrae ist deutlich genug denen die mit Salpeter umbzugehen wissen."

captain demanded one thousand pounds for his process, which Worsley rejected as containing nothing new. Worsley then relates that he offered his own process for making saltpeter to the auditor general and his commissioner without "demanding mechanically any summe of them" except his own costs. The army officials accepted this proposal, but Worsley was unable to act on it until the date of his letter to Hartlib. As he says,

> I am now at length actually undertaking [the project], my excesse of businesse hitherto having hindered it. And by this meanes, if the Lord please to blesse and assiste me in it, I hope not only to give a very good account of Peter and the nature of it, but something also of vegetation. And now I shall tell you a litle further, what my thoughts are on this subject, and what I have within my selfe instituted.[116]

This important passage reveals that Worsley was about to compose a treatise, or "good account," of saltpeter and vegetation if God should assist him. Although he then proceeds to recapitulate the doctrines found in his 1653 *Observations*, it is clear that the "good account" is something that he plans to do in the near future; it is not the letter of 16 May itself. We thus propose that the "good account" of niter and vegetation is the treatise *De nitro theses quaedam,* and that Worsley had not yet composed it.

In Worsley's *De nitro,* the theory and practice outlined above are fleshed out more fully in the form of numbered theses. Much of the tract is intended to prove that decaying organic matter produces saltpeter and that saltpeter acts as a fertilizer to plants. This was of course common knowledge by the 1650s, having been an integral part of the saltpeter discussion among chymical authors of the late sixteenth and early seventeenth centuries. But then, in his eighth thesis, Worsley introduces a new element to the discussion, and one that can be firmly located in the mid–seventeenth century:

> It is certaine that Salt-Peter hath Parts Volatill, inflammable and spirituous and parts fixed exceedingly causticke fiery and wonderfully detersive. It is found by 1000s of Experiments that all Plants likewise containe in them a Salt, which Salt hath parts inflammable Volatil and spirituous, which is the Subject of fire and combustion and parts that are fixed Caustic and detersive, which is that part which lieth in the ashes. This is plaine in every one, and will bee demonstrated in Sugar itselfe.[117]

116. HP 66/15/1A–4B.
117. HP 39/1/16A–20B.

What Worsley is referring to here is the famous discovery by Johann Rudolph Glauber that saltpeter can be "fixed" by burning it with charcoal. Later in his *De nitro* he makes this even more clear, saying that "it [niter] is easily fixed."[118] In terms of modern chemistry, potassium nitrate burns vividly with charcoal to produce potassium carbonate (K_2CO_3), a "fixed Caustic and detersive" salt. But Glauber also knew—as did every chymist of the seventeenth century—that ordinary saltpeter, unlike the "fixed niter," is flammable, since it is, after all, an essential ingredient of gunpowder. It had also been known for centuries that a "spirituous" substance could be obtained from niter by distilling it with fuller's earth, for this was the standard way of making *spirit of niter,* our nitric acid. Hence Glauber concluded that saltpeter was a twofold substance, a "hermaphroditic salt," containing both a volatile acid substance that he called volatile niter (nitric acid)and a solid caustic one that he called fixed niter (potassium carbonate). Because these two components of niter could between them dissolve a host of substances (nitric acid being a powerful corrosive and potassium carbonate a saponifying base), Glauber thought that he had discovered Van Helmont's alkahest, the universal dissolvent.[119] But Glauber also went on to recombine his volatile and fixed niter and recognized that the product was once again his twofold salt—simple niter—reconstituted. This was the same experiment upon which Robert Boyle built his famous "Essay on Nitre," published in 1661, to which we shall return.

In terms very similar to those used by Glauber, Worsley's *De nitro* refers to saltpeter as being "of a double Nature and different Parts, so it is constituted of a double Matter of a fixt and common salt mixed with an Aetherial heat and spirit."[120] But Worsley's eighth thesis also maintains that a salt like saltpeter can be found in plants, which is patently untrue if one thinks of niter only in terms of potassium nitrate. What Worsley certainly has in mind, however, is a philosophical *sal nitrum,* which reveals its flammable, volatile side in the burning of the plant and leaves behind its solid residue of "fixed niter" in the ashes. This fixed salt appears after the plant's ashes have been leached—it is mainly potassium carbonate obtained by lixiviating plant ashes. Worsley's claim that a sort of saltpeter is thus revealed in plants is once again pure Glauberian doctrine. Indeed, Glauber's published writings even use the example of sugar, to which Worsley rather cryptically al-

118. HP 39/1/19B.
119. Kathleen Ahonen, "Johann Rudolph Glauber: A Study of Animism in Seventeenth-Century Chemistry" (Ph.D. diss., University of Michigan, 1972), 107 n. 59.
120. HP 39/1/19B.

ludes, and contain arguments strikingly similar to Worsley's for the artificial production of saltpeter from grass and other plant matter.[121]

There is every reason to believe that Worsley knew Glauber's doctrine of fixed and volatile niter by the mid-1650s. First, Worsley had made an extended trip to the Netherlands in 1648–49, as a result of Hartlib's inquiries and urging, precisely in order to learn from Glauber; this mission has been richly documented by the recent work of John Young. Although Worsley was largely frustrated in these attempts, while living with Glauber's close friend Johann Moriaen, he got to know other associates of the German chymist, such as Johann Sibertus Kuffler. As is clear from Worsley's and Moriaen's letters, the two carried out agricultural experiments, exactly the sort of thing that would have engendered discussion about the *sal nitrum*. Indeed, Worsley tells Hartlib in a letter of 1649 that he will soon "write largely" about saltpeter. Furthermore, Moriaen wrote Worsley on 29 November 1651 that he was negotiating with Glauber to buy the latter's secret for making the alkahest. As we mentioned above, Glauber believed that he had located the alkahest in niter, so this is further evidence that Worsley may have learned of Glauber's views on the double nature of saltpeter through Moriaen.[122] But Young has also documented that copies of Glauber's works were being sent to Hartlib from the Netherlands as early as 1646. Hartlib did his best to have these translated, and discussed the works with his English associates.[123] It is not difficult, then, to see where and how Worsley could have acquired a general knowledge of Glauber's chymistry. But it is unlikely that he knew of the Glauberian doctrine of niter as a "hermaphroditic" or composite salt before his trip to Amsterdam in 1648–49, for Glauber had not yet published his findings about fixed niter, which only began to make their way into print in 1653, nor did Worsley have access to Glauber's discovery in unpublished form in 1646, since his association with Glauber's Dutch circle did not commence until 1648.[124]

121. Johann Rudolph Glauber, *Continuatio operum chymicorum* (Frankfurt, 1659), 2:370–91 (pt. 1, chap. 3, of *Dess Teutschlands Wohlfahrt*).

122. Young, *Faith, Medical Alchemy*, 217–20. Worsley to Hartlib, 22 June 1649, HP 26/33/1A–3B; see 2B. For Glauber's offer to sell Moriaen his secret of the alkahest see the extract of a letter from Moriaen to Worsley, 29 November 1651, HP 63/14/13A.

123. Young, *Faith, Medical Alchemy*, 198–99. See Heinrich Appelius to Hartlib, 16 October 1646, HP 45/1/28A.

124. Glauber began printing his books in 1646, with the *De auri tinctura* of that year and the *Furni novi philosophici* of 1646–49; neither of these works contains the doctrine of the fixed and volatile niter. The doctrine of *sal nitrum* as a hermaphroditic salt is spelled out in the *Miraculum mundi*, whose first part was not published until 1653. See Ahonen, "Glauber," vi–x, 102–3, 107, 115. See also Ahonen's entry in the *Dictionary of Scientific Biography* (New York: Scribner's, 1972), s.v. "Glauber, Johann Rudolph."

Thus it is very unlikely that Worsley's *De nitro* could date from 1646, and the references to the fixed and volatile parts of niter in it point to a date of composition no earlier than the mid-1650s. This dating is corroborated by the fact that on 16 May 1654 Worsley admitted that he had not yet composed his "very good account" of saltpeter and vegetation. And we also know from Worsley's attempts to buy Joyner's process from his "mistress" in 1646 that he was not confident of his own knowledge about niter at the time of making his parliamentary petition.

All of this evidence combines to form a picture of Worsley before his trip to the Netherlands as an enthusiastic amateur with great hopes but little concrete knowledge of technical chymistry. The letters of Boyle from around the same time do not dispel that image. Can we say anything more concrete about Worsley's chymical knowledge prior to the composition of Boyle's 1650 "Memorialls Philosophicall"? Fortunately, Worsley engaged in copious correspondence during his extended visit to the Netherlands. Another close friend of Samuel Hartlib's, John Dury, was keenly interested in techniques of distillation. Dury's wife, Dorothy, was prepared to set up as a distiller of *aqua vitae* and aromatics, but neither she nor Dury had the skill to do so. Although Starkey would later be enlisted to help Dury and his wife, a number of letters from 1649 show that Worsley was initially their main source of information about distilling.[125] Worsley's letters exhibit considerable knowledge of distillation techniques as far as they pertain to alcoholic beverages, but there is little evidence in them of metallic or pharmaceutical chymistry (beyond the medicinal virtue to be had from cordials). Consistent with our new analysis and dating of *De nitro* is Worsley's mention of "Glauberus" on distillation, possibly a reference to the latter's *Furni novi philosophici* of 1646–49, but also perhaps based on oral communication.[126]

WORSLEY AND TRANSMUTATIONAL CHYMISTRY

Worsley's knowledge of the *aqua vitae* trade as exhibited in his 1649 letters conforms very nicely with the nature of the recipes that he imparted to Boyle in 1650, since aside from surgical and dental material and a few botanical medicaments, those concern only fermentation and brewing. Yet we do know that Worsley had originally hoped to learn more than the technology

125. For Mrs. Dury's interest, see Worsley to Hartlib, 22 June 1649, HP 26/33/1A–3B. For Starkey's role, see Newman, *Gehennical Fire*, 78–80.
126. Worsley to Dury, 27 July [1649?], HP 33/2/18A–19B; see 19B; and the copy extract from this letter, undated, HP 26/33/7A–8B. At 33/2/19B Worsley mentions "our metallicke busines," apparently a reference to his abortive venture with Glauber and his associates.

of distillation from his Dutch voyage. As Young has shown in fascinating detail, Worsley was involved in an extended metallurgical venture with Moriaen, Kuffler, and a goldsmith or refiner *(Aurifaber)* named Antony Grill. The original goal of this enterprise was the extraction of gold from tin slag, possibly according to a process bought from Glauber.[127] Numerous letters from Moriaen to Worsley, written after the latter's return to England in fall 1649, reveal that Worsley began conveying additional information to Moriaen about two further substances of metallurgical, indeed specifically transmutational import. The first of these was a *luna fixa* (fixed silver) having the specific gravity and resistance to corrosion of gold but lacking its color, and the second was a mysterious "Sophic Mercury" involving martial regulus of antimony. The first extant references to these products are found in Moriaen's letter to Worsley of 19 May 1651, where he asks for more information about "your mercury" and "your luna."[128] In subsequent letters to Worsley, Moriaen repeatedly makes the same request, even asking that Worsley send him a sample of the Sophic Mercury.[129] Finally, after making such inquiries repeatedly over a period of three months, Moriaen loses all patience and berates Worsley, saying that there can be no excuse for his failure to impart further information about the *luna fixa*. Henceforth, Moriaen says, he will urge Worsley no further, but will resign himself to the will of God.[130]

Worsley's failure to respond to Moriaen's repeated solicitations raises some interesting questions. Why did Worsley bring up the issue of the *luna fixa* and Sophic Mercury with Moriaen if he could not deliver them, or at least provide some information on their manufacture? An answer to this question emerges straightforwardly. George Starkey had immigrated to England in the fall of 1650, and Worsley had observed him "extract silver out of Antimony," as Hartlib relates, in late April or early May of 1651. Around that same time, Starkey had developed his process for making his Philosophical Mercury with martial regulus of antimony and silver, which he divulged to Boyle in his famous "Key," probably written around the time of Hartlib's entry. As Starkey himself pointed out in the letter to Boyle, Worsley was then urging him to join the venture with Moriaen, Kuffler, and Grill, but Starkey declined, saying that this would put him in the menial position of a "Millhorse," whereas he was a master rather than

127. Young, *Faith, Medical Alchemy*, 226–32, 256–57.
128. Moriaen to Worsley, 19 May 1651, HP 9/16/5A–B.
129. Moriaen to Worsley, 2 July 1651, HP 9/16/10A–B.
130. Moriaen to Worsley, 4 August 1651, HP 9/16/13A–B.

an amanuensis and in no need of partners.[131] In a letter that Starkey then wrote to Moriaen on 30 May 1651, the young chymist politely reiterated this position, saying that he had "no secrets for sale."[132] The evidence, then, points to the following explanation for Worsley's failure to deliver. Since Worsley's earliest mention of "his" *luna fixa* and Sophic Mercury occurs very shortly after he observed Starkey's chymical performances in late April or early May 1651, it is highly likely that the products Worsley was advertising were actually Starkey's. Nonetheless, none of Moriaen's letters ever link Starkey's name to "Worsley's" *luna fixa* and Sophic Mercury; instead, Moriaen refers to an anonymous "nobleman" *(nobilis)* about whom Worsley had told him.[133] The absence of Starkey's name in this context suggests strongly that Worsley hoped to insert himself as a middleman between Moriaen and Starkey in the matter of the *luna fixa* and Sophic Mercury.

Starkey did write Moriaen directly on 30 May 1651. This was not at Worsley's urging, however, but probably at John Dury's, since the latter sent a "letter of recommendation," almost certainly to Moriaen, on Starkey's behalf.[134] Thus when Starkey established direct communications with Mori-

131. Starkey to Boyle, April/May 1651, *Notebooks and Correspondence*, document 3; also Boyle, *Correspondence*, 1:90–103. For the significance of this rejection, see Newman, *Gehennical Fire*, 69–72.

132. Starkey to Moriaen, 30 May 1651, *Notebooks and Correspondence*, document 4; HP 17/7/1A. Although this is the only extant letter to Moriaen, it is probable that Starkey continued corresponding with the Amsterdam chymist. He may even have sent a sample of his Sophic Mercury to the latter, for in another letter to Worsley (4 August 1651, HP 9/16/13A), Moriaen refers to the fact that a Sophic Mercury that has been sent to him leaves a golden spot when evaporated from silver, and most importantly, that it incalesces when gold is dissolved in it. These were important properties of Starkey's Sophic Mercury; see Principe, *Aspiring Adept*, 160–62, 171. Yet on 16 June Moriaen explicitly says that he was being sent antimonial mercuries from Cologne, making a sure identification with Starkey's mercury impossible (see HP 9/16/8A).

133. Young correctly suspects the presence of Starkey behind this *nobilis*. This does not mean that Starkey was willingly involved in the venture, however. Worsley's evasive silence, to the contrary, suggests that he was advancing Starkey's products to Moriaen without the American chymist's acquiescence. See Young, *Faith, Medical Alchemy*, 227–30.

134. Dury to Moriaen [?], 5 June 1651, HP 17/7/5A–6B. The editors of the Hartlib CD suggest that the letter might be to Hartlib, but that is excluded by the fact that "Domino Hartlibio" is referred to in the letter itself. Dury's letter begins by saying that he has not written his unnamed correspondent recently, because Worsley has already been in communication with the latter. Dury then begins a sort of encomium to Starkey *(Domino Stirkio)*, and passes to the fact that Starkey has also written a few days earlier to Dury's correspondent. We know, of course, that Starkey had written to Moriaen on 30 May, making it extremely likely that the addressee of Dury's letter is also Moriaen. We quote a passage from Dury's letter (17/7/5B):

> Velim autem scias illum [i.e., Starkey] sua sponte, nec me id petente Epistolam hanc tibi scripsisse; monstravit enim mihi illam exaratam ante aliquot dies cum ego necdum cogitabam de illo ad id officii genus provocando: quamvis enim mihi obversabatur hoc

aen, this must have placed Worsley in a somewhat embarrassing situation, for Starkey's letter describes in some detail the very *luna fixa* and Sophic Mercury that Worsley had mentioned but was unable to deliver. Although Moriaen explicitly mentions Starkey's letter of 30 May to Worsley in his own letter of 30 June, Moriaen's later letters show that Worsley did not take the opportunity to reveal that it was Starkey who had prepared the *luna fixa* and Sophic Mercury, suggesting that Worsley had adopted a strategy of silence.[135]

Since we do not have Worsley's letters to Moriaen from 1651, the cause of Worsley's peculiar reticence must remain partly conjectural. All the same, Moriaen's letters reveal one fact unambiguously: Worsley himself was not in a position to make either the Sophic Mercury or *luna fixa* himself. This again supports our contention that Worsley's chymical expertise was quite limited and that he was dependent on others for the technical procedures of metallic chymistry. Worsley himself seems to have realized this, and after his failure to extract Starkey's secret of the Sophic Mercury and *luna fixa,* he underwent a period of deep disillusionment with efforts at transmutation. As Moriaen related in a letter of 3 May 1652 to Hartlib, "that Mr. W[orsley] no longer wants to believe in any transmutation is a sign to me of his irresolute character."[136] Yet Worsley's disavowal of transmutation did not last—or was restricted to certain techniques he judged unsatisfactory—for by late February 1654 he believed that he was on the track to the great secret. In a much quoted letter of 28 February 1654 from Samuel Hartlib to Boyle, the German intelligencer includes a long passage of a letter from Worsley. Written in a tone of considerable excitement, this passage reveals that Worsley had contacted Hartlib's son-in-law Frederick Clodius about the proper way to make metals putrefy—such metallic degeneration was traditionally considered the necessary first step to making the Philosophers' Stone. Worsley also reveals a newfound contempt for the "vulgar chymistry" of strong-water distillers and other "laborants" who operate without the "key" of metallic putrefaction.

> For the truth is, I have laid all considerations in chemistry aside, as things not reaching much above common laborants, or strong-water distillers, unless we

illi prope diem proponere; tamen necdum oppertinatem mihi videbatur nactus commodam id faciendi; verum quando videbam Deum ita praeparasse animum ipsius ad id quod optabam, antequam ego quicquam ea de re dixissem illi, agnovi providentiam supremam nos manuducere ad hanc correspondentiam, tecum ineundam: quam utrisque vobis et mihi utrique conjunctissimo ausspicatissimam fore augurer.

135. Moriaen to Worsley, 30 June 1651, HP 9/16/9A–B; see 9B.
136. Extract of a letter from Moriaen to Hartlib, HP 63/14/20A. See also Young, *Faith, Medical Alchemy,* 232–33.

can arrive at this key, clearly and perfectly to know, how to open, ferment, putrify, corrupt and destroy (if we please) any mineral, or metal. . . . A man once
knowing how to corrupt or destroy a metal, it is then nothing near so difficult, for him to proceed further, and to make a metal, as it was at first, or is
now generally to men, to understand, how they should begin to destroy a
metal.[137]

We do not know what immediate response Worsley got from Clodius,
but his insistence on the need for a transmutational "key" soon developed
into a full-blown claim on Worsley's part that he had discovered this very
secret. Worsley believed that this key lay in the same substance that fur-
nished his other chymical hopes during this same period—the philosophical *sal nitrum*. By June 1654 Worsley had circulated what seems to have
been a substantial "Discourse" announcing his discovery concerning the
magnum opus. Although no copy of Worsley's *Discourse* has been located, it
excited great interest among Hartlib's chymical correspondents. The earliest response seems to have come from Clodius on 4 July, and this soon led
to an extended and fascinating debate between Worsley and Hartlib's
"chymical son."[138] By 10 July, Cheney Culpeper had written a long reply to
Worsley in the form of queries.[139] This was followed on 25 July by an
anonymous German discussion of the letter, sent from Hamburg. The German letter, interestingly, refers to both Clodius and Worsley and attempts
to adjudicate their argument.[140] Finally, on 7 November, Dury entered the
discussion about Worsley's "Confidence of the knowledge of the great secret."[141] But the excitement raised by Worsley's claims did not end there.
Further exchanges between Worsley and Clodius extended at least into November, and as late as February 1656 we find Worsley pronouncing on the
best books to read for knowledge of the "great secret."[142] We will now
briefly consider the nature of the argument between Worsley and Clodius,
as it illuminates an often neglected facet of early modern chymistry, namely
the disparate schools, with strong differences of opinion, that coexisted
even within one branch of the discipline—in this case, within chrysopoeia.

137. Quoted in Hartlib to Boyle, 28 February 1654, in Boyle, *Correspondence,* 1:154–63.
138. [Clodius] to [Worsley], 4 July 1654, HP 16/1/7A–B. The letter must be from Clodius, since it refers to Hartlib as "socer meus." The identification of the recipient as Worsley is
verified by its content. Note that this supplants the tentative identification of the recipient as
Boyle made previously in Newman, *Gehennical Fire,* 83.
139. Braddick and Greengrass, *Letters,* 363–67.
140. HP 39/12/131A–134B.
141. Dury to Hartlib, 7 November 1654, HP 4/3/55A–56B.
142. Copy extracts of letters from [Worsley] to [Hartlib?], 31 October 1654, 29 November
1654, HP 42/1/3A–4B; copy letter of [Worsley] to [?], 14 February 1655/6, HP 42/1/5A–
6B.

In addition, this correspondence reveals a Worsley resolutely adhering to the theory of a universal *sal nitrum* several years after Robert Boyle had abandoned this belief in favor of the Mercurialist approach of George Starkey.

Worsley versus Clodius on the "True Matter": Two Schools of Transmutational Alchemy

In his 4 July 1654 response to Worsley's discourse, Clodius accuses Worsley of speaking like a rhetorician, since he describes only the materials from which the Sophic Mercury can *not* be derived, rather than giving any positive information about it. Worsley had evidently excluded "salt, May-dew, rain, urine, and vulgar mercury, etc," and Clodius replies that he himself had never claimed the Sophic Mercury to be derivable from salts, but had always maintained that it must come from the mineral kingdom. Indeed, Clodius continues,

> With Geber I believe that the mercury of the wise is quicksilver, not the vulgar sort, but extracted from that, for this is not mercury in its own nature, nor in its whole substance, but the medial and pure essence of that, which draws its origin from it [i.e., from quicksilver], and is created therefrom.[143]

This passage clearly reveals that Clodius was an adherent of the Geberian school of alchemy, which had maintained since the High Middle Ages that the starting point of the Philosophers' Stone must be quicksilver. Despite the insouciant dismissals of historians who have not made a special study of chymistry, the influence of the supposed Arabian sage Geber was still at the heart of the discipline in the mid–seventeenth century.[144] Clodius goes on to claim that even the alkahest itself is to be made with the aid of common mercury, which acts upon unspecified salts and converts them into a ponderous liquor. He concludes the letter by saying that Worsley may yet convince him, but only if he can write more compelling and clearer arguments.

Although a number of the letters from this exchange are lost, it is clear that Worsley responded to Clodius's challenge. On 31 October, Worsley

---

143. [Clodius] to [Worsley], 4 July 1654, HP 16/1/7A.

144. For the dismissal of Geberian influence as "special pleading," see Scott Mandelbrote, review of Newman, *Gehennical Fire,* in *British Journal for the History of Science* 30 (1997): 109–11. For Geber's very real influence in the mid–seventeenth century, see Newman, "L'influence de la *Summa perfectionis* du pseudo-Geber," in *Alchimie et philosophie à la renaissance,* ed. Jean-Claude Margolin and Sylvain Matton (Paris: Vrin, 1993), 65–77; Newman, "Arabic Forgeries in the Seventeenth Century: The Case of the *Summa Perfectionis,*" in *The "Arabick" Interest of the Natural Philosophers in Seventeenth-Century England,* ed. G. A. Russell (Leiden: Brill, 1994), 278–96; and Newman, "Boyle's Debt to Corpuscular Alchemy," in *Robert Boyle Reconsidered,* ed. Michael Hunter (Cambridge: Cambridge University Press, 1994), 116–17.

sent a letter to Hartlib in which he reiterates his denial that quicksilver is the material of the Philosophical Mercury. Then Worsley issues a disclaimer:

> For the Philosophers stone I neyther pretend much, to it, nor have any fixed purpose ever much to pursue it, at least not anxiously yet well know that if there be any truth in that art, or in their writings, the foundation & nature of the worke is thoroughly, understood by me (I will not say so of every Enchiresis of it) so that if I should apply my selfe to it my error can not be great in it.[145]

It is hard to imagine a more eloquent testimony to the fact that Worsley's knowledge of the great secret was of the armchair variety. Despite the fact that he had announced his full confidence of the *magnum opus* only five months earlier, Worsley was now backing away from this claim. All the same, in another letter Worsley takes up the argument with Clodius himself. This exchange is particularly interesting, for it explicitly condemns the metallic chymistry of Moriaen and others, presumably the 1651 attempts at extraction and transmutation that included Starkey's Sophic Mercury and *luna fixa*.

> To Reduce all this discourse now to some argument If any man (like our good freind Moriaen or some other Phylosophers among my owne Contrimen) should upon reading the Phylosopher's discourses (of transmutation by the helpe of theire ☿ [mercury] of the Mettalls) Labour by I knowe not how many tedious & operose preparations to torture out of the mettalls a ☿ [mercury] currens (though this I am of opinion without more or lesse of Sophistication is not easye to bee done, yett admit itt) hee would find but a Delusion in the name, and would bee as farre from the ☿ [mercury] the Phylosopher's meane or from beeing able to make any such transmutation as they discribe & intend as hee was before having onely the comon ☿ [mercury].[146]

Despite his confidence that Moriaen and others who "torture" a mercury out of the metals are on the wrong track, Worsley is woefully ambiguous about the real matter from which the Sophic Mercury should be extracted. His rhetoric, however, places him again squarely in the school of the *sal nitrum* theorists. In his *Euphrates* of 1655, Worsley's contemporary Thomas Vaughan, a committed devotee of *sal nitrum*, would also complain

145. [Worsley] to [Hartlib], 31 October 1654, HP 42/1/3A-4B. The letter refers to a *cinnabaris nativa*, a cinnabar mineral, mentioned in "your Sonnes letter." Worsley then requests some of this ore from "Monsieur Clodius" and says that "Monsieur Clodius" does well to call it *Limosa Minera*. Clearly the "sonne" (or rather son-in-law) is Clodius, and the addressee therefore Hartlib.

146. Worsley to [Clodius], undated, 42/1/26A-27B.

of the *"torture* of *metalls"* by ignorant metallic alchemists.[147] This was a common refrain among chymists of the *sal nitrum* school; Michael Sendivogius had even written a humorous dialogue between an alchemist and mercury, in which the latter complains of being "tortured" by excessive heating and treatment with disgusting materials such as pig excrements.[148]

Worsley's allegiances did not go undetected by Clodius. In an undated response, possibly to the letter just quoted, Clodius tries to pin Worsley down. Clodius first clearly restates his own view—quicksilver is the immediate matter out of which metals are produced by nature, and by implication the material from which the Sophic Mercury should be made. Clodius then asserts that "you seem to me to be of the opinion that a certain central salt *[Sal quiddam Centrale]* is the first matter of all the metals."[149] The correctness of Clodius's assessment of Worsley's allegiance is proven by another letter from Worsley to Hartlib, probably of later date, where Worsley reiterates the necessity of putrefying the metals if one is to arrive at the Philosophers' Stone. This putrefaction, Worsley asserts, must be carried out by a menstruum that he seems to equate with the Sophic Mercury. But where should we look for this menstruum or Sophic Mercury? Worsley replies clearly that "you must know the meaning of *Sal Centri terrae.* I say you are to study to gett *Sal centri terrae.* For in salt is all energy."[150] This "salt of the center of the earth" is of course the philosophical *sal nitrum* of Sendivogius and his school. It is described at length in Sendivogius's *Novum lumen chemicum,* though the material from which it should be derived and the means of its extraction are no more clearly stated there by the Polish adept than they are here by his English epigone.

Worsley's debate with Clodius once again clearly positions the former in the camp of Sendivogius and the other exponents of a central *sal nitrum.* What is perhaps most interesting about Worsley is that his revelation concerning the putrefaction of metals, somehow involving the central *sal nitrum,* occurred at the same time as he was making his proposals for the production of common niter. These two simultaneous projects were running on parallel tracks—indeed, it is perhaps best to view them as two parts of a single endeavor. While Worsley was using the theories and practical processes of Glauber in his attempt to multiply common saltpeter, he was also employing the more "elevated" and theoretical chymistry of Sendi-

---

147. Thomas Vaughan, *Euphrates* (London, 1655), A3v.

148. Michael Sendivogius, *Novum lumen chemicum,* in Manget, *BCC,* 2:477–78.

149. Clodius to [Worsley], undated, HP 42/1/36A–37B; a scribal copy of this letter exists at BL, vol. 2, fols. 88–89, implying that the Worsley-Clodius debate was widely circulated through the Hartlib circle.

150. [Worsley] to [Hartlib], undated, HP 42/1/38–39B.

vogius to seek out its ultimate nature. What is clear, finally, is that neither project carried Worsley into the realm of intense metallic and pharmaceutical experimentation performed by Starkey. As we have seen, even during the high point of his own interest in transmutation, Worsley ridiculed the metallic alchemy of Moriaen and others. Worsley was content with the knowledge that if he should ever descend to actual practice, his "error can not be great in it." Although Worsley would claim many years later, just before his death, to have finally succeeded at making a green vitriolic oil with marvelous properties, his descent to manual experiment seems to have been desultory at best.[151]

WORSLEY AND BOYLE

Despite the rather negative conclusions that we have reached regarding Worsley's own expertise in practical, experimental chymistry, it does not follow that he had an insignificant influence on the developing natural philosophical interests of Robert Boyle. There are two areas, in particular, where a Worsleian influence is likely. First, as we have pointed out, Worsley's *De nitro theses quaedam* of the mid-1650s recapitulates Glauber's finding that saltpeter is composed of fixed and volatile parts. As we have also mentioned, this discovery forms the basis of Boyle's own "redintegration" of niter experiment described in his "Essay on Nitre," published in 1661. The preface to Boyle's "Essay" in fact refers to Glauber's discovery, though only to deny that Boyle had prior knowledge of Glauber's work before the composition of his own.

> I might perhaps venture to adde, that though I could not justifie my self by so convincing a proof of my Innocence, yet he, that shall take the pains to consider, that I could not borrow of *Glauber* the various Phaenomena I have particularly set down, and much less the Reflections on them, & shall compare in what differing manners, and to what differing purposes, we two propose the making of Salt-petre out of its own Spirit, and fixt Salt (He but prescribing as a bare Chymical Purification of Nitre, what I teach as a Philosophical Redintegration of it;) He, I say, who shall compare these things together, will, perchance think, that I was as likely to find this last nam'd Experiment as another.[152]

Although Boyle may have found a few passages to support his vaguely documented claim that Glauber intended his combination of fixed and volatile niter merely as a purification, the idea that this is all that Glauber

---

151. [Worsley] to [Boyle?], 25 August 1677, in Boyle, *Correspondence*, 4:452–54.
152. Boyle, "Essay on Nitre," in *Certain Physiological Essays*, in *Works*, 2:89.

had in mind is wishful thinking. In fact, Glauber refers explicitly to the re-combination of fixed and volatile niter to produce saltpeter years before Boyle's publication of the *Essay*.[153] We know that Boyle read some of Glau-ber's writings on saltpeter several years earlier, for Hartlib noted around September 1656 that Boyle had commended "the annexed discourse of salpeeter De Nitro" found in Glauber's *Tractatus de Prosperitate Germa-niae,* which had just been published at Amsterdam. Boyle then noted to Hartlib that the secrets of Glauber's treatise he "himself had before."[154] If we take this to mean that the fact of niter's composition—the central idea of Boyle's "Essay on Nitre"—was already known to him before reading of it in Glauber's *Prosperitas Germaniae,* it seems probable that Boyle may have learned of fixed and volatile niter from Benjamin Worsley. While Wors-ley did not make this discovery about niter himself, it is quite possible that he transmitted the information about niter as a hermaphroditic or twofold salt to the young Boyle, along with any number of other Glauberian ideas. We noted above how the chymical content of Boyle's c. 1650 "Booke of Na-ture" is manifestly Glauberian, and so we have a clear precedent for Glau-berian chymistry coming early to Boyle, most likely through the mediation of Worsley.

Worsley's commitment to the idea of the philosophical *sal nitrum* ubiq-uitous in natural things may also explain a puzzling passage at the start of the "Essay on Nitre," where Boyle declares niter to be "one of the most Catholick of Salts" because it is found in so many bodies "Vegetable, Ani-mal, and even Mineral." If Boyle were speaking simply about potassium ni-trate, this claim would be incomprehensible (and also would have rendered the perpetual and highly disruptive search for saltpeter before the modern age unnecessary), but Boyle is clear that such saltpeter is found "either in its rudiments, or under several disguises."[155] Similarly, an important section of Worsley's *De nitro* is devoted to demonstrating how "the Salt that is found in all Vegetables and in all Animals nourished by Vegetables is re-ducible or convertible into Salt-Peter."[156] Indeed, such ubiquity is an at-tribute of the philosophical *sal nitrum,* which Worsley himself located in such places as grass gone to seed, leaves, and blood.

The redating of Worsley's *De nitro* to c. 1654 also juxtaposes it with a

153. For example, in Glauber, *Continuatio operum chymicorum,* 2:388.
154. Hartlib, *Ephemerides* 1656, HP 29/5/92B. For remarks on Boyle's fear of being seen as a plagiarist of Glauber, see *Works,* 2:xii; and Michael Hunter, "Self-Definition through Self-Defense: Interpreting the Apologies of Robert Boyle," in *Robert Boyle (1627–1691): Scrupulos-ity and Science* (Woodbridge: Boydell, 2000), 146–48.
155. Boyle, "Essay on Nitre," in *Works,* 2:93.
156. Worsley, *De nitro,* HP 39/1/17B–18B.

sudden general interest in saltpeter that developed among the Hartlibians at that time. In January 1655, Boyle communicated to Hartlib a secret on making saltpeter that he had been given by Starkey, and Clodius was set to try it out.[157] Boyle had in fact recorded this process in his "Collections" on 12 January 1655.[158] At the same time, Clodius was also trying a method for producing saltpeter from sea salt.[159] It is possible that this renewed interest may have been sparked by Worsley's treatise. The same may be true of Boyle's wider interest in saltpeter which first appeared at this time, as witnessed by the appearance of a tract (now lost) entitled "Of Seeds, Dung & and Salt-peter" in a list of his "Philosophicall Essays" drawn up in 1655–56.[160] Thus while acknowledging that a key importance of Boyle's "Essay on Nitre" lies in its linkage of the niter experiment to "Notions of the Corpuscular Philosophy"—an issue absent from the niter speculations of Glauber, Worsley, Starkey, and Clodius—we may nonetheless also recognize the background and possible motivation of Boyle's interest in niter among his chymical associates of the mid-1650s.

A second possible Worsleian influence on Boyle's natural philosophy may be found in a more theoretical area. Rose-Mary Sargent has recently emphasized Boyle's use of Baconian "intermediate" explanations, that is, descriptions of natural phenomena that do not make recourse to the ultimate principles of nature—for Boyle, these were matter and motion.[161] Boyle in fact distinguished himself from the more reductionist corpuscularians such as Descartes by his own willingness to engage in these lower-level explanations. In his preface to *Certain Physiological Essays*, for example, Boyle tries to save physicians from the opprobrium of ignorance just because they do not necessarily employ corpuscularian principles:

157. Hartlib, *Ephemerides* 1655, HP 29/5/6A–6B. On Boyle, the Hartlibians, and niter, see Robert G. Frank Jr., *Harvey and the Oxford Physiologists: A Study of Scientific Ideas and Social Interaction* (Berkeley and Los Angeles: University of California Press, 1980), 121–28.

158. BP, vol. 8, fol. 142.

159. Mentioned in Boyle's "Collections" for 1655, BP, vol. 8, fol. 142v. The same process was being tried by others in London, and apparently this desirable goal was a project of long duration, for John Ward records in his diary around June 1662 that Starkey told him that "there were some in London who would undertake to make nitre of sea water, a ridiculous thing it being a contraire nature" (Ward Diaries, Folger Shakespeare Library, V.a.292, fol. 58r).

160. BP, vol. 36, fol. 70; printed in *Works*, 14:332 (on the dating of this list to no earlier than 1655, see above). The "Dung" section of this tract may have been carried over from an earlier essay (also lost) entitled simply "Of Dungs," which Boyle recorded on a list of his writings on 25 January 1650; this list is published in *Works*, 14:331.

161. Rose-Mary Sargent, *The Diffident Naturalist: Robert Boyle and the Philosophy of Experiment* (Chicago: University of Chicago Press, 1995), 40, 52, 54, 58, 69, 133–35, 206.

The Physitian that has observ'd the Medicinal vertues of Treacle, without knowing so much as the names, much less the Nature of each of the sixty and odd Ingredients whereof it is compounded, may cure many Patients with it. And though it must not be deny'd, that it is an advantage as well as a satisfaction, to know in general, how the Qualities of things are deducible from the primitive Affections of the smallest parts of Matter, yet whether we know that or no, if we know the Qualities of this or that body they compose, and how 'tis dispos'd to work upon other Bodies, or be wrought on by them, we may, without ascending to the Top in the series of Causes, perform things of great Moment.[162]

Thus Boyle argues that a physician's explanations have explanatory value even if they fail to ascend to the top in the series of natural causes, the *prima naturalia* or smallest corpuscles and their properties. Similarly, Boyle elsewhere argues that brewers need not understand the natural philosophy of fermentation in order to make beer.[163]

Boyle's doctrine of intermediate causes, although Baconian in origin, is strikingly similar to an argument raised by Benjamin Worsley in his debate with Frederick Clodius. In a hitherto unnoticed letter (probably addressed to Hartlib), Worsley thanks his correspondent for sending him a previous epistle written (evidently by Clodius) to the German natural philosopher Joachim Jungius.[164] Apparently Clodius had mentioned Van Helmont's theory, described in the *Ortus medicinae*, that there is an underlying sand called *Quellem* that provides the material structure to the earth. Van Helmont thought that this *Quellem* was itself composed of water that had been acted upon by *semina*, according to the theories that we discussed in chapter 2. Worsley, however, was skeptical of the view that such explanations were always necessary, as his response shows:

> But though I verie well remember the *Quellem* of Helmondt, & that it is his opinion that water is the *primum principium* of all things, though allso I thinke it doth much become a Philosopher to consider things *per analysin simplicissimam*, & though we must according to this analysis at length determine our thoughts into deCartes principles of not onely of water, but of atomes. yet as these thinges have theire commendation, so the knowledge of

162. Boyle, *Certain Physiological Essays,* in *Works,* 2:24. We have here chosen the first edition's reading of "wrought" rather than the later editions' "brought."
163. Principe, *Aspiring Adept,* 255.
164. [Worsley] to [Hartlib], undated, HP 42/1/38A–39B. The transcribers of the Hartlib papers CD have misread "Monsieur Iungius" as "Monsieur Iungivi."

other bodyes which may (if I have leave so to speake) be *principia subalterna;* though Atomes or water maye be said to be *principia generalissima,* are often times very usefull. For example if I would advance the art of Brewing, it may suffice if I take notice onely of the malt not of the fundation of what I at present aime at in my Instruction, though I say not any thing of the manner of making malt of graine or of producing that graine, by sowing or though I doe not proceed further analytically.[165]

The similarity between Worsley's position and Boyle's is striking. First, Worsley admits that it is a desideratum of natural philosophy to find the first principles of nature, by means of "the most simple analysis" *(analysis simplicissima).* He then points out that by this logic, one should not terminate one's analysis with water, but proceed all the way to Cartesian atoms. All the same, there are subordinate principles *(principia subalterna)* beneath these highly abstract first principles that are often "very usefull." Having made this claim, Worsley then gives one of the very examples eventually used by Boyle—that of brewing. When a brewer malts his grain or when he plants the seed that will produce his grain, he does not need to consider the most abstract principles of matter, but only the immediate causes at hand. One does not need to be a Cartesian, or for that matter a Helmontian, in order to make a good beer. Hence, although atomic principles are indeed the simplest in nature, and the most desirable to a natural philosopher, it does not follow that *principia subalterna* are without value.

The similarity between Worsley's *principia subalterna* and Boyle's intermediate causes suggests that the young aristocrat may well have derived this important part of his natural philosophy from the Baconian ethos of his older compatriot. On the basis of this previously unremarked letter, one must even consider the possibility that Worsley had an influence on Boyle's early corpuscularian philosophy. Yet, as we have shown, the matter is otherwise for Boyle's exposure to the theory and practice of pharmaceutical and metallic chymistry in the early 1650s. When it came to chymistry, Worsley's allegiance to the *sal nitrum* school led him to condemn the very theory and practice that Starkey was imparting to the eager young Boyle. Moreover, as Worsley himself admitted, the "tedious & operose preparations" of laboratory practice that consumed the greater part of Starkey's life were not for him. Unlike Starkey, the Worsley of Boyle's youth was content to rest assured in the superiority of theory and let others descend to the harsh world of laboratory practice.

---

165. [Worsley] to [Hartlib], undated, HP 42/1/38A–39B.

HARTLIB'S "CHYMICAL SON" FREDERICK CLODIUS AND BOYLE

Another early chymical associate for Boyle was Frederick Clodius, a person to whom we have already referred repeatedly. In spite of considerable data in the Hartlib Papers, very little attention has been paid to Clodius's life.[166] He came to England in mid-1652, having been recommended to Hartlib by Johann Moriaen, with whom he had lived and worked for a while in Amsterdam.[167] Prior to that time he was employed "into several parts of Europe, by a rich and powerful Prince, to purchase Rarities."[168] This rich and powerful prince was Frederick III, duke of Holstein-Gottorp (1609–70); Clodius's father Johann was also in contact with the duke. Clodius was himself probably a native of Holstein—Hartlib's *Ephemerides* are full of information about Holstein and its duke sent from Clodius, and Clodius claimed to have been offered a position in the duke's court in 1656, whereupon he referred to Holstein as his "homeland."[169] These sources also make it clear that the duke of Holstein was a devotee of Van Helmont and that he sent Adam Olearius (1600–71) to Van Helmont's widow in order to acquire his unpublished manuscripts. Although Clodius's involvement in this operation remains somewhat uncertain, he himself visited the widow Van Helmont, probably in mid-1651, who told him a story about the arson of the chymist's home and apparently gave him some manuscripts (or copies of them). Francis Mercurius Van Helmont (1618–99), Joan Baptista's youngest son, also had lent a portion of his father's manuscripts to the court of the duke of Holstein, some of which Clodius borrowed and neglected to return.[170] Clodius showed one of these manuscripts to Starkey in July 1653, who found it new to him.[171] Clodius also sent a copy

166. On Clodius see Young, *Faith, Medical Alchemy*, 80–82 and passim, and Dobbs, *Foundations of Newton's Alchemy*, 74–79.

167. He was still with Moriaen in Amsterdam on 19 April 1652; see HP 63/14/19A–B. Nonetheless, Hartlib had been receiving communications from Clodius, which he recorded in his *Ephemerides*, since mid-1651.

168. Boyle makes this description of a "Dr. C." in *Usefulnesse* (3:341), and the reference is undoubtedly to Clodius. Of course, Boyle's information about Clodius's previous activities came from Clodius himself and so may not be entirely unbiased.

169. On Clodius's father see Hartlib, *Ephemerides* 1651, HP 28/2/25B, and the letter to his son at HP 26/34 (cf. HP 68/3/14), which although in German is addressed "Allo mio Carissimo figliolo adesso con Signore Hartlib." On Clodius's delivery of information about Holstein and its duke, see, for example, HP 28/2/19A, 22A, 28B, 29B, and 32A. On Clodius's reference to his "homeland," see Clodius to Boyle, 3 March 1656; Boyle, *Correspondence*, 1:195–98, on 197.

170. Hartlib, *Ephemerides* 1653, HP 28/2/63A; Hartlib records euphemistically that Clodius "reserved" some of the manuscripts to himself.

171. On the duke of Holstein's attempt to publish Helmontian manuscripts, see Hartlib, *Ephemerides* 1651, HP 28/2/25A; on Clodius showing Starkey a manuscript, see *Ephemerides*

of the otherwise unknown Helmontian tract "Magna virtus verborum et rerum" to Johann Moriaen in 1654.[172] Henry Oldenburg himself tried to swap a secret process with Clodius for a Van Helmont manuscript in 1659.[173] Indeed, one reason for Clodius's quick rise among the Hartlibians may have been his ownership of these manuscripts, in much the same way that part of Starkey's fame rested upon his ownership of manuscripts by the great adept Eirenaeus Philalethes.

Upon Clodius's arrival in England he lived with Hartlib, set up a laboratory in his house, and in September 1653 married Hartlib's daughter Mary. In early 1658 Clodius was "again in . . . a labyrinth" and bringing upon himself "a new kind of undoing"; he moved out of Hartlib's house, seemingly on rather poor terms with his father-in-law.[174] Moriaen makes reference to Clodius's departure and apparent ingratitude to Hartlib, as does John Dury more fully in 1661 when he refers to "the wrong he doth you" and how "he makes himself incapable of the love of honest friends who will not bee able or willing to Trust or assist him."[175] Thereafter little is heard of Clodius save his wife's miscarriage in 1658 and some references to him in letters. His last extant exchange with Boyle occurs in December 1663, in a letter where he tells Boyle of his dire financial circumstances and asks for money.[176] Several later letters from Clodius to Boyle survived until the eighteenth century, as witnessed by the list made by William Wotton shortly after Boyle's death, but are no longer extant.[177] The last of these was dated 12 March 1670, and Thomas Birch, who made a list of Boyle's

---

1653 (July), HP 28/2/69A. The story about the arson is attributed to Clodius and recorded in Hartlib, *Ephemerides* 1651, HP 28/2/24B, while Clodius's conversation with the widow Van Helmont ("Helmontiana") is mentioned at *Ephemerides* 1651, HP 28/2/19A; Clodius is described as having been "with her [Van Helmont's widow] not long ago" (bey Ihr unlängst gewesen) in Moriaen to Hartlib, 19 April 1652, HP 63/14/19B. For an account of the loss of Helmontian manuscripts in the Great Fire of London and Boyle's interest in them, see Clericuzio, "From van Helmont to Boyle," 311–12.

172. Hartlib to Boyle, 15 May 1654; Boyle, *Correspondence*, 1:179–84. The title is reminiscent of the published but "imperfect" tract in the *Ortus* entitled "In verbis, herbis, et lapidibus est magna virtus," 575–84.

173. Hartlib to Boyle, 17 May 1659; Boyle, *Correspondence*, 1:352–54.

174. Hartlib to Boyle, 7 January and 2 February 1658; Boyle, *Correspondence*, 1:247–54.

175. Moriaen to Hartlib, 1 January 1658, HP 31/18/25A; George H. Turnbull, *Hartlib, Dury, and Comenius: Gleanings from Hartlib's Papers* (London: University Press of Liverpool, 1947), 8. Young, *Faith, Medical Alchemy*, 80, interprets Moriaen's criticisms to refer to Starkey, but it seems more likely that Moriaen's general disappointment with *der Mensch* ("mankind," not "the man," as Young renders it) is sparked by Clodius's recent breakup with Hartlib than by Starkey's "degeneration" five years earlier. Moreover, Moriaen mentions continued good relations with Starkey. Dury to Hartlib, 2/12 August 1661, HP 4/4/30A.

176. Clodius to Boyle, 12 December 1663; Boyle, *Correspondence*, 2:229–30.

177. On Wotton's list, see Boyle, *Correspondence*, 1:xxvii, 6:397–414.

surviving letters in the early 1740s, described this item as "Fr. Clodius . . . Represents his low Circumstances; intends to dedicate his Works to Mr B. of whose Resentment against him he complains."[178] After this date, Clodius seems to have sunk into obscurity, although there are references in the papers of the Welsh landowner Edward Lloyd to one Dr. Clodius or Claudius who came to Llanvorda for chymical purposes in 1680–81.[179]

Of particular interest is the close relationship that developed between Clodius and Sir Kenelm Digby in 1654, when the Catholic expatriate returned briefly from his exile. Much has already been written on Digby (who was Boyle's kinsman), including his chymical studies.[180] Digby offered to finance Clodius's laboratory and family, exchanged secrets with him, and claimed that "in all his travels and converses with the choicest arts, both in *Italy* and *France*, he hath not met so much of theoretical solidity and practical dexterity both together" as he saw in Clodius. Hartlib further remarked that "both their judgements and their experiences agree mightily together, to the very amazement of each other" and asserted that Digby was sending for all his arcana and papers out of France, to give them all to Clodius.[181] Digby even had a book dedicated to Clodius in 1658, whose dedication cites Clodius's "worthiness, wisdom, and deep Learning."[182] It was Digby and Clodius who planned a "universal laboratory" of which Hartlib informed Boyle—what is probably a manifesto of this project survives—and Digby offered to contribute six to seven hundred pounds to it.[183] It was these two again who constituted the core of the "general chemical council," a group whose full membership and activities remain somewhat unclear.[184]

While Hartlib was enthusiastic about his "chymical son," and Digby equally so, contemporary opinions about Clodius vary widely. John Evelyn, for example, accused Clodius of having "Insinuated himselfe into the Acquaintance of his Father-in-Law" using a "*Methodus Mendicandi* [regimen

178. Ibid., 4:172.

179. Brynley F. Roberts, *Edward Lhuyd: The Making of a Scientist* (Cardiff: University of Wales Press, 1980), 15.

180. Dobbs, "Studies in the Natural Philosophy of Sir Kenelm Digby," *Ambix* 18 (1971): 1–25; 20 (1973): 143–163; 21 (1974): 1–28; on Digby more generally, see R. T. Petersson, *Sir Kenelm Digby: The Ornament of England, 1603–1665* (Cambridge: Harvard University Press, 1956), and E. W. Bligh, *Sir Kenelm Digby and His Venetia* (London: S. Low, Marston, 1932).

181. Hartlib to Boyle, 8 or 9 May 1654; Boyle, *Correspondence*, 1:169–79, on 175.

182. Pierre Borel, *A New Treatise, Proving the Multiplicity of Worlds*, tr. D. Sashott (London, 1658), Epistle Dedicatory, [iii].

183. Hartlib to Boyle, 8 or 9 May 1654; Boyle, *Correspondence*, 1:169–79, on 175; "Laboratorium Clodianum," HP 16/1/79A-80B; on the "chemical council," see Dobbs, *Foundations of Newton's Alchemy*, 75–78.

184. See Webster, *Great Instauration*, 303–5.

of lying] (& pretence of Extraordinary *Arcana*)."[185] But Evelyn's dispar-
agement pales in comparison to the scorn and bile heaped upon Clodius by
Henry More in his letters to Lady Anne Conway. More apparently introduced Clodius to Lady Conway in late 1653, no doubt hoping that he might
be able to cure her chronic headaches. Besides delivering medicine, Clodius
was also engaged to convey a portrait of Lady Conway to her brother John
Finch in Padua. It is unclear whether the portrait ever arrived.[186] In regard
to the medicines for Lady Conway's headaches, it is equally unclear
whether any of these actually materialized either. But Lady Conway's com-
plaints and Henry More's outrage are clear enough:

> This Clodius has moved my indignation above all measure. . . . He is so
> mainly like a cheat, that I utterly suspect his skill. For if he had any it were his
> advantage to be honest. The thinges in your [Conway's] letter you alleadge
> against him are so foule and grosse that in my own judgement, I know not
> how to trust him in any thing, that makes no more consciences of what he
> speakes.[187]

More also mentions "the money he has couzen'd you of" and further
counsels his friend to take legal action against Clodius "that he may be
made an example. For if thinges holds as ill as they represent themselves for
the present, he is as accurs'd a Raskall as ever trod on English ground, and
I am sorry Mr Hartlib should have such a wretch for a son-in-law."[188] Evidently, things continued to appear very bad indeed, as More persisted in
pressing for legal measures, and wrote to a "Mr. Newbury," possibly a mag-
istrate, about it. More had in fact paid Clodius for some "service," for he recalls that Clodius "look'd as pale as ashes . . . when he receiv'd the sixteen
peeces of gold of me from you, which makes me suspect that he did never

185. John Evelyn to William Wotton, 12 September 1703; published in Michael Hunter, *Robert Boyle by Himself and His Friends* (London: Pickering, 1994), 91–99, on 92. "*Methodus Mendicandi*" is a pun on the *methodus medicandi* that Clodius should have been practicing instead. Evelyn's comments are in response to questions asked by Wotton, who was writing his
aborted "Life of Boyle." On this endeavor, see Hunter, *Robert Boyle*, xxxvi–liv, 84–148.
186. John Finch to Anne Conway, 6 November 1653, in Sarah Hutton, ed., *The Conway Letters: The Correspondence of Anne, Viscountess Conway, Henry More, and their Friends, 1642–1684*, rev. ed. edited by Marjorie Hope Nicolson (Oxford: Clarendon, 1992), 88: "I make no
doubt that Mr Frederick will find some meanes of conveying your picture to me." Finch to
Conway, 9/19 November 1653; ibid., 89: "I heare no news of your picture though many Dutch
from Flanders and Holland are lately come hither." Finch to Conway, 30 November/10 December 1653, in *Conway Letters*, 90: "Your picture which I so much desire is not come to my
hands yet. I hope Mr Frederick convey it safe."
187. Henry More to Anne Conway, April 24 [1654], *Conway Letters*, 94–95. Unfortunately, Lady Conway's letter is not extant.
188. Ibid.

think in his own confidence that he could ever deserve any such reward. . . . What a trouble it is to deal with men dishonest."[189] Whether the four pounds was for delivery of a portrait that never arrived or a medicine that finally materialized (and which More counseled Conway not to take, as it was either worthless or dangerous) is not entirely clear.[190] The last we hear of Clodius's "gross foul villany" is in the summer of 1654.

> I can not think of that misshapen monster Clodius, but I am ready to vomitt at his gross foul villany, and pitty many a poor soul, that I think on my conscience, our lawes have knitt up at Tiburne for less then the twentieth part of that wickedness this wretch stands gilty of . . . in a course of Physick, [I] intend to vomitt tomorrow, with possett drink, and the fulsome remembrance of that foul wretch Clodius, it will serve me instead of oxymel of squills, to move my Stomack.[191]

Surviving documents do not indicate that Boyle considered Clodius a "foul wretch," much less a "misshapen monster" or an efficient emetic. Whether or not the two met before Boyle's departure for Ireland in June 1652 is unknown, but rather unlikely. But they certainly met in London the following summer, when Boyle was back in England for about three months. Judging from the very familiar tone of Boyle's 27 September 1653 letter to Clodius, the two became close immediately.[192] Indeed, the tone of many parts of Boyle's letters to Clodius borders on the obsequious. In terms of what Boyle obtained from Clodius, the "Collections" that Boyle made in 1654 show a few recipes from Clodius, but in 1655 Clodius's contributions to Boyle's "Collections" become substantial. This is also the time at which entries from Kenelm Digby first appear, and their juxtaposition may reflect the collaboration between the two. Digby himself undoubtedly encouraged Boyle's chymical studies.[193] Additionally, Digby's best-known

189. Henry More to Anne Conway, May 8 [1654], *Conway Letters*, 97–98.

190. Ibid. In a letter of 7 June 1654, More again rails against Clodius, noting that "I can not think of Clodius without some nauseating. I did diverse times think with my self, that he has cheated you of the picture you trusted him with." *Conway Letters*, 102.

191. More to Conway, 18 June 1654, *Conway Letters*, 104.

192. Boyle, *Correspondence*, 1:148–50.

193. Besides the items in the "Collections" that came from Digby, Boyle also owned a handsomely bound collection of chrysopoetic manuscripts bearing both Digby's "KD" monogram and marginalia in his hand; this may well have been a gift directly from Digby. See Principe, "Newly Discovered Boyle Documents in the Royal Society Archive: Alchemical Tracts and His Student Notebook," *Notes and Records of the Royal Society* 49 (1995): 57–70.

contribution to natural philosophy—the atomistic system contained in his *Two Treatises*—may well have been an important influence on the young Boyle, who was writing his "Atomicall Philosophy" around this time, and in which he refers to his "deservedly famous Countryman Sir Kenelme Digby."[194] Its content bears some affinities with Digby's own interests, particularly in its emphasis on effluvia and their particulate nature.[195]

Clodius also prepared a copy of the *Epistolae philosophicae* and *Statuta philosophorum incognitorum* (attributed to Michael Sendivogius) for Boyle; the manuscript in Clodius's hand survives among the Boyle Papers.[196] It is worth pointing out that another member of the Clodius family—possibly a nephew or cousin to Frederick—lived with Hartlib in 1659, for mention is made of "young Clodius" being in Hartlib's chamber working on translations of German alchemical treatises, while the elder Clodius was "gone to *Gravesend*."[197]

Several of the items that Clodius shared with Boyle—like items that Starkey contributed—found their way into Boyle's later publications. Most of these items appear in *Usefulnesse* (1663). In one place, Boyle cites "our Ingenious Friend Dr. *C*" in relation to a story from Van Helmont's widow.[198] In the same work, when Boyle tells of the three "Chymical medicines, that I do the most familiarly employ," the first is Starkey's *ens veneris* while the third is an "Essentia Cornu cervini" (essence of hartshorn), which can be linked to Clodius through Boyle's 1655 "Collections," as well as through a letter from Moriaen.[199] Indeed, after treating the simplest preparation of this spirit of hartshorn, Boyle notes that there is "a more elaborate and costly" way of making it used by "an excellent Chymist who makes great advantage of it," and this is almost certainly Clodius. Indeed, since Boyle notes that the use of this medicine is "principally in Affections of the Brain," this spirit may have been what attracted More to Clodius in hopes of relieving Lady Conway's excruciating headaches. It is probably Clodius again who is referred to under the initials of "Dr. *N-N*" later in the work.[200]

194. Boyle, "Atomicall Philosophy," in *Works*, 13:227.

195. Digby, *Two Treatises;* on Digby's atomism, and also on the clear influence of Walter Charleton on Boyle's atomism, see Robert H. Kargon, *Atomism from Hariot to Newton* (Cambridge: Cambridge University Press, 1966), 70–73, 97–99.

196. BP, vol. 34, pp. 238–323.

197. Hartlib to Boyle, 10 May 1659, Boyle, *Correspondence*, 1:350–51.

198. Boyle, *Usefulnesse*, in *Works*, 3:348.

199. BP, vol. 8, fol. 144; Moriaen to Hartlib, 1 January 1658; HP 31/18/1–3, on 2A; Boyle, *Usefulnesse*, in *Works*, 3:394–95.

200. Boyle, *Usefulnesse*, in *Works*, 3:395, 530.

STARKEY AND CLODIUS

Starkey and Clodius were in direct communication very soon after the latter's arrival in England in 1652, and a cordial letter from Starkey to Clodius is still extant.[201] Throughout late summer 1652, Clodius was Hartlib's chief informant on Starkey's activities.[202] On 26 August 1652 Clodius witnessed a demonstration of Starkey's alkahest, presumably the material he had been working on since his January dream and that he recalled wistfully (after having lost it when a digesting flask broke) in *Pyrotechny Asserted*.[203] This is very probably the same event retold by Boyle in *Usefulnesse*, involving an unnamed "Chymist" endeavoring to make the alkahest and "Our Ingenious Friend Dr. *C*" who agreed to give the chymist "Two hundred Crowns for a Pint of this *Menstruum*."[204] If, as seems likely, this is an account of Starkey and Clodius, it is curious to note that Starkey, the inventor of the wonderful menstruum worth two hundred crowns a pint, goes wholly unnamed, while Clodius is at least given an initial and some identifying attributes. Additionally, it seems likely that Clodius attempted to take credit with Hartlib for preparations made with Starkey's alkahest. On 9 August 1652, Hartlib recorded that Clodius had brought him a medicine made from antimony using the alkahest, with the implication that it was Clodius who had prepared it.[205] There is little indication that Clodius had his own alkahestical solvent at this time, and Boyle explicitly refers to Starkey's preparation of a medicine from antimony using his alkahest, saying that the "sweet crystals" made thereby were used to cure Cheney Culpeper. Thus, it seems quite possible that Clodius was attempting to pass off Starkey's preparation as his own, at least to Hartlib. Indeed, it is noteworthy that many of the items that Hartlib's summer 1652 *Ephemerides* attribute to Clodius bear a striking resemblance to Starkey's ideas and activities.[206] It

201. See *Notebooks and Correspondence*, document 9; HP 16/1/72A–72B.

202. For example, see Hartlib, *Ephemerides* 1652, HP 28/2/32A.

203. Hartlib, *Ephemerides* 1652 (26 August), HP 28/2/33A; Starkey, *Pyrotechny Asserted*, 34.

204. Boyle, *Usefulnesse*, in *Works*, 3:341. Hartlib tells of Clodius having seen mercury "brought over the Helm by the Alkahest of Stirk," (*Ephemerides* 1652 [August 26], HP 28/2/33A), and here Boyle recounts that "Dr. *C*." saw common sulphur thus dissolved and distilled, as well as antimony reduced to sweet crystals, which were later used to cure "Sir C. C." (presumably Cheney Culpeper) "of a very radicated and desperate disease."

205. Hartlib, *Ephemerides* 1652, HP 28/2/29B.

206. For example, Clodius emphasizes the difference between the preparations of the alkahest and the Philosophers' Stone (HP 28/2/30A), a topic that Starkey often stressed; Clodius speaks of better distillations of oil of cinnamon and how the common article is adulterated with oil of almonds (HP 28/2/32A)—this was one of Starkey's projects, and he notes this adulteration in a letter to Moriaen (30 May 1651; see *Notebooks and Correspondence*, document 4); Clodius refers to the "7 or 8 noble medicines" including the *ens veneris* that are pre-

seems that the newly arrived Clodius was borrowing rather heavily from Starkey.

Whatever relationship existed between Starkey and Clodius soon deteriorated. We have no testimony of this from Starkey's end, but Clodius's recorded remarks about Starkey suddenly turn sour in 1654. Whereas Clodius was willing to pay handsomely for Starkey's alkahest in summer 1652 (and may have been trying to claim it as his own), in 1654 he dismissed the product that Starkey had "dreamt up" *(somniavit)* and promoted his own process instead.[207] Similarly, in 1653, Clodius was enthusiastic about the manuscripts of Eirenaeus Philalethes, claiming that they would "uncover all" and "clearly discover that Mystery [of the Philosophers' Stone]," but in 1655 he complained on more than one occasion that these same manuscripts were faulty and unhelpful.[208] It is probably no mere coincidence that Clodius's statements of dissatisfaction with Starkey fall very close to the time of Hartlib's widely cited disaffection with Starkey and of Clodius's push to gain Boyle's patronage.

It is probably also no coincidence that Hartlib's and Clodius's turn against Starkey also aligns closely with the arrival of Digby in England and his commencement of a close relationship with Clodius. Although no extant document directly records Digby's opinions of Starkey, there is good reason to believe that he held a long-standing grudge against the young American. In 1651, Starkey had rebuffed—in no uncertain language—the offers of Digby's close associate Dr. Richard Farrar to buy the secret of the Philosophical Mercury and *luna fixa*.[209] After word of Starkey's successes began to circulate among the Hartlibians, "Farrar in two or 3 days after Came Gaping, & he would give so much, viz. 30s per ounce for the *luna* or 5000li for the secret." Starkey was appalled by this attempt "in such a way of lucre [to] prostrate so great a secret," and told Farrar to his face how little he thought of him:

> I told him I had had tryal of some of the world & in that point I had foun[d]
> no smal basenes, that unlesse a man wil stoope to some mens humors in some

---

pared on the way to Butler's stone (HP 28/2/29B)—these multiple arcana were clearly laid out in Starkey's "Vitriologia" (*Notebooks and Correspondence,* document 11; Sloane MS 3750, fols. 9r

207. [Clodius] to [Worsley], 4 July 1654; HP 16/1/7. Clodius's "somniavit" may also refer literally to Starkey's dream about the alkahest, related to Boyle on 26 January 1652. See *Notebooks and Correspondence,* document 6; also Boyle, *Correspondence,* 1:121.

208. Ibid.; see Hartlib, *Ephemerides* 1653 (May), HP 28/2/61B and 63A; cf. *Ephemerides,* 1655 (February), HP 29/5/12A.

209. For Farrar, see Newman, *Gehennical Fire,* 67, 74–75, 300 n. 94. Farrar's "epitaph" for Digby is noted in [Thomas Longueville], *The Life of Sir Kenelm Digby* (London: Longmans, Green, 1896), 6 n.

things which they who Court Nature not Sophistry, Count sordid, they will speak of them what not, & one of that farina I had found him & therefore for his mony I advised him to keepe it.[210]

Given Starkey's fiery temperament and his indignation at this attempted "venality of Natures Secrets," we may assume that the account of this scene that he gave to Boyle was more sedate than the browbeating Farrar received. As a result of the episode, Starkey refused to have "the least familiarity . . . with [Farrar] nor any in whom he had influence," which presumably included Digby. And Starkey notes explicitly that the rebuff now "makes the Earl & Farrar his Parasite to revile me basely"; this "Earl" is probably Digby himself, for the latter sometimes styled himself with that title in the 1650s.[211] Given this history of bad blood between Starkey and Digby, it is possible that the alienation of Starkey from the Hartlib circle in early 1654 was motivated as much by the machinations of Clodius and the ill-will of Digby as it was by the young American's financial problems.

Despite Clodius's turn against Starkey, the Holsatian chymist continued to benefit from Starkey's work, although it seems without Starkey's permission. This brings up another interesting facet of Boyle's treatment of Starkey, namely his management of the secrets entrusted to him. Boyle appears to have been a constant "leak" of privileged information. A clear example of this involves Starkey's development of a process for making wine and spirit of wine from fermented legumes like peas and beans as well as from "corn" (cereals). Such a process was promised (but not delivered upon) by Glauber. This preparation was potentially quite lucrative, for it would allow the production of wines and, perhaps more importantly, the production of the expensive spirit of wine (necessary for many chymical and medicinal preparations) on a large scale in a country so unfavorable to viniculture as England.[212] Hartlib recorded that Starkey was at work on the process in March 1653 and had hired an assistant. When Boyle returned briefly from Ireland in summer 1653, Starkey had apparently been successful in the pro-

---

210. Starkey to Boyle, April/May 1651, *Notebooks and Correspondence,* document 3; also in Boyle, *Correspondence,* 1:90–103, 94.

211. Johann Hiskias Cardilucius, *Magnalia medico-chymica* (Nuremberg, 1676), 298–300. Cardilucius met Digby in Germany, where the latter was traveling under the assumed title of "Earl" or "Count," translated as *Graf* by Cardilucius.

212. Since wine is only 8–12 percent ethanol, a large quantity must be distilled to produce only a small yield of alcohol. In our own trials using seventeenth-century implements, a liter of good wine gave only about 100–125 ml of "common spirits of wine" (about 50 percent ethanol), which upon repeated rectifications gave only about 40 ml of "dephlegmed spirits" (more than 90 percent ethanol). In a practical sense, this means that Starkey's contemporaries needed ten gallons of wine (imported) to produce every gallon of common spirit, and over twenty gallons to prepare each gallon of the purer spirits.

cess and imparted it to him. But Boyle, before returning to Ireland, made sure to pass on the process to Clodius. Writing on 27 September 1653 from Bristol, as he was awaiting transport back to Ireland (some of his baggage was already loaded on board), Boyle informed Clodius that he would send him the process in his next letter, adding the somewhat curious direction that since he "received this Processe from a freind as a Secret; I shall beg you would not let it loose that Name."[213] Starkey is nowhere mentioned in the letter by name. It is curious that Boyle should function as an intermediary between Starkey and Clodius, who already knew one another well. One explanation is that Boyle was giving Clodius what Starkey would not impart of his own accord. The same kind of transmission by Boyle occurred with Starkey's process for multiplying saltpeter in early 1655. Hartlib even tags the passage about it in his *Ephemerides* with the word "secret."

It might be argued that the corn wine and saltpeter projects were cooperative Hartlibian endeavors, and thus Boyle was merely acting as a conduit for information that would have been transmitted anyway. Yet this would not explain why Boyle omits Starkey's name from the secret about wines when passing it on to Clodius by letter. If Starkey and Clodius had been cooperating, would Boyle have attributed the wine process merely to "a freind"? In the actions of Boyle and Clodius we see the same pattern as in the case of the letters between Moriaen and Worsley regarding the *luna fixa* and Sophic Mercury—two members of the Hartlib circle were eking out the results of Starkey's laboratory practice and using them as commodities of exchange in their economy of "secrets."

These are not the only examples of Boyle's transmission of Starkey's secrets to Clodius. In a letter from Ireland in spring 1654, Boyle sent Clodius a recipe for extracting the "Sulphur of stella martis." This too is Starkey's work; Starkey had told Boyle of it with great excitement in a 1652 letter.[214] But the most striking example of Boyle's revelation of Starkey's secrets is the case of the Philosophical Mercury—the item that caused Starkey's explosion at Digby's "Parasite" Farrar. When Starkey first imparted this grand secret to Boyle in 1651, he wrote: "I account these things in your breast to be *tanquam in arca sigillata* [as if in a sealed chest]. Therefor I do not

213. Boyle to Clodius, 27 September 1653; Boyle, *Correspondence*, 1:148–50, on 149; Hartlib recorded the recipe at once in his *Ephemerides:* "Mr Boyle hath imparted to Clod making wine from Pease" (HP 28/2/74A).

214. Boyle to Clodius, April/May 1654, Boyle, *Correspondence*, 1:165–68; Starkey to Boyle, 3 January 1652, in *Notebooks and Correspondence*, document 6; also in Boyle, *Correspondence*, 1:107–11, on 108–9.

praeingage Secrecy confiding that it is your intimate genious so to be."[215] Yet in August 1653 Boyle opened up that "sealed chest" in which Starkey had such confidence and gave Starkey's "Key" to Clodius, allowing it to be transcribed and sent to Johann Moriaen, who distributed it yet further. Starkey himself had refused to send the recipe to Moriaen (or anyone else), and this must have been well known in the Hartlib circle.[216]

Boyle's free hand with Starkey's secrets, coupled with the deferential tone of his letters, implies that he was trying to win Clodius's favor. Despite the dubious nature of Clodius's charms, it is clear that Boyle was willing to transgress the implicit (and sometimes explicit) terms of his established relationship with Starkey in order to win the affections of the German chymist. At the same time that Boyle was extracting and transmitting Starkey's secrets, however, Clodius himself was equally eagerly seeking Boyle's patronage. When Digby pressed his offers to support Clodius's laboratory, Clodius—at least according to Hartlib's report—demurred, replying darkly "Timeo Danaos et dona ferentes" and implying none too subtly that he would prefer Boyle's patronage:

> for whether his [Boyle's] estate will suffer him to contribute little or much for the carrying on of all our physical and chemical affairs and designs, he alone is to be entrusted with a full and entire communication of them all and others, as we shall advise amongst ourselves.[217]

Clodius noted also—perhaps acutely—that in Digby "he could discern rather gallantry than goodness" and averred that he saw both those qualities in Boyle "eminently and superlatively." In the event, it was Digby who entered into the closer relationship with Clodius.

Boyle maintained a relationship of sorts with Clodius at least until 1663. But at no time did that relationship pay off to the extent of his relationship with Starkey. While we have identified items in Boyle's corpus attributable to Clodius, these are few in comparison to those derived from Starkey. Clodius, however, seems to have tried to take credit for training Boyle in chymistry; at least that is the report tendered to John Ward by William Welden, who had lived for a time with Clodius. It is possibly the echoes of this rumor—and/or the letters to Clodius in Boyle's *Nachlass*—that provoked William Wotton to ask John Evelyn if Clodius had in fact initiated

---

215. Starkey to Boyle, April/May 1651, in *Notebooks and Correspondence,* document 3; also in Boyle, *Correspondence,* 1:90–103, on 99.

216. Hartlib, *Ephemerides* 1653 (August 8), HP 28/2/70B; Newman, *Gehennical Fire,* 172–73. This may have been the origin of the German version of Starkey's letter published in 1722; see *Notebooks and Correspondence,* document 3.

217. Hartlib to Boyle, 15 May 1654; Boyle, *Correspondence,* 1:179–84, on 180.

Boyle "among the Spagyrists."[218] It should, however, be clear from this study that not only did Starkey precede Clodius in his relationship with Boyle, but that Clodius himself was partly parasitical off of Starkey's work. While Clodius's affections were certainly the object of Boyle's desires for a while, and while he did in fact give Boyle some chymical information, the Holsatian cannot claim to have been Boyle's "first teacher" in the ways of chymistry. If that title belongs to anyone, it is to George Starkey.

CONCLUSIONS

Various portrayals of the Hartlib circle have emphasized their common goals and motivations, the spirit of free communication, cooperation, and irenicism.[219] But much of the material presented in this chapter casts a rather different light on the members of the group. Factionalism, manipulation, and appropriation seem notable attributes of many of the Hartlibians. The Clodius-Digby alliance may well have been in part responsible for turning Hartlib against Starkey. Various members attempted to appropriate Starkey's work: Worsley claimed Starkey's Philosophical Mercury and *luna fixa* as his own to Moriaen, Clodius may have misrepresented Starkey's early alkahest as his own to Hartlib, and Boyle not only divulged secrets entrusted to him by Starkey, but often gave the impression—sometimes implicitly and sometimes explicitly—that they were his own discoveries. The whole spectacle that this represents—of Clodius, Worsley, Boyle, Digby, and Farrar competing to extract Starkey's secrets, each behind the other's back—seems almost farcical. The comedy becomes even more burlesque when we consider that none of these figures divined the true origin or nature of Starkey's mysterious New England adept, the supposed source of his transmutational arcana, himself a product of the young American's fecund imagination. Driven by the seventeenth-century's passion for "projecting," the overheated world of the Hartlib circle was scarcely a model of harmony or disinterest.

It was largely within this context that Boyle's chymical interests and expertise developed. While the cogent arguments that have been made for the *cause* behind Boyle's turn toward natural philosophy and away from devotional writing have rightly sought factors outside the confines of the Hartlib circle, this does not mean that the primary *content* of Boyle's nat-

218. Ward Diaries, V.a.295; on Evelyn to Wotton, see below, n. 222.

219. For example, Webster, *Great Instauration*, passim (note, however, that the bitter Worsley-Petty debate over the surveying of Ireland is also covered herein); Dobbs, *Foundations of Newton's Alchemy*, 62–80; Ronald S. Wilkinson, "The Hartlib Papers and Seventeenth-Century Chemistry, pt. 2," *Ambix* 17 (1970): 85–110.

ural philosophical training is to be sought beyond the Hartlibians.[220] Among these associates of Hartlib, George Starkey clearly stands out—the methodology, expertise, and originality of his chymical practice and theory outstrips those of any other early associate of Boyle. While Boyle clearly benefited from chymical contributions drawn from a wide array of associates, his relationship with Starkey was not only the earliest and most productive, but also apparently unique in the way that Starkey "tutored" Boyle during his training in chymistry. Even though the closest terms between Boyle and Starkey lasted for only a few years, Starkey's influence permeates through Boyle's life—perhaps most dramatically in Starkey's initiation of Boyle's forty-year quest for the correct means of transforming the Philosophical Mercury Starkey taught him how to make into the Philosophers' Stone.[221]

Yet in spite of Starkey's significant contributions to the young Boyle, there is little acknowledgment of this by Boyle himself; indeed, if anything, there is rather the opposite tendency, as when Boyle attributes his chymical training to "Illiterates" rather than to the young Harvard graduate, as we noted in chapter 1.[222] We also noted there his appropriation of Starkey's laudanum in *Usefulnesse* (1663), of Starkey's discovery of the freezing power of sal ammoniac in "Mechanical Origine of Heat and Cold" (1675), and even of Starkey's prized secret, the preparation of the Philosophical Mercury, in the *Philosophical Transactions* (1676). Some early works studied in this chapter, notably the "Essay on Poisons," show a similar pattern. While Boyle did have a fairly general policy of not citing living authors by name (in praise or blame) he seems yet more indisposed to mention Starkey either by name (he never did in print), or by initials (as he did with Clodius and others), or even by more allusive, nameless references (which he did only occasionally with Starkey).

Interestingly, Boyle began asserting a degree of ownership over Starkey's

220. Hunter, "How Boyle Became a Scientist."
221. On the Philosophical Mercury, see Principe, *Aspiring Adept*, 153–79. Even Boyle's late works bear information collected from Starkey; for example, *Christian Virtuoso II* (first published by Thomas Birch and Henry Miles in 1744 from fragments written in the 1680s), uses testimony regarding cochineal from "a physician, whose writings are not unknown, that was born in America"; *Works*, 12:441. For Starkey's experiments with cochineal, see his letter to Hartlib, in *Notebooks and Correspondence*, document 8.
222. Indeed, William Wotton, in writing his aborted "Life of Boyle" made inquiries to discover who Boyle's chymical teacher was; when asked, John Evelyn pled his own nescience of "who it was Innitiated Mr. Boyle among the Spagyrists, before I had the honor to know him" and hypothesized that this occurred during his years at Oxford; see Evelyn to Wotton, 12 September 1703, in Hunter, *Robert Boyle by Himself*, 92.

work very early—even while the two were still on close terms. The earliest example is in the "Essay of the Holy Scriptures" (c. 1652), where Boyle makes his first known reference to the *ens veneris*. There Boyle writes that a "(secret but an) Easy Sublimation yeeld[s] me a venerall Body,"[223] using the first person singular, even though we know that the process was actually executed by Starkey. Likewise in the 1654 "Collections," Boyle refers to a variation on "our Ens," and similarly in a set of "C[hymical] Notes" dating from the second half of the 1650s, he refers to "our Ens Veneris" using the same potentially ambiguous first person plural that he does in *Useful-nesse*.[224] It is likely, however, that Boyle felt a level of ownership over the *ens veneris* since he apparently underwrote its original preparation.[225] For his part, on the other hand, Starkey seems to have resisted Boyle's attempts to categorize the young American as a chymical "operator" or technician, a status that would have legitimized such proprietorship.[226]

Experiences from Starkey also proved useful for undergirding Boyle's own expertise. In "Essay of the Holy Scriptures," Boyle makes offhanded comments about his own experience to the effect that "I have seen an Alkali Volatile," referring to what he presumably saw of Starkey's work on vola-tilizing alkalies, as he does in "Atomicall Philosophy" by citing "my owne Experience" in distilling a "Mercury incomparably more subtle & volatile then is made by any of our knowne chymicall Processes"—a glancing refer-ence to Starkey's Philosophical Mercury.[227] Boyle tends to use the vague term "we," which leaves the reader unsure whether this is only an authorial plural or a reference to an actual plurality of workers. In "On Poisons," Boyle writes vaguely that Van Helmont's "Hint" directed "some of us" to correct opium with alkali, and Boyle writes similarly of "our Laudanums," and of how the correction of arsenic is "our preparation."[228]

Elsewhere we have noted instances of Boyle's use of the allusive, "hint-ing" style so common in the more secretive chymical works—a practice that displays the author's knowledge and authority in a measured way. Thus

223. Boyle, "Holy Scriptures," in *Works*, 13:204.

224. BP, vol. 25, pp. 347, 72; the latter collection is dated to "Sept. 29" of an unspecified year.

225. Boyle, *Usefulnesse*, in *Works*, 3:501; Hartlib apparently did not see things this way, as he records the *ens veneris* simply as Starkey's, as when Boyle informed him in September 1653 that Starkey has prepared large quantities of several medicaments, and Hartlib recorded that "Stirk has great store of his Laudanum and ens veneris and haematina" (Hartlib, *Ephemerides* 1653, HP 28/2/72B).

226. Newman, *Gehennical Fire*, 62–78.

227. Boyle, "Holy Scriptures," in *Works*, 13:204; Boyle, "Atomicall Philosophy," in *Works*, 13:233.

228. Boyle, "Poisons," in *Works*, 13:247, 254–55.

it is worth noting that Boyle hints at Starkey's own secrets as well as his own. One previously mentioned example of this concerns Starkey's Philosophical Mercury, but Boyle does the same with Starkey's volatilization of alkalies. Recall that Starkey's prized secret was the use of the air as a medium for volatilizing salt of tartar, which he identified as the "artificial and hidden circulation" hinted at by Van Helmont. While *Nature's Explication* and *Pyrotechny*, written in the aftermath of this discovery, frequently allude to the need to know this "hidden circulation," Starkey never mentions that it consists of exposure to the air for extended periods, save making a single allusive, "dispersed" mention of the air.[229] But Boyle himself hints several times at Starkey's secret. In a section of *Usefulnesse* written in the 1650s, Boyle mentions "the hidden ways of making fix'd Bodies volatile, and volatile fix'd" and refers to "the power of the open Air in promoting the former of these Operations."[230] Similarly, in the "Essay on Nitre" (written at about the same time as *Usefulnesse*) and Starkey's works, Boyle notes somewhat mysteriously that

> we have known such changes (seemingly Chymical) made in some Saline Concretes, by the help chiefly of the volatilizing operations of the open air, as very few, save those that have attentively consider'd what *Van Helmont*, and one or two other Artists, have hinted on that subject, or have made tryals of that nature themselves, will be apt to imagine.[231]

Given what we have presented in the previous chapter, it is easy to recognize this passage as a reference to Starkey ("one or two other Artists") and his volatilization of salt of tartar (a "Saline Concrete") by means of air. Even much later, in the 1675 "Mechanical Origine of Volatility," Boyle refers to the "probable Relations of some Chymists" that "the Air does much contribute to the volatilization of some bodies that are barely, though indeed for no short time, exposed to it."[232] Again, this is an allusive

---

229. In one place Starkey refers to how "many Tunnes or never so little quantitie" of fixed salts "laid in any Field" will become volatile, and elsewhere he mentions "the heavenly influences, how by a mean they visit" (this "mean" may refer to the air). In both cases he immediately advises the reader to "imitate Nature" if he wishes to find the secret of volatilizing alkalies; *Pyrotechny*, 86, 143.

230. Boyle, *Usefulnesse*, in *Works*, 3:317. Note that the word "hidden," present in the manuscript (BP, vol. 8, fol. 22) was omitted from the published text.

231. Boyle, "Essay on Nitre," in *Works*, 2:107.

232. Boyle, "Mechanical Origine of Volatility," in *Mechanical Origine of Qualities*, in *Works*, 8:438.

reference to Starkey's multimonth exposure of salt of tartar and oil to the air to produce the Elixir of volatile salt.[233]

In chapter 1, we outlined possible causes for Boyle's very limited acknowledgment of Starkey. These included Boyle's interest in asserting his own independence and originality and enhancing his own status. But there is a further factor that should be mentioned. Starkey undoubtedly became something of a pariah toward the end of his life. His acerbic character, already in full evidence in his declamations against the medical establishment as a "Goose-quill tribe" of "piss-pot prophets" in *Natures Explication,* became even more corrosive through the late 1650s and early 1660s. Every year seemed to bring Starkey into another violent public controversy, whether in his constant struggles and pamphlet wars (like his *Smart Scourge for a Silly, Saucy Fool*) against the depredations of his work (or the raiding of his market) by "empirics," or his perennial conflicts with and (unanswered) challenges to the "Galenists" of the Royal College of Physicians. Starkey's high-profile status as a medical controversialist was far removed from Boyle's chosen public image as an irenic natural philosopher. Did Boyle wish to avoid being publicly associated with Starkey? This is undoubtedly the case. But this chariness toward public association does not mean that Boyle did not benefit from Starkey's tutelage or that he refrained from employing parts of Starkey's work. To reiterate, what is important for our study is the historiographical consequence—Boyle's silence about his sources has made it seem that he had none and has consequently given the impression of a greater discontinuity in the history of chemistry at Boyle's period than is really the case.

When Starkey's longtime friend Jeremiah Astell published the manuscript of Starkey's *Liquor Alchahest* ten years after the American chymist's death, he dedicated it to Boyle. One wonders if perhaps he might have been chiding his dedicatee when he wrote in his preface that "it would not, I believe, lessen the esteem of some eminent Practitioners, should they acknowledge with me, that they had from him [Starkey] those fundamentals of Art that hath rendred them thus famous."[234]

Starkey—the man behind the figure of Eirenaeus Philalethes, the mysterious cosmopolite and adept whom Boyle so wanted to meet and learn from—was in fact Boyle's key chymical teacher all along.

---

233. One might speculate that Starkey's discovery regarding the power of the air might in fact have contributed to Boyle's initial interest in pneumatics and the air pump. The timing is close—Starkey's perfection of the volatilization using the air occurred in 1656, and Boyle's work on the air pump commenced in 1657–58—but of course there is no clear evidence to support this speculation.

234. [Jeremiah Astell], preface to Starkey, *Liquor Alchahest* (London, 1675), [xi–xii].

SIX

The Legacy of Van Helmont's and Starkey's Chymistry

BOYLE, HOMBERG, AND THE CHEMICAL REVOLUTION

The foregoing chapters have explored a variety of issues of laboratory practice in chymistry. We have identified several themes—the emphasis on weights and measures dating back to the High Middle Ages, the synthesis of the metallurgical tradition of medieval alchemy with the Paracelsian notion of *Scheidung* or *spagyria* to give rise to early quantitative analysis in Alexander von Suchten and Joan Baptista Van Helmont, the reliance upon and deployment of "mass balance" in Van Helmont and Starkey, and Starkey's own highly formalized and rigorous laboratory methodology and its sources. Finally, we noted that a surprising amount of Boyle's early training in chymistry came from Starkey. We can now see that several features routinely designated as essential parts of the modern practice of chemistry had already become quite highly developed in mid-seventeenth-century chymistry, particularly in the realms of *spagyria* and chrysopoeia.

But it is certainly worth following the story further to examine how this tradition of practical chymistry endured and developed over time. While it is clear that Boyle's early chymistry of the 1650s was closely akin to that of Starkey, from whom he learned it, what of Boyle's later chymistry? After his years at Oxford, the initiation of his scientific publications, and his move to London, were the traditions he inherited through Starkey and his careful reading of other chymical authorities rejected? Did the English natural philosopher outgrow his American mentor and abandon his juvenilia in favor of a more "modern" chymistry? We now know that this is definitely not the case with Boyle's chrysopoetic preoccupations; he remained a dedicated believer in transmutation and seeker for the Philosophers' Stone until his death in 1691. Moreover, throughout his life, Boyle continued to pursue practical chrysopoeia along the specifically Mercurialist lines that he was taught originally by Starkey.[1] Yet setting aside for the moment Boyle's

---

1. Principe, *Aspiring Adept.*

transmutational endeavors, what is left in his mature chymistry that might be traceable to Van Helmont and his champion Starkey?

It is clear that Boyle's first set of publications, which appeared in the early 1660s and upon which so much of the traditional view of him has been drawn—are strongly Helmontian. The *dubia* that Boyle famously cast on the efficacy of fire analysis in *The Sceptical Chymist* (1661) have long been known to have had Helmontian origins.[2] Similarly, Boyle's water culture experiments are linked to Van Helmont's demonstration that a willow tree is "made of water," and, even though Boyle expresses reservations about certain specific aspects of Helmontian chymistry, he continues to display his admiration and indebtedness to the Belgian. Boyle's identity as a Helmontian chymist was, moreover, clearly perceived by contemporary readers. John Ward, one of Boyle's associates at Oxford, read the *Sceptical Chymist* very shortly after it was published, and recorded in his diary upon reading it that "Mr Boghil [Boyle] does mightily commend van Helmont."[3] Similarly, the Netherlandish chymist Willem Spannut, writing to Boyle specifically about the *Sceptical Chymist* in late 1664, exclaims that "I have seen no one hitherto who touches upon and penetrates Helmont so acutely."[4] These testimonies from contemporary readers who unambiguously identify Boyle as a Helmontian provide a different sense of the *Sceptical Chymist* from the common one linking him to the Royal Society and the Scientific Revolution—in short, they saw the book as part of an existing tradition, not as something radically new. *Usefulnesse of Experimental Naturall Philosophy* (1663) is likewise a strongly Helmontian work resting upon the Helmontian essays that Boyle wrote in the 1650s. This is particularly true of part 2, with its emphasis on medical and iatrochemical topics. Thus the Helmontian grounding that Starkey imparted to Boyle (and that was reinforced by other contacts) endured into the 1660s and undergirded the very works upon which Boyle's reputation was built. It is much less recog-

2. Allen G. Debus, "Fire Analysis and the Elements in the Sixteenth and Seventeenth Centuries," *Annals of Science* 23 (1967): 127–47; Antonio Clericuzio, "A Redefinition of Boyle's Chemistry and Corpuscular Philosophy," *Annals of Science* 47 (1990): 561–89 (esp. 564–68); Clericuzio, "From van Helmont to Boyle: A Study of the Transmission of Helmontian Chemical and Medical Theories in Seventeenth-Century England," *British Journal for the History of Science* 26 (1993): 303–34, esp. 314–19, 331.

3. John Ward Diaries, V.a.292, fol. 83 (July 1662).

4. Spannut to Boyle, 12 December 1664; Boyle, *Correspondence,* 2:436–39, on 436: "neque vidi hactenus qui Helmontium acutius attigit, et punxit." Spannut continues with a punning reference to Helmont's *Opuscula medica inaudita* that "inaudita vere plurimis detexit, donec ex mente V. D. Vidimus, ipsi parem, si non mage astutum" ("he [Helmont] hid the unheard-of things from many, until we saw them from your mind, equal to his own, if not more astute").

nized, however, that Boyle's chymistry maintained Helmontian elements even in his later work.

In the following section we present some of these less recognized Helmontian features of Boyle's mature chymistry by pointing out elements of Van Helmont's chymical system that remain even in the work published by Boyle during the last decade of his active life. We then pursue the story of the impact of Helmontian and Starkeian laboratory practice yet further into the eighteenth century and finally to the explicit and laudatory references to Van Helmont made by Antoine Laurent Lavoisier himself.

THE CHYMISTRY OF SALTS IN BOYLE AND VAN HELMONT

Salts were a major preoccupation of Helmontian chymistry. Van Helmont expended enormous labors on the investigation of alkali salts, and his famous alkahest itself was supposed to be a "circulated salt." Starkey, for his part, so strongly emphasized the importance of salts that he actually (and spuriously) derived the origin of the "al" at the beginning of "alchemy" from the Greek *hals*, or *salt*, thus defining the word to mean "the Art of Separating Salts."[5] Similarly, Boyle has traditionally been lauded for his experiment on the "redintegration" of niter and his development of color indicators, both of which fall within the chymistry of salts.[6] We will therefore consider the relationship between Boyle's work on salts and Van Helmont's—first by looking at the two authors' classification systems and then by considering the closely related theories of corrosivity and the possibility of a universal dissolvent that works without acidity or alkalinity—namely the alkahest.

Boyle has been heralded for his critical attitude toward earlier chymical "families," or classes of substances, and has also been credited with devising a "more comprehensive classification" of saline substances by distinguishing three classes of salts: acid, alkalizate or lixivial, and urinous or volatile.[7] This classification first appears in the early *Usefulnesse of Experimental Naturall Philosophy* (1663). In the midst of extolling the alkahest and volatile salt of tartar (replete with references to Van Helmont) Boyle draws a lengthy comparison between these Helmontian *arcana maiora,* which are supposed to dissolve other substances *sine repassione*—that is, without becoming exhausted (or exantlated)—and common corrosive "saline liquors."

5. George Starkey, *Pyrotechny Asserted* (London, 1658), 4; on other curious derivations see William R. Newman and Lawrence M. Principe, "Alchemy vs. Chemistry: The Etymological Origins of a Historiographic Mistake," *Early Science and Medicine* 3 (1998): 32–65, esp. 42–43.

6. Marie Boas Hall, *Robert Boyle and Seventeenth-Century Chemistry* (Cambridge: Cambridge University Press, 1958), 94, 126–41, 217–19.

7. Clericuzio, "Redefinition," 588; see also 564.

Here Boyle advises his reader not to be discouraged in his quest for Helmontian arcana or to be

> brought to despair of ever seeing any noble Menstruum, that is not sharpe to the taste, nor of any of the three peculiar kinds of Saline Liquor. (Acid as *Aquafortis,* Urinous, as the Spirits of Blood, Urine, and other Animal substances, nor Alcalizate, as Oyle of Tartar *Per deliquium*) . . . And whereas the vulgar Saline Menstruums, (which alone seem to have been known to *Sala* and *Billychius*) are so specificated, if I may so express it, that what an Acid Menstruum dissolves, an Alcalizate, or an Urinous will precipitate, & *è converso;* And whichsoever you choose of these three sorts of Menstruums, one of the other two will disarm, and destroy it.[8]

Thus Boyle classifies "vulgar" saline liquors into three types—acid ones (familiar enough in *"Aquafortis,"* or nitric acid), urinous ones (by which he means ammonia and its salts), and alkalizate ones (namely, the "fixed" alkalies, such as potassium and sodium carbonates). These "Saline Liquors" are liquids derived from their parent salts by a variety of practical processes, primarily distillation, deliquescence, lixiviation, and solution. While the three classes that Boyle lists have their own characteristic qualities, they also interact with one another—"disarming and destroying" each other—as witnessed by effervescence or precipitation.[9] Thus for Boyle, there is a clear, observable, practical means of distinguishing among the three kinds of salts and their cognate liquors.

Boyle also supported this classification with a set of practical chymical tests that could categorize a given salt. These tests are employed rather piecemeal in both *Usefulnesse* and the *Sceptical Chymist,* while the fullest version appears in *Experiments Touching Colours*.[10] There Boyle uses a series of two indicator tests to assign any salt into one of the three "Tribes." He first pours some of the saline liquid into an aqueous tincture of *lignum nephriticum;* if the tincture's blue color vanishes—as it does upon the addition of vinegar or spirit of vitriol—the salt is acid. (Boyle notes that a similar test can be carried out with syrup of violets, which, like our litmus paper, turns red in acid.) If the salt does not change the color of the extract, Boyle can then assign it to one of the two remaining "Tribes" by using a solution

---

8. Robert Boyle, *Usefulnesse of Experimental Naturall Philosophy,* in *Works,* 3:413.

9. At the end of this passage (3:413), Boyle also claims to have discovered a "Menstruum that was but a degree to" the alkahest, which can draw a red tincture from glass of antimony. But since this tincture is not precipitated by any of the three kinds of salts, it lies outside the threefold classification of "vulgar Saline Liquors"—this anomalous menstruum was quite possibly the succedaneum to the alkahest that Starkey prepared in 1652.

10. For example, Boyle, *Sceptical Chymist,* in *Works,* 2:310, and *Usefulnesse,* in *Works,* 4:370.

of "Venetian sublimate" (corrosive sublimate, or mercuric chloride) as a second indicator. If the salt produces an orange precipitate when added to the solution of sublimate, it is concluded to be of a "Lixiviate Nature," like a calcined alkali. Salts of a "Urinous Nature," on the other hand, precipitate out a white deposit in the second test.[11] Thus by combining these two solution tests, Boyle not only devises a practical, experimental scheme for identifying specific salts, but also undergirds his division of all salts and their derivative liquors into three classes.

Boyle's threefold division of salts elicited interest from readers and recurred in many of his books throughout his maturity. In 1668, Michael Behm, a colleague of the astronomer Johann Hevelius at Danzig, wrote to Boyle urging him to publish more of his "chemical experiments on the differences between salts," since he had heard from Timothy Clarke, physician to Charles II and fellow of the Royal Society that Boyle was considering doing so.[12] In later years, the threefold division formed the basis for Boyle's section divisions in the appendix to the 1680 second edition of *The Sceptical Chymist*, entitled *The Producibleness of Chymicall Principles*, which is arguably Boyle's most developed chymical work.[13] The same division is used in other works, for example, the "Mechanical Origine of Corrosiveness," published in 1675/76.[14]

The reader will note that the first appearance of this threefold division of salts (in *Usefulnesse*) occurs in a Helmontian context—during a discussion of the alkahest and its distinctness from "vulgar" saline corrosives. This juxtaposition should alert the reader to the possible origin of the classification itself. Boyle's scheme is in fact developed ultimately from *De lithiasi*, one of his (and Starkey's) favorite Helmontian tracts. In the third chapter of that text—entitled "The Content of Urine"—Van Helmont describes at length his laboratory operations on human and animal urines, experiments that he made in the hope of finding the cause, and hence the cure, of kidney and bladder stones. Here Van Helmont notes how he "examined salts by every analysis."[15] Like Boyle, Van Helmont points out that saline liquors congeal when their solutions are deprived of "fretting" power. Repeated observa-

---

11. Boyle, *Experiments Touching Colours*, in *Works*, 4:154–55.

12. Michael Behm to Boyle, 2 October 1668, in Boyle, *Correspondence*, 4:103–14, on 103–4. In a similar letter written almost a year earlier to Hevelius (and forwarded to Oldenburg), Behm connected his desire for further elucidation to having read *Colours* and mentioned explicitly the three types of salts outlined by Boyle: lixivial, acid, and volatile; see Behm to Hevelius, 1 November 1667; Oldenburg, *Correspondence*, 3:572–77 on 572.

13. Boyle, *Producibleness of Chymicall Principles*, in *Works*, 9:36–47. Here the terms "volatile salt" and "urinous salt" are used interchangeably.

14. Boyle, "Mechanical Origine of Corrosiveness," in *Mechanical Origine of Qualities*, in *Works*, 8:463.

15. Van Helmont, *De lithiasi*, in *Opuscula*, no. 11, p. 23.

tions taught him that "the acid spirit of sulphurs or of salts would become earthy with an alkalized body," meaning that these two types—acid and alkalizate—oppose one another, resulting in the formation of an inactive, "earthy" body.[16] Having observed this result, Van Helmont then performed analyses on salts and learned that "the spirits of all salts are acid with the exception of the alkalized ones and the essential sulphurs in vegetables." Presumably Van Helmont is thinking of the fact that marine salt, saltpeter, vitriol, and several other salts provide the mineral acids when subjected to destructive distillation. Now at first glance it may look as if Van Helmont is merely describing a bifurcation into acids and alkalies, a system that would later be developed by some of his followers and criticized by Boyle.[17] But in fact things are not so simple. After stating the antagonism of acids and alkalies, Van Helmont proceeds to describe a product obtained from the distillation of putrefied human urine and notes that "the spirit of human urine is neither acid, nor alkaline, but merely salty *(salsus)*, just as is that of beasts of burden." The salt of urine thus forms a third class of saline substances. Van Helmont then uses practical laboratory tests to verify and indicate the distinctness of these three classes of saline substances. For example, unlike acid saline spirits, the saline spirit of human urine will not coagulate blood or milk. Also, unlike either acids or alkalies, it can coagulate with spirit of wine to form the famous *offa alba* (actually ammonium carbonate precipitated out of the spirit of urine by the alcohol).[18]

Thus a foundation for Boyle's division of salts into three families or "Tribes" can be found in Van Helmont's distinction of acid and alkaline salts from the volatile salt of urine. As is common in the works of Van Helmont, however, the Belgian chymist chose to write in the form of hints and paradoxes rather than to develop his doctrine into a systematic whole. Nor does Van Helmont seem to have regarded the salt of urine as the primary exemplar in an eponymous class of "urinous salts," as Boyle did, even though Van Helmont did note the similarity of urinous spirit to spirit of human blood, a substance Boyle later classified among "urinous" bodies.[19] Yet the terminology Boyle uses to describe his three classes of salts argues for the Helmontian origin of the system. For example, in the *Experiments Touching Colours* of 1664, Boyle refers to urinous salts as "*Salsuginous* (if I

16. Ibid., 22–23.

17. Hall, *Boyle and Seventeenth-Century Chemistry*, 59, 88–89, 134–36, 148–54; Marie Boas Hall, "Acid and Alkali in Seventeenth-Century Chemistry," *Archives internationales d'histoire des sciences* 9 (1956): 13–28.

18. Newman, *Gehennical Fire*, 182–88, for the significance of the *offa alba* in Helmontian chymistry.

19. Van Helmont, *Ortus*, "Aura vitalis," p. 726.

may for Distinction sake so call the Fugitive Salts of Animal Substances.)"[20] The term "salsuginous" seems modeled on Van Helmont's term *spiritus salsus* for the volatile salt of urine, itself clearly one of the "Fugitive Salts of Animal Substances."

The Helmontian affiliations of Boyle's scheme become clearer when we examine his terminology for the two types of salt that are not acid. In both *Usefulnesse* and *Colours* Boyle sets up a dichotomy between acid salts and "Sulphureous" ones. In *Usefulnesse* he observes how "the contrariety of acid and sulphureous Salts makes them sometimes disarm, sometimes after some ebullition, precipitate each other," and he also notes "that sulphureous Salts, such as Oyl of Tartar, made *per Deliquium*, being drop'd into the expressed Juices of divers Vegetables, will, in a moment, turn them into a lovely green."[21] At several points in *Colours* he again uses the term "Sulphureous Salts" explicitly to describe together both "the Urinous and Volatile Salts of Animal Substances, and the Alcalisate or fixed Salts that are made by Incineration" in opposition to acid salts.[22] Finally, a similar terminology is employed in *The Sceptical Chymist*, where Boyle speaks of "the three chief sorts of Salts, the Acid, the Alcalizate, and the Sulphureous."[23]

This notion that the two "Tribes" of nonacidic salts—those that turn Boyle's vegetable indicators green rather than red—are somehow sulphureous does not immediately make sense on its own. If we consider the properties of sulphur—for example, its characteristic taste, smell, and color—there is no obvious way in which these salts would ever seem "sulphureous." Indeed, this usage is comprehensible only in the light of *Van Helmont's theory of alkalies*. As described in chapter 2, Van Helmont argues that the process of burning combustible materials produces alkalies. During the cremation, a volatile Salt in these substances "seizes upon" a portion of the nearby Sulphur, resulting in a partial or total fixation of two into an alkali.

Now one might think that Boyle's labeling of "urinous" salts as sulphureous does not agree with Van Helmont's theory of alkalies, because urinous salts are volatile whereas Van Helmont asserts that the combination of the Salt and Sulphur during combustion should yield a *fixed* alkali salt. But in fact Van Helmont emphasizes in "Blas humanum" that the fixation is *relative* rather than absolute. For example, when a coal is made by destructively distilling honey, the entire coal burns away when heated in the open air, demonstrating that the alkali formed in the honey's cremation re-

20. Boyle, *Colours*, in *Works*, 4:154.
21. Boyle, *Usefulnesse*, in *Works*, 3:317, 370.
22. Boyle, *Colours*, 4:109, 112.
23. Boyle, *Sceptical Chymist*, in *Works*, 2:289.

mains relatively volatile—less volatile than the initial volatile Salt and Sulphur in the honey, to be sure, but more so than the fixed alkali salt of tartar. Van Helmont clearly means that the alkali is only "more fixed" than the initial volatile Sulphur and Salt and not absolutely fixed.[24] The same action of a volatile Salt seizing on Sulphur accounts for the partial fixation of arsenic (our arsenic sulphide) and saltpeter when the two are burned together. Likewise, Van Helmont believed that one could even remove the flammability from spirit of wine by refluxing it with salt of tartar; the oily, sulphureous component of the spirit of wine responsible for its flammability would be seized and made more fixed by the salt of tartar. The departure of the Sulphur from the spirit of wine was supposed to result on the one hand in the formation of more salt; on the other hand, the "desulphurized" spirit of wine was itself supposed to become mere water.[25]

These examples show that Van Helmont thought that the combination of Sulphur and a volatile Salt yields an alkaline product whose fixity could vary considerably. In the case of the honey coal, the "alkalized" salt is still volatile enough to sublime away without residue, while in the case of salt of tartar, the "alkalized" salt is not volatile at all. Indeed, this theory emerges in "Blas humanum," where the immediate subject is human blood, for Van Helmont wants to argue that the salt produced during the destructive distillation of blood is actually an artifact of the fire rather than a constitutive ingredient of blood in a living being. Thus, he obviously cannot mean that the salt produced from blood is *absolutely* fixed, for he recognizes that this salt of blood (mostly our ammonium chloride) is volatile. This artificial salt, produced de novo by the combination of Sulphur and volatile Salt, is fixed only in relation to the much more volatile blood that is constantly sublimed away invisibly at low temperature through the pores of the living body. Van Helmont's concept of relative fixity thus allows for the presence of Sulphur "fixed" with Salt even in volatile "urinous" salts (such as salt of human blood) as well as completely nonvolatile alkalies (such as salt of tartar).

Returning now to Boyle, it is clear that all alkaline salts, which Boyle can distinguish from acidic ones using a decoction of *lignum nephriticum* or syrup of violets, must—by Helmontian principles—contain a "hidden" Sulphur. Indeed, the "Oyl of Tartar" (salt of tartar that has liquefied by deliquescence) that Boyle uses as his example of a "sulphureous Salt" is the paradigmatic substance that Van Helmont uses to illustrate the fixation of Sulphur in the formation of an alkali. Boyle's use of the term "Sulphure-

24. Van Helmont, *Ortus,* "Blas humanum," no. 39, p. 187: "totum sal, in concreto erat volatile, non in forma fixioris alkali."

25. Van Helmont, *Ortus,* "Complexionum et mistionum elementalium figmentum," nos. 9–11, p. 105.

ous" for two of the classes of salts therefore argues for an unequivocally Helmontian origin for his thinking on salts and their classifications; Boyle's term "sulphureous Salt" implicitly relies upon the Helmontian theory that all alkaline substances contain a Sulphur married to a salt.[26]

The approbation of the Helmontian theory of alkalies that is implicit in Boyle's coining of the term "sulphureous Salt" is also coupled with his explicit affirmation of the Helmontian belief that salt of tartar could "seize" the Sulphur in spirit of wine and thereby transmute that substance from a flammable liquid into a watery phlegm. In *Usefulnesse* and *The Sceptical Chymist* Boyle rather hesitatingly approves of this operation and its underlying theory, saying that the combination of the sulphureous component in spirit of wine with the salt of tartar makes good sense, but pointing out that he himself has not been able to effect the process. Yet in the later *Origine of Formes and Qualities* (1666), a work sometimes viewed as a capstone of Boyle's scientific oeuvre, the English natural philosopher accepts the conversion of spirit of wine into water without reservation. Here Boyle relates that certain "Ingenious Persons," working by his directions, and ignorant of each others' labors, "did both of them reduce considerable quantities of high rectify'd Spirit of Wine" into an insipid phlegm by means of a "duely prepared Salt of Tartar."[27]

STARKEY'S ROLE IN CHYMICAL CLASSIFICATIONS

We noted above that although the third chapter of Van Helmont's *De lithiasi* contains the rudiments of Boyle's "Tribes" of salts, Van Helmont's material is by no means as thoroughly developed or as clearly expressed as Boyle's similar threefold division. Indeed, Boyle apparently wrote a treatise on the "Various Chymicall Distinctions of Salts," to which he refers in the *Sceptical Chymist* and *Usefulnesse*, and from which—as he implies—much of the material in those works on the "Tribes" of salts may be derived.[28] While no manuscript of that essay survives—probably because its contents were largely subsumed into printed works—it is mentioned in a mid-1650s list of his essays.[29] This early work may have been a platform for Boyle's further development of the ideas in *De lithiasi*. At the same time, however,

---

26. Discussions of Van Helmont's theory of alkalies—along with much else of Helmontian character—occur in the correspondence between Boyle and the chymist Daniel Coxe in 1665–66; see esp. Coxe to Boyle, 19 January 1666, in Boyle, *Correspondence*, 3:30–44, esp. 34–36.

27. Boyle, *Origine of Formes and Qualities*, in *Works*, 5:438–439. *Usefulnesse*, in *Works*, 3:530; *Sceptical Chymist*, in *Works*, 2:263.

28. Boyle, *Sceptical Chymist*, in *Works*, 2:310; *Usefulnesse*, *Works*, 3:405.

29. For the listing of Boyle's tract "Observations on some Chymicall Distinctions of Salts," see *Works*, 14:332.

there is good evidence that Boyle may have derived the threefold classification of salts from George Starkey, or at least collaborated with him in its formulation, for Starkey was already employing this distinction by October 1657. In a dated notebook fragment presenting a Scholastic discussion about the making of the alkahest, Starkey presents the following *quaestio:* "Whether [the alkahest exists as] a liquor capable of undergoing distillation, or rather as a liquor impregnated with the immortal salt." He responds in the fashion of a trained schoolman, by bifurcating the question into two parts, corresponding to three classes of substances.

> If it should be desired in the form of a liquor, whether an acid or a saline liquor should be sought? If an acid one, the acidity of the same will be destroyed both in all alkalies as well as in all urinous spirits . . . If such a spirit be purely saline, however, it will be nothing except urinous spirit, spirit of alkalies, or the fat and sulphureous spirit of vegetables.[30]

By means of this Scholastic analysis—whose content is itself based on the third chapter of *De lithiasi*—Starkey first excludes acid saline liquors as candidates for the alkahest. Acids do not satisfy the desideratum of working *sine repassione*, since they are destroyed by alkalies and "urinous spirits." Having excluded acids, Starkey passes to the remaining candidates—nonacidic substances, namely, urinous spirits, spirit of alkalies, and sulphureous vegetable spirits. We need not follow him further, for it is clear that he has explicitly treated "urinous spirits" as a category distinct from "spirit of alkalies" and noted that both of them have the effect of "destroying" acids. This is the same approach Boyle takes in *Usefulnesse, The Sceptical Chymist,* and other works of the 1660s. Indeed, Boyle's early comment about the "three sorts of saline liquors" made in *Usefulnesse*—that "whichsoever you choose of these three sorts of Menstruums, one of the other two will disarm, and destroy it"—is closely akin to Starkey's notebook passage.

Starkey's tripartite division of salts into acid, alkali, and urinous classes is not restricted to an isolated manuscript, but appears in print as well. His *Brief Examination and Censure of Several Medicines,* devoted to the analysis and exposure of several commercial medicines, refers unequivocally to the spirits of hartshorn, soot, and bones as "urinous." This shows that Starkey, unlike Van Helmont, had plainly turned spirit of urine into an eponymous exemplar under which a whole class of substances could be grouped.[31] Additionally, Starkey's posthumously published *Liquor Alcha-*

---

30. *Notebooks and Correspondence,* document 14; Sloane 631, fols. 198v–199r.
31. George Starkey, *A Brief Examination and Censure of Several Medicines* (London, 1664), 38–39.

hest explicitly lists the same three categories of salts and uses the same names for them—acid, alkalizate, and urinous.[32] *Liquor Alchahest* also speaks of "a volatile Salt in the soot" from burnt wood, which Starkey classifies as "plainly and truly Urinous"—as Boyle also did.[33] Although *Liquor Alchahest* was not published until ten years after Starkey's death in 1665, these comments clearly reflect the same tripartite division that had already surfaced in the 1657 manuscript. The fact that Boyle's tract on salts was also written in the 1650s, at a time when he was very close to Starkey, and indeed often heavily indebted to him, strengthens the linkage.

If we return to Boyle's 1680 *Producibleness*, it is revealing that immediately after the section on salts, the treatise also gives a tripartite division of spirits as well, classifying them as acid, urinous, or vinous. Boyle uses this division to organize the entire second part of the book. It is certain that this second tripartite division, like the division of salts, also appears earlier in Starkey's writings. In *Pyrotechny Asserted*, when Starkey describes the various medicaments that he plans to make, he refers to *"acid,* and *vinous,* and *urinous spirits"* as three distinct classes.[34] Elsewhere in the same work he also declares explicitly that "there are three distinct kind of Spirits, *Acetous, Urinous,* and *Vinous,*" using "Acetous" as a synonym for "acidic."[35] Starkey's threefold distinction of spirits and his tripartite division of salts both devolve from the same kind of exhaustive investigation that we have found to be characteristic of the American chymist's laboratory notebooks. Starkey is combing through the chymical realm for the *arcana maiora;* in order to facilitate his search he carefully divides the vast array of chymical substances into manageable classes made up of substances that show analogous properties in the laboratory. As we saw in his 1657 manuscript, it is not at all far-fetched to see in this classification system the same Scholastic elements that Starkey deployed in guiding his laboratory practice.

To conclude this part of our argument, we find that the important work on the classification of salts that remained a key part of Boyle's chymistry well into his later years was (although obviously and importantly elaborated by his use of color indicators) actually built upon a Helmontian foundation that was probably developed and transmitted to him by Starkey. This transmission through Starkey is further corroborated by the analogous division of spirits found in both Boyle's and Starkey's publications.

32. George Starkey, *Liquor Alchahest* (London, 1675), 7–9. For a fuller analysis of the dating of this text see *Notebooks and Correspondence.*

33. See, for example, Boyle, *Sceptical Chymist,* in *Works,* 2:310, 359; *Producibleness,* in *Works,* 9:35–36.

34. Starkey, *Pyrotechny Asserted,* 171.

35. Ibid., 127.

OTHER HELMONTIAN NOTIONS IN BOYLE'S LATER CHYMISTRY
While Boyle's threefold division of salts appears to have lasted throughout his life, what of other Helmontian notions about salts? Although Boyle in the 1660s supports (to varying degrees) the Helmontian theory of Salts fixing Sulphurs into alkalies, his 1680 *Producibleness* devotes a long section to the production of "Alcali's or Lixiviate Salts" in which there is a sustained dialogue with this Helmontian theory. Boyle adduces several pieces of experimental evidence designed to cast doubt on Van Helmont's idea. For example, Boyle points out that common brimstone must contain a Sulphureous principle (as shown by its oiliness and flammability) as well as a Salt (as shown by the acidic saline spirit collected when it is burnt under a bell jar). Yet when brimstone, composed of a Salt and a Sulphur "combin'd by nature," is put into the fire, it can be easily sublimed; hence there is in brimstone an apparently unfixed combination of a Salt and a Sulphur. Now the force of this objection rests solely on the assumption that Van Helmont claimed that all combinations of Salt and Sulphur give a fixed product, but as noted above, Van Helmont actually held a principle of *relative* fixity. Van Helmont could simply have replied that the brimstone is indeed more fixed than its putative components (which have already "seized" one another), even though it is still capable of being sublimed. So Boyle may here be arguing against a simplified version of Helmontian thought postulating that the mutual "seizing" of Salt and Sulphur leads inevitably to a state of total fixation, as in the case of salt of tartar. Boyle himself presumably did not interpret Van Helmont in this fashion, for as we saw above, he considered volatile, urinous salts to be "sulphureous" in his works of the 1660s. Indeed, what may amount to a critique of "vulgar Helmontianism" here seems akin to Boyle's comments on the simplified acid-alkali duality proposed by Otto Tachenius in the 1660s, which the English natural philosopher assailed in his "Reflections upon the Hypothesis of Alcali and Acidum" of 1675.[36]

Thus while Boyle expresses some disagreement with Van Helmont's theory of alkali production, it is important to note how he does so, for it displays his intellectual commitments and rhetorical methodology. Boyle often adopts and hybridizes several (sometimes competing) theories, accepting some parts of each and rejecting others. He also occasionally uses notions from a system that he rejects in order to cast doubts on yet another system. In *Producibleness*, for example, he temporarily adopts a position

---

36. Boyle, "Reflections," in *Mechanical Origine of Qualities*, in *Works*, 8:407–19; see Hall, "Acid and Alkali."

from his opponents the "vulgar chymists" against Van Helmont.[37] Clearly, Boyle is not suddenly adopting the view of those whom the entire work is intent upon refuting. Boyle's "polyphonic" style should not obscure the fact that the whole point of his essay is to argue in favor of the production of alkali salts de novo against the "common opinion of the Chymists" who believed them to be preexistent in mixed bodies. In this argument, Van Helmont was obviously an ally, and indeed, Boyle is explicit about how he dissents from only one aspect of the Belgian chymist's system. He writes how he finds Van Helmont's theory of alkalies somewhat less than satisfactory "especially since 'tis applied to *all* fixt *Alcalies*"; that is, "it seems not, at least universally true."[38] This is an expression of Boyle's well-known "diffidence" and his great dislike of framing universal claims about the natural world. Indeed, in the same section of the work, Boyle claims that alkalies may in fact "be produc'd in multitudes of mixt Bodies, especially in a good number of Vegetables, after the way proposed by *Helmont,* or by some such like."[39] Boyle is not being inconsistent. It is the universal extension of Van Helmont's theory to the production of *all* alkalies by *all* methods, or the seizing of *all* Sulphurs by *all* Salts in *all* cremations to produce absolutely fixed alkalies that Boyle denies. Thus, we should not read this section of *Producibleness* as a wholesale rejection of Helmontian ideas in Boyle's later chymistry.

Indeed, Boyle's attitude toward Van Helmont is often similar to his attitude toward writers on metallic transmutation. While Boyle does not hesitate to voice criticisms in the *Sceptical Chymist* and elsewhere regarding the chrysopoeians' annoying secretive language or certain defects in or overextensions of their theories, he nonetheless continues to maintain several of their notions and to pursue their goals. Moreover, chrysopoetic theories are often very useful in arguing against Boyle's real chymical foes, the "vulgar chymists." Boyle deploys Van Helmont similarly; he is willing to criticize aspects of Helmontian theory—particularly when it makes universal claims or is extended by "vulgar" interpreters to do so—but at the same time he retains much of Van Helmont's thought and often finds him useful in opposing the "common opinion of the Chymists." What is key in all of this is the need for sensitivity to the divisions within seventeenth-century chymistry and to the finesse and direction of Boyle's arguments. Boyle did in fact "admit no man's opinions in the whole lump," and so we must be

37. Boyle, *Producibleness,* in *Works,* 9:39–41.
38. Ibid., 9:39, 41.
39. Ibid., 9:43.

careful when reading his criticisms (or praise) of various authors or schools not to overextend them.

The fact that Boyle continued to be inspired by Helmontian chymical practice becomes immediately apparent in the following section of *Producibleness*, which describes Boyle's study of the salts in urine. Boyle notes that besides the volatile salt extractable from urine it also contains two fixed salts. These are distinguishable by their different crystal shapes, which Boyle describes. One of these is "like common salt," and upon "a light suspition" that the table salt ingested with our food appears in the urine, he attempts to compare human urine with horse urine to test the idea. Finally, Boyle notes that a fixed alkaline salt is never found in urine, even if the distillation residue is cremated and extracted.[40] All of these observations are strikingly similar to those expressed in the third book of Van Helmont's *De lithiasi*— the background text for Boyle's classification of salts.[41] There Van Helmont describes how he "thought to anatomize the salt of urine" and found in it "two more fixed salts" besides the volatile salt. Like Boyle, Van Helmont separates these fixed salts by careful fractional crystallization, and describes their different crystal shapes, remarking that one of them is a "marine salt" taken in with our food, while the other is "born in our digestion." Indeed, Van Helmont observes, perhaps on the basis of an experiment on himself, that if one abstains from the use of table salt for a few days, the digestion-generated salt, which is the "truly urinous" salt—presumably unlike "marine" salt—still continues to appear in the urine. Van Helmont also notes the different tastes of the two salts and their crystallization in different regions of the evaporating vessel. Finally, Van Helmont, like Boyle, compares human and horse urine and also makes a point of noting that no alkali salt is ever found in urine. Both conclude that there are three salts— one volatile and two fixed—in putrefied urine. The linkage of Boyle's practice to Van Helmont's is made clear by Boyle's quotation from the third chapter of *De lithiasi* at the end of the *Producibleness* essay: "Wisdom despises those that despise the indagation of Urine." Curiously, while Boyle states that he is "not altogether of Helmonts mind" in this particular, "yet I think that those who understand the mystical writings of some of the best Chymicall Philosophers of former times, will look upon it as a more tolerable Hyperbole, than other Men or even Vulgar Chymists imagine it to be."[42] This sentiment about the secrets of urine may refer to Starkey's belief that the long sought-after alkahest is to be found there.

40. Ibid., 9:49–51
41. Van Helmont, *De lithiasi*, in *Opuscula*, nos. 19–25, 25–26.
42. Boyle, *Producibleness*, in *Works*, 9:51; the original is Van Helmont, *De lithiasi*, in *Opuscula*, no. 33, p. 30.

THE FATE OF QUANTITATIVE ANALYSIS IN BOYLE'S
MECHANICAL CHYMISTRY

Another element in Boyle's critique of the Helmontian alkali theory not only illuminates Boyle's commitment to mechanical explanations of change in general, but highlights a characteristic feature of his own chymistry. Expressing his allegiance to the mechanical philosophy, Boyle suggests that pace Van Helmont, merely "Mechanicall changes" may suffice to explain the production of alkalies in the fire. That is, it is not necessary for the chymical species of Salt and Sulphur to combine in order for an alkali to be formed. The gist of his approach is that even though most fixed alkalies (excepting "Egyptian Nitre," a naturally occurring soda) are indeed productions of the fire—as Van Helmont claimed—it is unnecessary to suppose that alkalies *must* arise specifically from Salt and Sulphur fixed together by fire. Instead, Boyle argues, "the size, shape, and solidity or weight of the saline Corpuscules" suffice to make a fixed and alkaline salt.[43] As an example, Boyle claims that by heating niter with pipe clay, that is, without the addition of anything discernibly Sulphureous, he is still able to produce an "*Alcali* of Nitre," thus providing evidence that "a congruous change of Texture may suffice" to produce an alkali from a nonalkali.[44] The combination of discrete chymical species to form a particular product is not necessary within Boyle's mechanical framework, which claims that the mere alteration of the texture of a body can transform part or all of it into a different one. This is clearly contrary to Van Helmont's assertion that if either of the ingredients of the alkali is lacking (as in the case of rotten wood) no alkali can be formed.[45]

Boyle goes on to note that while fire does in fact often produce alkalies, it can also destroy them. By keeping salt of tartar in fusion "for a good while" and then dissolving the cooled salt in water and filtering the solution, Boyle obtained an earthy substance, despite the fact that the purified salt had initially dissolved in its entirety without leaving any residue in the filter. Boyle performed the same series of fusion, dissolution, and filtration at least sixteen times on the same portion of salt of tartar, and at each repetition always found newly produced "slime or mudd" in the filter paper. This "slime" was "of a nature very differing from Salt of Tartar"; it was insipid and earthy, and thus no longer alkaline.[46]

43. Boyle, *Producibleness,* in *Works,* 9:41.
44. Ibid., 9:46.
45. Van Helmont, *Ortus,* "Blas humanum," no. 41, p. 188; see above, chapter 2.
46. Boyle, *Producibleness,* in *Works,* 9:47. Boyle was actually isolating silicates that the fused alkali had dissolved from the walls of his crucibles.

Boyle's argument here has a subtle consequence, but one of great significance. It has sometimes been noted that Boyle's mechanical philosophy, taken to its logical conclusion, works against the idea of fixed chemical species (for example, elements) that can combine and separate without undergoing fundamental change.[47] According to Boyle's favored mechanical scheme, anything can be produced from anything by a change of corpuscular texture alone without the mediation of specific chymical species. This notion, when carried to its logical extreme, undercuts the possibility of quantitative analysis altogether, because if specific chymical species are not necessary components of a given "concrete," then they are not necessary products of analysis either. Thus no true analysis could ever be possible. We can see, then, that while both Van Helmont and Boyle criticize fire analysis on the same general practical grounds—namely that the fire produces substances de novo rather than merely isolating them—Boyle is actually more radical. While Van Helmont asserts that certain specific substances *are* necessary in order to compose certain other substances (as in the case of the requirement that both Salt and Sulphur must be present to produce an alkali), and therefore only certain substances can be isolated from the decomposition of a given substance, Boyle's mechanical system must reject even this. In this case, according to Boyle, alkalies can be made by various agents from anything and converted by various agents into anything. Thus Boyle denies the notion of "constant composition" implicit in Helmontian theory, whereby Sulphur and Salt must combine in order to produce an alkali (and conversely, an alkali must be composed of Sulphur and Salt).[48] Without such an underlying notion of "constant composition," quantitative analysis is futile. We will return to this point below after a further consideration of the importance of analysis.

To sum up what we have seen thus far, then, the Boyle of the 1660s supported several key aspects of Van Helmont's chymistry. Helmontian features are explicitly and implicitly present in the important works that Boyle

---

47. This point was made long ago by (among others) Thomas Kuhn, "Robert Boyle and Structural Chemistry in the Seventeenth Century," *Isis* 43 (1952): 12–36. It has recently been reaffirmed by Ursula Klein, "Robert Boyle: Der Begründer der neuzeitlichen Chemie?" *Philosophia naturalis* 31 (1994): 63–106. Although Kuhn and Klein exaggerate the degree of Boyle's opposition to fixed chemical species, the position held by him in *Producibleness* suggests that Antonio Clericuzio's rebuttal of the Kuhnian position is itself overstated; see Clericuzio, "Redefinition," 563–64.

48. Note that the term "constant composition" we employ is *not* intended to refer to the early nineteenth-century discovery of "constant proportions," meaning that a specific weight of one substance always combines with a given weight—neither more nor less—of another substance in a constant ratio. By "constant composition" we mean only that a given substance must be formed by the union of other specific substances, in contradistinction to Boyle's notion of "anything from anything."

published at that time. These Helmontian notions include the threefold division of salts—which apparently came to Boyle through the mediation of Starkey—and the theory that alkaline substances are produced by a combination of Sulphur with a volatile Salt. In the later phase of Boyle's career he explicitly promoted a mechanical alternative to Van Helmont's theory of alkalies, even while retaining the three "Tribes" of salts and the possibility that Van Helmont's alkali formation theory was true in some instances.[49] As can easily happen, a casual reading of Boyle brings to light criticisms of his predecessors' theories but not the origins of his own, nor the degree to which he continues to adopt other notions from the thinkers he criticizes. In this way, we can see that some Helmontian foundations of Boyle's edifice remain even after the superstructure was remodeled along the lines of a stricter mechanical philosophy. As we shall now show, the same pattern may be found in Boyle's treatment of acids and their relationship to the alkahest.

EXANTLATION AND THE ALKAHEST IN BOYLE AND VAN HELMONT
In chapter 2 we described the Helmontian theory of exantlation, whereby corrosives are thought to lose their fretting power and become exhausted— "exantlated" in Helmontian language—not only by going into combination with another substance, but also because of an internal loss of vigor. One of the things that distinguishes the alkahest from ordinary corrosives is that the latter are affected by the substances upon which they act, even when they do not physically combine. The alkahest, on the other hand, works *sine repassione*—it does not undergo any change as a result of dissolving another substance, but rather can be recovered intact from the dissolved material, retaining its powers of dissolution unimpaired indefinitely. One bulwark of Van Helmont's exantlation theory is his illustrative experiment of making a "vitriolic alum" from oil of vitriol and quicksilver. Van Helmont acknowledges that the mercury initially goes into combination with the oil of vitriol to form a *larva* (literally, "mask") or superficial compound, but when he washed this compound and evaporated the wash water, he found aluminous crystals. Upon reducing the washed compound back into mercury, Van Helmont thought that he could regain the full weight of his original quicksilver and therefore concluded that the aluminous crystals were produced solely from the oil of vitriol, by the "radial activity" of the

49. Already in the *Sceptical Chymist*, in *Works*, 2:358–64, Boyle speaks of Van Helmont's theory in mechanical terms, but this comes directly after his acceptance of the Helmontian notion that Mercury, Sulphur, and Salt can exist in bodies as "heterogeneities" without being their ultimate principles (at 2:357). Hence Boyle is not denying the Helmontian theory that a volatile Salt is fixed by Sulphur, but only providing a parallel mechanical explanation.

mercury. In other words, these crystals contained no mercury, but were merely the exantlated acid.

Boyle used Van Helmont's theory of exantlation in a number of his more mature works. Yet in the case of exantlation Boyle does not set up a parallel mechanical explanation as an alternative as he did with the alkali formation theory. On the contrary, Boyle uses this strange Helmontian idea to illustrate and to support his own mechanical hypothesis, even though Van Helmont is clear on the point that the radial activity he describes is nonmechanical. Given the Helmontian character of *Usefulnesse* (as mentioned above), it is not surprising that Boyle uses the exantlation theory in that text, where he refers to "the generality of Corrosives, and the like Acid or Saline Liquors, which work but upon few kinds of bodies, and soon coagulate, or exantlate themselves by working."[50] The appearance of exantlation, along with Van Helmont's illustrative vitriolic alum experiment in *Certain Physiological Essays,* however, is more interesting, for there it is used to support Boyle's mechanical explanation of the firmness and fluidity of bodies. Boyle particularly wants to refute those "Eminent Modern Philosophers" who argue that a fluid must be composed of parts that are themselves fluid. So Boyle describes the treatment of quicksilver with common oil of vitriol to provide a "ponderous Calx or Powder" in the retort after the liquid has been distilled off. Boyle, following Van Helmont's experiment (even though the Belgian chymist is not mentioned), washes this powder and finds only part of it soluble. What is of particular interest to Boyle, as to Van Helmont, is the residue remaining after the evaporation of the wash water because that residue is composed of "a store of saline and brittle bodies." Recapitulating Van Helmont's explanation of the process, Boyle notes that these crystals "proceeded rather from the *Menstruum* [the oil of vitriol] than the metal." Therefore, the isolation of this brittle salt from the wash water indicates to Boyle that "the saline Corpuscles that chiefly compose it [the oil of vitriol], do retain their stiffness" even when in the state of "the very liquid Oyl of Vitriol."[51] Thus since the "saline and brittle bodies" proceed from the oil of vitriol, which was entirely liquid before, Boyle's unnamed opponents are forced to abandon their claim that all fluids are composed of parts that are themselves fluid. The force of Boyle's demonstration lies in the supposed fact that the "alum" contains no mercury and comes entirely from the oil of vitriol, which is Van Helmont's claim.

50. Boyle, *Usefulnesse,* in *Works,* 3:411.
51. Boyle, *Certain Physiological Essays,* in *Works,* 2:186.

How does Boyle know that this aluminous salt is not a combination of the oil of vitriol with the quicksilver? Van Helmont had determined initial and final weights of the quicksilver, and finding no loss concluded that since no mercury was unaccounted for, the crystals could not contain any. Boyle, however, is not so circumspect, and it seems that he relies on Van Helmont's testimony, for his own reasoning is rather weak:

> And that these [crystals] proceeded from the *Menstruum* rather than the metal, we were induc'd to think, by observing the dry Calx, before any water was pour'd on it: for though the saline part of the Mixture did not weigh (perhaps anything near) so much as the Mercurial distinctly did, yet the Aggregate or Mixture did weigh a great deal more than the Quicksilver did when it was put in; and the Oyl of Vitriol that was abstracted, a great deal less than it did before it was committed to distillation.[52]

Boyle's own quantitative attempts seem rather halfhearted. Content with such expressions as "a great deal more" and "a great deal less," Boyle does not give either absolute or relative quantities of his ingredients. His reasoning seems to be that the "Aggregate" calx produced from the mercury and oil of vitriol weighed more than the initial mercury, and the oil of vitriol that was distilled off weighed considerably less than the oil of vitriol that was used. Hence there was a loss of weight in the oil of vitriol and a gain in the calx. But all that this tells us is that some oil of vitriol combined with the quicksilver to produce the "dry Calx"; it does not demonstrate that all of the quicksilver remained in the calx after it was washed or that only the originally "vitriolate" parts dissolved in the wash water. This is exactly the point that Van Helmont endeavors to prove by his more explicitly quantitative study of the process. There are two possible explanations for Boyle's laissez-faire approach. He either felt that he could rely upon the account provided by (the here unreferenced) Van Helmont, and so did not need to carry out exact measurements himself, or having found that Van Helmont's measurements were wrong, he chose to elide the issue by avoiding exact weights altogether.[53] At any rate, we will see shortly that Boyle's relative indifference to precise measurements of initial and final weights is not uncharacteristic of his approach, and has implications of its own.

Boyle reuses Van Helmont's experiment to illustrate the mechanical philosophy in the *Sceptical Chymist*. Here Boyle wants to argue that the dry distillation of vitriol produces a liquid "salt," that is, oil of vitriol, because

52. Ibid.; see also 2:145.
53. For some evidence that Boyle found Van Helmont's experiment less than "fit to be rely'd on," see Boyle, *Formes and Qualities*, in *Works*, 5:371.

the fire of the distillation causes various particles of the vitriol to be "re-duc'd to such a shape and bignesse as is requisite to compose such a Liquor as Chymists are wont to call Phlegme or Water" which phlegm then distills over with the saline parts of the liquid. In order to support this viewpoint, Boyle alludes to the fact that he has "divers times separated a salt from Oyle of Vitriol itself (though a very ponderous Liquor and drawn from a saline body) by boyling it with a just quantity of Mercury, and then washing the newly coagulated salt from the Precipitate with fair Water."[54]

Here again Boyle implicitly relies on Van Helmont's concept of exantla-tion and radial activity to support this mechanical hypothesis. Since the "newly coagulated salt" contains no mercury, it must be a product of the vitriol itself, but in a different physical state than it was while liquid. The tex-ture of the vitriol corpuscles has been altered—first by being distilled into the liquid oil of vitriol and then by reverting to a solid state under the influ-ence of the mercury—even though the mercury added nothing of its sub-stance to transform the liquid into a solid; the mercury acted without corporeal aggregation or material interchange.

Exantlation occurs only with "vulgar" corrosives and not with the alka-hest; hence, the theory behind this "immortal solvent," which can work *sine repassione*, is closely related to the Helmontian theory of exantlation. It is now well-known that Boyle eagerly sought the Helmontian alkahest as a potential agent of analysis far superior to fire and even wrote an entire tract on the subject. Yet there is an interesting relationship between Van Hel-mont's theory of the alkahest and Boyle's own mechanical philosophy that has gone largely unnoticed.[55] This Helmontian arcanum provided a sur-prising and powerful support for Boyle's reductionist corpuscular theory. Van Helmont's own explanation of the alkahest's unusual ability to dissolve substances without combining with them was couched in expressly corpus-cular terms.[56] As he says in the *Ortus medicinae*, the alkahest is a "liquid which, reduced to the smallest atoms possible in nature, would chastely spurn the marriages of any ferment."[57] Because of the small size of its par-ticles, the alkahest had "attained the final limit of subtlety in nature."[58] It was the minuscule particle size of the alkahest that allowed it to penetrate readily into the pores of even the densest substances, such as gold, and cut

54. Boyle, *Sceptical Chymist*, in *Works*, 2:359. The experiment is also mentioned in *Colours*, *Works*, 4:158–59, but without the background of exantlation.
55. Principe, *Aspiring Adept*, 183–84; Boyle mentions his authorship of this treatise in the preface to *Producibleness*, in *Works*, 9:27, but no trace of it now remains.
56. Newman, *Gehennical Fire*, 141–51.
57. Van Helmont, *Ortus*, "Imago fermentis impraegnat massam semine," no. 28, p. 116.
58. Van Helmont, *Ortus*, "Potestas medicaminum," no. 24, p. 474.

them apart. By dividing the particles of substances into such minute portions that they could no longer support the action of the *semina* that supplied their qualitative characteristics, the alkahest would ultimately reduce
all substances into their primordial water.[59] Relying upon medieval discussions of the philosophical mercury, Van Helmont also asserted that the
alkahest's corpuscles were *uniformly* tiny—this homogeneity of particle
size allowed the alkahest to be distilled away from whatever it had dissolved
without leaving the slightest residue.

Although it is certainly true that Van Helmont invested this medieval
corpuscular theory with vitalistic concerns of his own devising, it was an
easy matter for Boyle to dismiss these while still using the Helmontian theory for his mechanical philosophy. In his "Mechanical Origine of Volatility," Boyle describes the alkahest in explicitly mechanical terms:

> And I see not how those that admit the Truth of this strange Alkahestical op-
> eration, can well deny, that Volatility depends upon the Mechanical affections
> of matter, since it appears not, that the Alkahest does, at least in our case,
> work upon bodies other than Mechanically. And it must be confest, that the
> same material parts of a portion of corporeal substance, which, when they
> were associated and contexed (whether by an *Archeus,* seed, form, or what
> else you please,) after such a determinate manner, constituted a solid and fixt
> body, as Flint or a lump of Gold; by having their texture dissolved, and (per-
> haps after being subtilized) by being freed from their former implications or
> firm cohesions, may become the parts of a fluid body totally Volatile.[60]

The agreement between the "mechanist" Boyle's explanation of the
alkahest and that of the "vitalist" Van Helmont is, to say the least, striking.
Like Van Helmont, Boyle argues that the alkahest works by cutting bodies
into such small particles that they form a liquid and even become volatile.
Indeed, Boyle's explanation of the alkahest is used in "Mechanical Origine
of Volatility" precisely to support the English natural philosopher's contention that volatility is related to small particle size. In a similar vein, Boyle
had already argued in *Certain Physiological Essays* that the ability of the
alkahest to dissolve bodies by cutting them apart supported his corpuscular
explanation of fluidity because it showed that "the most solid Body by being divided into parts small enough to be put into motion by the causes that
keep those of water and other Liquors in agitation, may become fluid Bodies."[61] Boyle's appeal to the alkahest in these mechanical contexts was not

59. Van Helmont, *Ortus,* "Progymnasma meteori," p. 68, #7.
60. Boyle, "Mechanical Origine of Volatility," in *Mechanical Origine of Qualities,* in
Works, 8:432.
61. Boyle, *Certain Physiological Essays,* in *Works,* 2:128.

merely inserted for the sake of its exotic novelty—the fact that the Helmontian alkahest was capable of being withdrawn intact from the substances it had dissolved meant that it did not combine with the corpuscles of the dissolved body. As a result, the characteristics of volatility and fluidity induced in the dissolved body could not be attributed to some portion of the alkahest remaining with it. Hence the induced volatility and fluidity had to be due to a mere change in the texture of the dissolved body itself, which was precisely the point that Boyle desired to prove.

If we now compare Boyle's employment of the alkahest as a prop for the mechanical philosophy with his treatment of Helmontian exantlation, an interesting fact emerges. Boyle uses the production of "alum" from oil of vitriol and the action of the alkahest as evidence for a change of state that is induced in a substance without the addition of any other substances that go into combination with it. Since the Helmontian "alum" was congealed from oil of vitriol by the "radial activity" of the mercury alone, it contains nothing but oil of vitriol, just as a substance dissolved by the alkahest contains nothing but the dissolved substance in a changed state. In general terms, then, both examples serve as compelling demonstrations of the fact that changes of state need not stem from the addition or subtraction of some specific ingredient with particular qualities, but can derive merely from the dissociation, association, or rearrangement of corpuscles whose material identity remains unchanged. Boyle's eventual dissatisfaction with the Helmontian theory of alkalies falls into the same pattern, for there Boyle questions the Helmontian requirement that two specific substances (Sulphur and Salt) are required to combine in order to form an alkali in favor of a mechanical process whereby the agency of fire alters the texture of a single substance into or out of an alkaline form without the need for the addition or combination of specific material ingredients. These examples clearly indicate Boyle's tendency to deemphasize chemical combination in favor of the mechanical alteration of a single substance.

If we now return to the thought world of Van Helmont, a surprisingly parallel situation can be discerned. Both Van Helmont and Boyle believed in a "uniform catholic matter" from which all things are made. For Van Helmont, this substance is water; for Boyle it is the homogeneous, quality-less matter comprising the *prima naturalia*—the miniscule corpuscles that lie at the basis of material existence. Despite this difference in the identity of the ultimate form of matter, both Van Helmont's and Boyle's views of changes in matter have a curious, unexpected resonance. Van Helmont posited the origin of real change in the action of *semina* upon the universal substance, thereby converting water into the multifarious substances of the world of sensible phenomena. In Helmontian chymistry—especially in the form of it passed on to Boyle by Starkey—it is these deep internal workings of the *semina* that constitute the ultimate object of practical laboratory studies. The chymist needs to know about the *larvae*, or disguises, that substances assume either when they undergo a spontaneous change of state like water frozen into ice or when they go into superficial combination—as when mercury assumes the *larvae* of cinnabar, corrosive sublimate, calomel, and so forth—but he must recognize these superficial products of his practice primarily for the purpose of directing his real study away from them and toward the world of the *semina* and their more profound activity.

Ordinary chymistry operates at the level of gross corpuscular interaction, while the production of Helmontian *arcana maiora* such as the alkahest and the Philosophers' Stone requires operations at the deeper level of subtle particles that can combine in an irreversible "union *per minima*."[62] Boyle, like Van Helmont, recognized that the reversible chemical reactions of "vulgar chymistry"—not to mention changes of state like freezing—do not represent the most intimate workings of nature. He explicitly argued in the *Sceptical Chymist*, for example, that metallic transmutation presupposes an analysis of the metallic corpuscles into smaller, more primitive component particles that can then be reassembled to give a new metal.[63] Operating on the newly formed metal by means of ordinary "vulgar" chymical processes will not allow one to disassemble it again into its components. It is only by working at a more intimate level that one can effect such radical change.

Additionally, Boyle's mechanical interpretations of exantlated acids and the dissolution of materials by the alkahest—like Van Helmont's explanations of these phenomena—focus on the changes induced in a single substance that retains its material identity while being divided into smaller corpuscles or otherwise undergoing a change in texture.[64] It was an easy matter for Boyle to "mechanize" Van Helmont's explanations by explicitly limiting the ensuing alteration to one in the size, shape, or association of corpuscles. In this emphasis on the possibility of qualitative change without

---

62. For this theory in Starkey's work, see Newman, *Gehennical Fire*, 141–43, 177–78.

63. Boyle, *Sceptical Chymist*, in *Works*, 2:274. Boyle uses this argument to cast doubt on Daniel Sennert's theory that the metals are composed of more or less permanent atoms, each of which must always "retain its own Nature." This is actually an uncharitable reading of Sennert's position, for the latter was himself a proponent of metallic transmutation.

64. It is true that Van Helmont thinks that continued exposure to the alkahest will make a substance revert to water, but what we have in mind here is merely the initial dissolution of a substance into its liquid *primum ens*, which retains its characteristic chymical qualities. In effect, the dissolved substance has merely undergone a division into smaller particles, hence keeping its other qualities while becoming a liquid at the level of sense perception.

chymical combination, and in the fact that both men postulate a single, primordial material from which all things spring, the Helmontian and the Boylean explanations are surprisingly similar.

Thus, despite the often extravagant flavor of Van Helmont's system, we can detect a deep affinity between the thought of the Belgian chymist and his English reformulator. Notwithstanding Boyle's overt subordination of chymistry to natural philosophy that we described in the first chapter, his mechanical ruminations mapped closely onto the more vitalistic corpuscular chymistry of his forebear. The transformation of the young Boyle, learning expressly Helmontian chymistry at the hands of George Starkey, to the militant reductionist of the *Origine of Formes and Qualities,* did not require the metamorphosis that later historians—or Boyle himself—have led us to believe. Boyle's long career witnessed a continual engagement with Helmontian theory and practice, as his systematization of three classes of salts and his continued deployment of Helmontian experiments demonstrate. Indeed, this should occasion little surprise when we realize that Van Helmont's writings constituted what was probably the most wide-ranging and influential chymical theory of the second half of the seventeenth century. Although parts of Van Helmont's original theory, such as the production of alkalies exclusively from the combination of Salt with Sulphur, were called into question in Boyle's later works, the overall structure of Boyle's thought remained in many ways Helmontian. Of course, it would be fatuous to argue that Boyle remained an acolyte of Van Helmont in the manner of Starkey. Boyle's ultimate aspirations lay in the realm of natural philosophy, not in that of chymistry, which he viewed as subordinate to natural philosophy. As a result, most of the chymistry appearing in Boyle's printed works served his greater philosophical goal—the establishment of the mechanical philosophy and the debunking of the Scholastic doctrine of qualities.

A HELMONTIAN BACKGROUND
TO THE CHEMICAL REVOLUTION

Part of our study of seventeenth-century laboratory practice has dealt with the Helmontian introduction of mass balance into the conceptual armory of practicing chymists. As we saw in chapter 2, Van Helmont openly attacked the Aristotelian principle that bodies are composed of four elements, two of which have weight and two of which are imponderable. Indeed, Van Helmont was opposed to the concept of imponderable substances in general. Even his mysterious "Gas," the forerunner of modern gases, was only relatively light. For example, Van Helmont noted that when sixty-two pounds of oak charcoal burn, only one pound of ash remains, and therefore the Gas or *spiritus sylvestris* produced, even though invisible and uncapturable, weighs sixty-one pounds.[65] The sublimation of snow—that is, its conversion into Gas—occurs by the threefold process of dividing and extenuating the corpuscles of a heavy substance (water) and inverting the shells of that substance's corpuscles. Gas could be lighter than the atmosphere only because the particles out of which it was made occupied more space per unit of matter. In other words, if Gas was lighter than the ambient air, this was because the specific gravity of Gas was less. According to Van Helmont, all matter has weight, and that weight is conserved even in the "final tortures" of the most striking physical and chemical change. Van Helmont's insistence on what we now call conservation of weight was implicitly adhered to by many an earlier alchemist; Geber, to name but one, assumed the constancy of weight in his metallurgical analyses.[66] But it seems to have been Van Helmont who explicitly combined the new Paracelsian emphasis on *spagyria*—analysis and synthesis—with the practical gravimetric techniques of earlier alchemists to arrive at the principle that "in our furnaces we read that there is no more certain genus of acquiring knowledge for the understanding of things through their root and constitutive causes, than when one knows what is contained in a thing, and how much of it there is."[67] Using such principles, Van Helmont devised experiments that challenged the Aristotelian system at its most basic level. His synthesis and analysis of glass, for example, in which he showed that the weight of the initial sand can be recovered in the final product, was a graphic illustration of the problems inherent in contemporary interpretations of Aristotelian mixture. Chapters 3 and 4 show, moreover, that Starkey's laboratory notebooks testify to his constant reliance on the principle of mass balance; they also show how the American chymist used weight determinations to follow the progress, measure the success, and suggest improvements to chymical processes as well as to provide quantitative measures of the relative amounts of constituents in mixed bodies.

Such deployment of and insistence upon initial and final weights cannot help but recall the justly celebrated developments of eighteenth-century chemistry usually known as the "chemical revolution"—meaning, in simplest terms, the reformulation of chemistry carried out largely by Antoine Laurent Lavoisier. This reformulation began with the savant's discovery in

65. Van Helmont, *Ortus,* "Complexionum atque mistionum elementalium figmentum," p. 106, #13–14.

66. Pace Emile Meyerson, *Identity and Reality* (New York: Gordon and Breach, 1989), 160.

67. Van Helmont, *De lithiasi,* in *Opuscula,* no. 1, p. 20: "In nostris furnibus legimus, non esse in natura certius sciendi genus, ad cognoscendum per causas radicales, ac constitutivas rerum; quam dum scitur quid, quantumque in re quaque, sit contentum."

the 1770s that combustion is due to the combination of a gaseous substance that he termed "oxygen" with the combustible material rather than to the emission of phlogiston (sometimes envisioned as an imponderable fluid) from the combustible material. Lavoisier's disclosure led to further research allowing him to prove in the 1780s that water—the fundamental Helmontian element—was itself a compound of oxygen and another substance that he termed "hydrogen." This in turn gave rise to the reformation of chemical nomenclature that provided chemistry with an ever growing number of constant elements instead of the four interconvertible ones that Aristotle had proposed, or the universal "catholic matter" of the seventeenth-century mechanists. As is well-known, Lavoisier's work began with the observation that the weight of combustible or calcinable substances—whether nonmetals such as phosphorus and sulphur or metals like tin and lead—increases upon their exposure to the fire. Although there were numerous references to this phenomenon for at least a hundred years before Lavoisier, it was Lavoisier's attention to this detail and his careful experimentation built on comparing initial and final weights of the substances undergoing reactions that allowed him to give a convincing demonstration of the role of oxygen in combustion.

An entire literary industry has emerged out of the quest to determine the sources that led Lavoisier to his bold new hypothesis and the techniques that allowed him to prove it. It has become a commonplace in the literature that Lavoisier ushered in a new age of quantitative chemistry that superseded the predominantly qualitative enterprise of the chemists before him. Beyond the inherent improbability of such a claim, we wish to point to the tradition of Helmontian chymical practice outlined in this book as one ultimate source of Lavoisier's quantitative methods. We do not, of course, wish to make the naïve claim that a direct reading of Van Helmont led Lavoisier to his emphasis on gravimetric methods. Rather, we argue that the practical approaches that Van Helmont pioneered over a century earlier became an engrained feature of the chemical traditions to which Lavoisier was heir. At the same time, we must also point out that even as a young man Lavoisier was an admirer of Van Helmont. Lavoisier purchased Van Helmont's writings in Strasbourg in 1767, discussed them in his 1770 *Mémoires sur la nature de l'eau,* and praised them at length in his *Opuscules* of 1774.[68] Indeed, Lavoisier's direct comments about Van Helmont in the *Opuscules* are quite

68. For Lavoisier's purchase of "Helmontii opera" in the Strasbourg bookstore of Amand Koenig, see *Oeuvres de Lavoisier: Correspondence,* ed. René Fric (Paris: Albin Michel, 1955), 1:96. The list provided by Koenig is surprisingly rich in works that discuss or profess transmutational alchemy. See Lavoisier, *Oeuvres,* 2:2–5, for a discussion of Van Helmont's willow tree experiment.

perceptive and reveal a close reading of the Flemish chymist. One of his most remarkable comments deals with Van Helmont's discussion of Gas, a forerunner of the "airs" so important to the pneumatic chemistry of the eighteenth century. Lavoisier wrote that

> we are astonished, in reading this Treatise [the *Ortus*], to find an infinite number of facts, which we are accustomed to consider as more modern, and we cannot forebear to acknowledge, that Van Helmont has related, at that period, almost every thing, which we are now acquainted with, on this subject . . . It is easy to see that almost all the discoveries of this kind, which we have usually attributed to Mr. Boyle, really belong to Van Helmont, and that the latter has even carried his theory much farther.[69]

Regardless of how well this praise would withstand modern historical scrutiny, the passage reveals that Lavoisier—at a distance of over a century—saw Van Helmont as a very important, even pioneering, figure in the development of the chemical methods and knowledge with which Lavoisier was working. In the following section, therefore, we will look first at the emergence of Lavoisier's quantitative methods and then make some suggestions about his sources.

LAVOISIER'S BALANCE SHEET METHOD

In his *Traité elementaire de chimie* of 1789, Lavoisier, at the apex of his scientific career, wrote:

> We may lay it down as an incontestible axiom, that, in all the operations of art and nature, nothing is created; an equal quantity of matter exists both before and after the experiment; . . . Upon this principle the whole art of performing chemical experiments depends: we must always suppose an exact equality between the elements of the body examined and those of the products of its analysis. Hence, since from must of grapes we procure alkohol and carbonic acid, I have an undoubted right to suppose that must consists of carbonic acid and alkohol.[70]

These were words that would have warmed the heart of Van Helmont. Yet they were not mere words. Lavoisier supported his conclusions by a technique that has subsequently come to be called his "balance sheet"

69. Antoine Laurent Lavoisier, *Essays Physical and Chemical*, tr. Thomas Henry (London: Joseph Johnson, 1776; reprint, London: Frank Cass, 1970), 7–11.

70. Antoine Laurent Lavoisier, *Elements of Chemistry*, tr. Robert Kerr (Edinburgh: William Creech, 1790; reprint New York: Dover, 1965; reprint of 1790), 130–31. For an alternative translation, see Douglas McKie, *Antoine Lavoisier* (New York: Henry Schuman, 1952), 283.

method. In the example of vinous fermentation referred to above, Lavoisier fermented known quantities of sugar and yeast in a known quantity of water. He then weighed the products of the fermentation—carbonic acid gas (i.e., carbon dioxide), water, and the residual liquor. Careful analyses subsequently allowed Lavoisier to determine the elementary constitution of the products, and by means of a series of tables he showed that the weight of the final products equaled that of the initial ingredients and that each of the individual elements also retained its initial weight. Hence Douglas McKie asserted in 1952 that Lavoisier's tables "constitute the first chemical 'balance sheet' and illustrate, for the first time, the principle that in a chemical reaction there is no loss of matter, but only changes in its forms of aggregation."[71]

Since McKie's work, Lavoisier's balance sheet method has been the subject of increasing historical study. It formed a central theme of Charles Coulston Gillispie's *The Edge of Objectivity*, where it was seen as exemplifying the "rationalization of matter." Gillispie argued vigorously against the notion that Lavoisier's concern with initial and final products was adapted from his work in agriculture and finance, saying that the roots of this method lay instead in the work of the British chemist Joseph Black.[72] Gillispie's assertion of an initial influence from Black was then drawn into question by Henry Guerlac, who pointed out that Lavoisier's early work on combustion (1772) predated his knowledge of Black's research.[73] More recently, Frederic L. Holmes has argued against the interpretation of Arthur Donovan and others who see Lavoisier's emphasis on gravimetric methods as an importation from physics; Holmes's work has reopened the possibility that Lavoisier's balance sheet method derived from earlier chemical sources.[74]

Holmes's position has been corroborated by Louise Palmer, who has found the nascent balance sheet method even in Lavoisier's earliest scientific notebooks. In the summer of 1764, Lavoisier began a comprehensive program of study centering on the mineral gypsum. Motivated by broad concerns in the realm of natural history as well as in chemistry, the young Lavoisier made numerous mineralogical observations in the field and alter-

71. McKie, *Antoine Lavoisier*, 284.

72. Charles Coulston Gillispie, *The Edge of Objectivity* (Princeton: Princeton University Press, 1961), 202–59, esp. 231–32.

73. Henry Guerlac, *Lavoisier—The Crucial Year: The Background and Origin of His First Experiments on Combustion in 1772* (New York: Gordon and Breach, 1961), 11–24.

74. Frederic L. Holmes, *Antoine Lavoisier: The Next Crucial Year, Or The Sources of His Quantitative Method in Chemistry* (Princeton: Princeton University Press, 1998), 9–11. See Arthur Donovan, *Antoine Lavoisier: Science, Administration, and Revolution* (Oxford: Blackwell, 1993), 45–73.

nated these with practical work in his laboratory. In the course of this research—whose culminating feature was the analysis and synthesis of gypsum—Lavoisier explained a peculiar characteristic of that mineral. When gypsum is calcined at high temperature, it becomes powdery "plaster of Paris"; upon exposure to water, this powder expands and becomes a solid, brittle mass. As Lavoisier showed, the formation of the solid mass from calcined gypsum is merely a reacquisition of the water of crystallization that the mineral lost upon calcination. What is remarkable about Lavoisier's conclusion is the form into which he cast it in his notes. As Palmer states, he was already employing his celebrated balance sheet method there in the early 1760s. During several trials, Lavoisier first weighed a sample of gypsum crystals, heated them to distill the water out of them, and then weighed all the products. Finding that the initial and final products differed by only a trifling amount (16 grains), Lavoisier concluded that slaked plaster is chemically the same as gypsum, and that these differ from unslaked plaster of Paris only by the fact that the former exist in a crystalline form combined with a fixed amount of water.[75]

As Palmer also points out, Lavoisier's gravimetric techniques seem to be related to the methods of contemporary chemists whom he read, such as Simon and Gilles-François Boulduc and Andreas Marggraf. But we would like to suggest the importance of another link—one that has been proposed by Holmes, but not with the Helmontian connection that we would like to make. In a paper on the development of chemistry in the early Académie Royale des Sciences, Holmes cites Wilhelm Homberg (1652–1715), who in 1691 had become one of the chief chemists at the Académie.[76] Indeed, Homberg was recognized as one of the most important chemists in France until his death in 1715, and his writings had great influence on the following generations of chemists, particularly Hermann Boerhaave. Homberg was also the mentor of Etienne-François Geoffroy, who would become famous in the subsequent literature for having published the first table of chemical affinities. Homberg's works were unquestionably well-known to Lavoisier at an early period, since already in 1765 the young savant was employing methods of determining the specific gravity of liquids with a special hydrometer devised by Homberg and described in the *Mémoires* of the Académie.[77]

75. Louise Palmer, "The Early Scientific Work of Antoine Laurent Lavoisier: In the Field and in the Laboratory, 1763–1767" (Ph.D. diss., Yale University, 1998), 163–69, 202–4.

76. Frederic L. Holmes, "The Communal Context for Etienne-François Geoffroy's *Table des rapports*," *Science in Context* 9 (1996): 289–311.

77. Palmer, "Early Scientific Work," 174.

Holmes points to a paper that Homberg published in 1703 involving a fairly complicated analysis of common sulphur. Knowing that sulphur burned under a bell jar produces an acid spirit *(oleum sulphuris per campanam)*, Homberg—like most of his contemporaries—concluded that sulphur contains an "acid salt" that is released during combustion. In order to test this theory and to determine how much of this salt was contained in common sulphur, Homberg burned a measured quantity of sulphur under a large glass balloon and collected the saline spirit that condensed on the walls of the vessel. From one pound of sulphur he obtained an ounce and a half of "acid spirit." Further quantitative analyses by other methods led Homberg to conclude that sulphur contained roughly equal weights of acid spirit, oily "earth," and "true sulphur" or Sulphur principle. Now Homberg was aware that his careful gravimetric treatment of initial and final products could not account for a loss of roughly a quarter of the weight of the common sulphur that had gone into his analyses.[78] His balance of input and output in this case was technically far inferior to the laboratory processes of Lavoisier. But even if Homberg found it "difficult to know precisely how much there is of the acid salt in a certain weight of common sulphur," he was still committed to the attempt to ascertain this value in spite of technical difficulties.[79] Thus, as Holmes points out, there is a strong similarity between Homberg's gravimetric techniques and those of his later compatriot at the Académie.

> Lavoisier did not invent the quantitative style that he made the hallmark of his contribution to chemistry. He was able to make the "balance sheet method" work in a manner that Homberg could not only because he was able, as Homberg was not, to carry out the combustion in an enclosed space.[80]

Like Lavoisier, Homberg was critically concerned with identifying, weighing, and recording the initial and final products of chemical processes. And like Lavoisier, Homberg compared the sum of the final weights to the sum of the weights of the ingredients, a practice that allowed him to ascertain when he had a problem of "missing mass." Indeed, the affinity between Homberg's chymical practice and Lavoisier's is even greater than Holmes has noted, for in the year following Homberg's analysis of sulphur, his stu-

78. Wilhelm Homberg, "Essay de l'analyse du souphre commun," *Mémoires de l'Académie Royale des Sciences* 5 (1703): 38. The *Histoire* and *Mémoires* appeared both in Paris quarto editions and Amsterdam duodecimos; here and throughout, the pagination refers to the Paris quarto editions.

79. Ibid., 32.

80. Holmes, "Communal Context," 301.

dent Geoffroy published a sequel paper on the resynthesis of sulphur from the principles into which his mentor had analyzed it. The account of this paper begins with words highly reminiscent of Lavoisier's dual emphasis on analysis and synthesis: "One is never so sure of having decomposed a mixed body into its true principles than when one can recompose it from the same principles."[81]

The similarity between the practical quantitative methods of Homberg and the later ones of Lavoisier raises an obvious concern with the conditions under which Homberg himself developed such techniques. We now suggest that Homberg represents a compelling link and conduit between the Helmontian chymistry to which this book has been devoted and the celebrated methods of Lavoisier. As we show below, a consideration of some of Homberg's other papers leads us straight back into the realm of Helmontian chymistry.

WILHELM HOMBERG AND HELMONTIAN CHYMISTRY

The topic of Wilhelm Homberg's chymistry is substantial in itself and would quickly lead well beyond the limits of this study. Homberg published extensively on a wide variety of topics—not all of them chymical—and had many contacts and much influence. Therefore, for this study, we will restrict ourselves to only a preliminary consideration of the roots of Homberg's influential laboratory practices and leave the fuller exposition of Homberg's chymistry for another time and place.[82] Of key importance for our present concern is the observation that Homberg's interest in final and initial weights and gravimetric analysis is a constant theme of his work, and that this interest extends well beyond his 1703 paper on the analysis of common sulphur cited above. For example, throughout the 1690s Homberg was involved in the Académie's long-term project toward the analysis of plants by distillation; part of this endeavor involved measuring the exact and relative quantities of the substances separated from plants.[83] One of

81. Etienne-François Geoffroy, "Maniere de recomposer le souphre commun par la réunion de ses principes," *Mémoires de l'Académie Royale des Sciences* 6 (1704): 278–86; the quotation is from the corresponding entry in the accompanying *Histoire*, 37; the annual *Histoire* was written by the perpetual secretary of the Académie, Bernard de Fontenelle.

82. One of us (LMP) is currently at work on an extensive and broad-based study of Wilhelm Homberg that will be published in due course. An early result of this investigation, upon which much of the following Homberg material is based, may be seen in Lawrence M. Principe, "Wilhelm Homberg: Chymical Corpuscularianism and Chrysopoeia in the Early Eighteenth Century," in *Late Medieval and Early Modern Corpuscular Matter Theories*, ed. Christoph Lüthy, John Murdoch, and William Newman (Leiden: Brill, 2001), 535–56.

83. Alice Stroup, "Wilhelm Homberg and the Search for the Constituents of Plants at the 17th-Century Académie Royale des Sciences," *Ambix* 26 (1979): 184–202.

Homberg's memoirs from this period records his careful comparison of the weights of lixivial salts before and after spirits of wine had been repeatedly distilled off of them. Using the principle of mass balance, Homberg concluded that the loss in weight that he recorded for the salt was due to the volatilization of the fixed salt by the action of the spirit of wine.[84] This volatilization of a fixed alkali by means of spirit of wine is reminiscent of Starkey's process of "alcoolization" described in *Pyrotechny Asserted*—a process, of course, with Helmontian roots.[85] Many of Homberg's other statements and activities likewise recall themes treated earlier in this book, such as the fixing of the Sulphureous principle out of spirit of wine with salt of tartar, the formation of alkalies during the combustion of vegetable matter, the threefold division of salts into acidic, lixivial, and urinous, and so forth.[86] But by far the most striking such item is Homberg's paper on the analysis of common mercury, which is analogous to his later paper on the analysis of common sulphur. While the emphasis on analysis again links this work with Lavoisier's later endeavors, both this analytical perspective and the details of Homberg's practical process return us to a surprising—and familiar—source.

In a 1700 paper, Homberg examines the solubilities of the various metals toward different acids. In the course of such studies, he found that certain dissolutions of mercury implied that "the composition of mercury is not uniform," and he promises "very convincing observations" of this fact.[87] These observations come in an appendix to the paper where Homberg describes the "long and tedious operation" that he had "done many times in order to purify mercury exactly."[88] This practical process involves repeated amalgamations and distillations of common mercury from the regulus of antimony. The reader of this book should immediately sit up with a smile of recognition. Common mercury treated "philosophically" with the regulus of antimony is of course the central secret encoded throughout the corpus of Eirenaeus Philalethes and is accordingly a practical process that fills many pages of Starkey's laboratory notebooks.[89] Indeed, the operation that Homberg goes on to describe is nothing less than the Mercurialist method

84. Memoir reproduced ibid., 194–96.

85. Starkey, *Pyrotechny Asserted*, 126–27.

86. For references to these endeavors see Stroup, "Homberg"; on Homberg's threefold division of salts, see Homberg, "Essays de Chimie," *Mémoires de l'Académie Royale des Sciences* 4 (1702): 36.

87. Homberg, "Observations sur les dissolvans du mercure," *Mémoires de l'Académie Royale des Sciences* 2 (1700): 192. Curiously, this article is missing from the first edition of the 1700 *Mémoires* (published in 1703), even though the appended "Suite des observations" does appear (190–195). Thus our references to this pair of papers cite the 1719 second ed.

88. Homberg, "Suite des observations sur les dissoluans du mercure," *Mémoires de l'Académie Royale des Sciences* 2 (1700): 197.

89. Principe, "Wilhelm Homberg," 546–49.

of preparing common quicksilver into the Sophic Mercury, the starting material for the Philosophers' Stone.[90]

Homberg's process proves identical to Starkey's in all details. For example, the exact proportions Homberg cites—nine parts of antimony to four parts of iron—are those provided in coded form in Philalethes' *Introitus apertus ad occlusum regis palatium* and in plaintext in Starkey's "Key."[91] After preparing a regulus using these proportions and purifying it "the third or fourth time with saltpeter" exactly as Starkey stipulated, Homberg fuses two parts of this regulus with one part of copper. Unlike other Mercurialists who used a similar process for making the Sophic Mercury, Starkey was highly unusual (perhaps unique) in employing copper at this point—an innovation he made in 1653 in his constant attempts to lessen the cost and difficulty of the process. Indeed, Starkey created the special name "net" for this alloy in honor of the net that Vulcan used to ensnare the fornicating Mars (iron) and Venus (copper), and in reference to the reticular crystal pattern produced on the alloy's surface as it cools. Homberg then amalgamates three pounds of common mercury with this venereal regulus of antimony, advising that the amalgam be ground in a hot mortar—again just as Starkey dictated—until the amalgam is soft and "ne paroisse plus de grumaux sous les doigts."[92] The amalgam must then be digested and washed repeatedly, according to Homberg, until the wash water ceases to be black (again as Starkey remarked), and then the mercury must be distilled from the amalgam. It should then be reamalgamated with fresh regulus in the same fashion as before, and these amalgamations must be repeated ten times. Starkey too advised that the amalgamation be repeated nine or ten times with successive washings and distillations.

Curiously, Homberg nowhere cites a source for his process, and so we must add him to the growing list of silent beneficiaries of Starkey's practical laboratory processes. Where did Homberg get such a detailed account of Starkey's laboratory processes? Recalling that Starkey's "Key" was originally written to Boyle in 1651, it is possible that Homberg was entrusted with this secret during 1677–78, when he was working in London with Boyle.[93] Although the "Key" was circulated on the Continent in manu-

90. For an account focusing on the long-term development of this practical process from von Suchten to Homberg, see Lawrence M. Principe, "Chacun à Son Goût: Experimental Continuity and Theoretical Diversity in Sixteenth- to Eighteenth-Century Chymistry," *Sudhoffs Archiv*, forthcoming.

91. Homberg, "Suite des observations," 197; on Starkey's method see chapter 3, above.

92. Homberg, "Suite des observations," 197.

93. At present, in spite of thorough searches of the Royal Society Boyle Papers and Boyle's printed works, there is no clear evidence to support Fontenelle's claim ("Eloge de M. Homberg," *Histoire de l'Académie Royale des Sciences* 17 (1715): 85) that Homberg actually

script, which might have provided a source for Homberg, the conjecture that he obtained it from a more privileged source is strengthened by the fact that Homberg's process employs a copper alloy rather than the silver alloy stipulated in the "Key." Starkey made this change around 1653, and Boyle adopted this modification in his own work—possibly on the basis of oral information from Starkey—but this change of metals is not found explicitly in any of the published sources.[94] Regardless of the exact avenue of transmission, it is clear that Homberg was heir to highly detailed accounts of Starkey's laboratory practices and processes.

Now as if the identity of Homberg's practical laboratory process with Starkey's were not enough, Homberg also deploys the process for the *same two purposes* as Starkey, namely to show the composition of common mercury and to conduct chrysopoetic experiments. In regard to analysis, Homberg's 1700 paper deals explicitly with analyzing common mercury in a way analogous to that of the 1703 analysis of common sulphur, which Holmes has seen as related to Lavoisier's balance sheet method. As we recounted in chapter 3, Starkey himself describes—both in his "Key" and the *Exposition upon Ripley's Epistle*—how the treatment of mercury with an antimony alloy can be used to perform an analysis of mercury in quantitative terms. Starkey likened the antimonial alloy to a "soap" that could "wash out" the heterogeneities of mercury, and so he explicitly advised that the fouled water used to wash the amalgam be saved in order to show "how the Heterogeneities of *Mercury* are discovered."[95] Strikingly, this is exactly the way Homberg employs the process in 1700. Homberg likewise collects the "dirty" water left from washing the amalgam and evaporates it to find "an earthy material, light, mouse-grey, without odor or taste." This powder, he argues, must be an impurity separated from the mercury rather than a part

---

worked in London with Boyle. The likelihood that Homberg got this recipe from Boyle is the clearest evidence of their collaboration at present. The date of this collaboration, if it occurred, had been set in 1674 (Hall, *Boyle and Seventeenth Century Chemistry*, 73–74, and *Dictionary of Scientific Biography*, s.v. "Homberg, Wilhelm"), but Stroup clearly indicates that this must be incorrect and that the date is more likely 1677–78 (see Stroup, "Homberg," 185–86). Interestingly, this time has now been shown to be Boyle's *annus mirabilis alchemicus* (Principe, *Aspiring Adept*), and if Homberg were with Boyle at that time, he would surely have encountered Boyle's chrysopoetic interests, which long centered around this very process.

94. Principe, *Aspiring Adept*, 169; Newman, *Gehennical Fire*, 133–34 n. 72. Starkey does make a reference to the substitution of "Venus" for the "Doves of Diana" in *Marrow of Alchemy*, but even if Homberg interpreted these metaphorical allusions correctly he could not have gotten the exact proportions and practical manipulations from that source; see Starkey, *Marrow of Alchemy*, pt. 2 (London, 1655), 15–16, and Principe, "Wilhelm Homberg," 548–49.

95. George Starkey [Eirenaeus Philalethes, pseud.], "Epistle to King Edward Unfolded" in *Ripley Reviv'd* (London, 1678), 14 (no. 27 in Newman, *Gehennical Fire*, 270).

of the regulus not only because it is not reducible back into a metallic form, but also because no more powder is expelled after the sixth amalgamation regardless of how much fresh regulus is added to the mercury.[96] Therefore, Homberg concludes that there is "a material that is found naturally in all common mercury, and which makes up an essential part of it, and which can be separated by this operation." Homberg weighs the powder and concludes from his analysis that three pounds of mercury contain five and a half scruples of this gray powder.[97]

Besides carrying out an analysis like that done by Starkey, Homberg also speculates on the means by which the regulus of antimony is capable of "purifying mercury exactly."

> We know that sulphur acts powerfully on mercury: this is what has led me to think that it could well be the sulphureous material of the regulus of antimony that acted as the dissolvent of this material [the gray powder] that is separated from the rest of the body of the mercury, and that this sulphur would have no action on the other parts of the mercury, because the gray powder being once separated by the first five or six amalgamations, the regulus no longer acts on the mercury, and all the amalgamations that one makes after the sixth separate nothing further; that is, the waters with which one does his washings are always clear, which agrees well with the idea that one has of the sulphur of antimony regulus; that is, that it is different from the burning sulphur of crude antimony, for the latter dissolves the entire body of mercury, while the former dissolves only the part that is separable by our operation.[98]

Here Homberg reiterates the explanation used by Starkey himself, which is based upon Helmontian principles. According to the theory expressed in the "Key" (and more fully in Starkey's *Exposition upon Ripley's Epistle* and implicitly in his laboratory notebooks), there are two antimonial sulphurs. Crude antimony (stibnite) contains an external, flammable mineral sulphur that makes the mineral an effective means of refining metals, for this sulphur can burn out and consume their base impurities. But Starkey also claims that when regulus is formed from crude antimony and iron, the antimony loses its mineral sulphur and the resultant regulus acquires an internal metallic Sulphur. It is this metallic Sulphur that acts upon the mercury,

---

96. The reader will recall that Starkey (like von Suchten) was able to reduce the black powder to regulus. Homberg, however, notes that if the wash water is hot, then a heavy black powder reducible to regulus is emitted, but if the water is just warm, then only the gray powder is expelled. Homberg, "Suite des observations," 198–99.

97. Ibid., 199–200.

98. Ibid., 197 ("pour purifier exactement le mercure") and 200.

cleansing it during its amalgamation and washing. As Starkey says in the "Key," the mercury acquires from the regulus "a spiritual seed that is a fire which shall thoroughly purge away all its superfluities."[99] In effect, Homberg has recapitulated this theory—the "burning sulphur of crude antimony" is different from the "sulphur of regulus," since the former acts upon the whole body of the mercury, while the latter acts specifically upon the heterogeneous matter and expels it.

Homberg used this specially prepared mercury to provide the central illustrative experiment in his important "Essais de chimie"—a kind of a serial textbook on the principles of chemistry that appeared in the *Mémoires* beginning in 1702. This deployment further displays Homberg's relationship to Van Helmont via the chrysopoeia of Eirenaeus Philalethes.[100] In 1705, Homberg states that if this specially prepared mercury is sealed in a glass egg with a long neck and heated, it will gradually thicken, and finally precipitate into a powder, first black, then white, and finally red.[101] Homberg explains this precipitation in corpuscularian terms. He notes that mercury's liquidity depends upon the shape of its corpuscles—they are smooth spheres that roll easily upon one another. When placed in the fire, these smooth corpuscles are struck by the rapidly moving particles of the "matter of light" contained in the fire. This matter of light is the keystone of Homberg's system, and is for him nothing other than the chymical Sulphur principle in the free state. The impacts of these particles scratch the surface of the spheres, so that the tiny corpuscles of the Sulphur principle can then stick in these scratches, "par son gluten naturel," as Homberg informs us, so that gradually these fine particles, transformed from the matter of light in the flame into a metallic Sulphur by virtue of their arrest by the mercury, coat the mercury corpuscles like a shell, but a rough, prickly shell, like smooth chestnuts encased within prickly husks, as Homberg expresses it.[102] Thus they can no longer roll on one another, and the liquid mercury becomes a powder; the powder also weighs more than the mercury, owing to its incorporation of the matter of light as a metallic Sulphur.

Homberg insists that this is not a significant change—not yet. For if the red powder is put into a stronger fire, where the matter of light is in greater

99. Starkey to Boyle, April/May 1651, *Notebooks and Correspondence,* document 3; Boyle, *Correspondence,* 1:97.

100. Homberg's chrysopoetic interests and his use of them in concert with the development of his corpuscular theories are detailed in Principe, "Wilhelm Homberg," to which the reader is referred for a fuller exposition of this process.

101. Homberg, "Suite des Essays de Chimie. Article Troisième, du Souphre Principe," *Mémoires de l'Académie Royale des Sciences* 7 (1705): 88–96.

102. Homberg, "Suite de l'article trois des Essais de Chimie," *Mémoires de l'Académie Royale des Sciences* 8 (1706): 262.

agitation, this more violent stream strips away the Sulphureous shells, liberating the mercurial kernels, and so liquid quicksilver distills over. But Homberg notes that even though most of the newly introduced Sulphur is swept away by a stronger fire, a very small portion of the red powder cannot be returned to its original mercurial state, but rather remains behind fixed as a solid metal, which as we might guess—and Homberg later tells us—is gold.

Homberg explains the twofold product of the precipitation of mercury. When the Sulphur (or matter of light) is attached only "superficially" to the mercury, it is easily removed; most of the red precipitate formed by the digestion of mercury contains such Sulphur attached only to the exterior of its corpuscles and is therefore readily returned to running mercury by a stronger fire. A small quantity, however, about one two-hundredth of the whole, has experienced a deeper penetration of the Sulphur. In this case the Sulphur "has entered the substance itself of the mercury" so fire cannot drive it out. When this happens, the stricter union of the mercury and Sulphur provides gold, for the precious metal has, according to Homberg, the fixed Sulphur "at its interior."

This distinction of exterior and interior Sulphurs comes from a long tradition of chrysopoetic thought. Building on Arabic sources that equated "interior" with potency and "exterior" with act, the medieval Geber had posited two different Sulphurs present in mercury, one "sealed up in its profundity" and the other "supervenient"; the superficial could be removed easily, but the interior was difficult or impossible to remove. Different substances result from different "positioning" of the principles. But it was again Van Helmont who transformed this from a quasi-metaphorical discussion of potency and act to a genuinely spatial language concerning kernels and surrounding shells of individual particles.[103] Homberg's distinction between a superficially attached and a profoundly penetrating Sulphur, complete with the language of kernels and shells, is closely akin to Van Helmont's own interpretation of sulphur and mercury. The explanation of how the Sulphur penetrates to the mercurial core by deeply scoring the mercurial corpuscles is Homberg's own ingenious additional layer of mechanical explanation built upon Van Helmont's theory.

CONCLUSIONS

We have thus outlined a clear legacy of Van Helmont's and Starkey's chymical theory and practice. Helmontianism, at least partly transmitted through and interpreted by Starkey, not only played an important role in Boyle's

103. Newman, *Gehennical Fire*, 92–114.

first set of natural philosophical publications but also persisted in somewhat different formats in his mature chymistry. But we have followed this legacy further. Wilhelm Homberg provides a discernible path of transmission for both experimental practices and theories from Van Helmont and Starkey (and undoubtedly others) into the French chemistry of the Académie Royale des Sciences that eventually nurtured Lavoisier; Homberg's reliance upon Starkey is astonishing and unquestionable. It is true that current research is revealing more and more evidence of the balance sheet method among chymists in the last decades of the seventeenth century, especially in the early Académie, such as the famous corpuscularian chymist Nicolas Lemery. But the traces of Van Helmont's influence are present even in the writings of the earliest Academicians—figures whom we have not been able to treat in this study, such as Samuel Cottereau Du Clos.[104]

The fact that we have found Homberg's nascent balance sheet method in Starkey's own work with antimonial amalgams is therefore suggestive, to say the least, of the relationship between seventeenth-century chymistry and the refocused discipline of the eighteenth century. As the major English follower of Van Helmont in the seventeenth century, Starkey was in an excellent position to refine and to pass on the gravimetric emphasis of his Belgian master to the subsequent generation, which included the *chimistes* of the Académie Royale des Sciences. These lines of transmission argue that the relationship between the thoughts and practices of the Helmontians and those of Lavoisier is not limited to Lavoisier's expressed admiration for his Flemish predecessor, nor are their practices merely isomorphic.

Moreover, Van Helmont's interest in weights and mass balance in chymistry, and the transmission of this emphasis as adopted and refined by his heirs—like Homberg—makes it yet more difficult to argue that Lavoisier had to borrow the gravimetric methods of physics in order to make his chemical breakthroughs of the 1770s.[105] What we have found, rather, is an independent tradition in chemistry in which an unsuspected degree of gravimetric emphasis was steadily developing over a long period, particularly as a tool for analyses and for the monitoring of practical processes. Al-

104. For Lemery and the balance sheet method, see Michel Bougard, *La chimie de Nicolas Lemery* (Turnhout: Brepols, 1999), 181–90. For the Helmontian tendencies of the Academician Samuel Cottereau Du Clos, see his "Sur les eaux minerales," in *Histoire et Mémoires de l'Académie Royale des Sciences*, 11 vols. (Paris, 1729–33), 4:46–48, where Du Clos speaks of the *première être* and *seminaires* of metals. Throughout the treatise, he refers to alkaline salts as Helmontian *sels sulphurés*.

105. See, for example, Donovan, *Antoine Lavoisier*, 48; John E. McEvoy, "Continuity and Discontinuity in the Chemical Revolution," *Osiris*, 2d ser., 4 (1988): 195–213, esp. 204–6; Arthur Donovan, "Lavoisier and the Origins of Modern Chemistry," in *Osiris*, 2d ser., 4 (1988): 214–31, esp. 230.

though Lavoisier certainly carried this tradition much further and to a far higher degree of precision than any of his predecessors had done, the difference between his practical efforts and theirs is one of degree and technical expertise rather than of type.

The importance of this tradition of weight determination and analysis for chemistry—encompassing Van Helmont, Starkey, Homberg, Lavoisier, and others—makes it imperative that we return briefly to the issue of Boyle's mechanical predilections and how they affected his relationship to this growing tradition. Curiously, Boyle seems not to have been particularly concerned in his chymistry about developing the Helmontian emphasis on "what is contained in a thing, and how much of it there is." But as noted above, a mechanical chymistry wherein anything could produce anything else via mechanical changes of corpuscular texture militated against the very possibility of meaningful quantitative analysis, because a purely mechanistic chymistry is incompatible with the notion of "constant composition" that undergirds the concept of analysis.[106]

Boyle's relative lack of interest in gravimetric studies appears in the very cases where he engages in the analysis and synthesis, or "redintegration" of substances. His "Essay on Nitre," a work at the vanguard of his program of placing chymistry in the service of natural philosophy, describes Boyle's famous experiment of burning saltpeter with charcoal to produce "fixed niter" (potassium carbonate), followed by the "redintegration" of the saltpeter by combining the fixed niter with "spirit of niter" (nitric acid). Boyle here shows relatively little interest in quantitative measures: instead, he gives qualitative arguments for the identity of his synthesized product with ordinary niter and then defers the project of determining whether "the whole body of the Salt-Petre, after it's having been sever'd into very differing parts by distillation, may be adequately re-united into Salt-Petre equiponderant to it's first self."[107] Later, in *Origine of Formes and Qualities*, a work that grew out of "Notes upon an Essay about Nitre," Boyle simply dismisses the possibility of making an "adaequate redintegration," that is, one having the same weight as the original materials.[108]

One might argue that Boyle's reluctance was due to experimental caution rather than lack of interest, and there would be some truth in this. Yet if we consider Boyle's goals for chymistry, his reticence makes perfect sense from a doctrinal position as well. Boyle's main target was the Aristotelian doctrine of qualities and related views among the "vulgar chymists." Thus his analyses and resyntheses of substances such as camphor, amber, and tur-

106. Kuhn, "Boyle and Structural Chemistry."
107. Boyle, *Certain Physiological Essays,* in *Works,* 2:108.
108. Boyle, *Formes and Qualities,* in *Works,* 5:372.

pentine—although their inspiration may derive from traditional *spagyria*—are primarily intended to demonstrate that "redintegration" is a purely mechanical process, like the taking apart and reassembling of a watch, rather than the removal and reimposition of a substantial form.[109] The intent of Boyle's redintegration experiments was not to learn about the nature of a substance by determining "what is contained . . . and how much of it there is," but rather to show that one could "Reproduce a Body, which has been depriv'd of its substantial Form."[110] Redintegration shows that qualities arise from texture, not from substantial form. Thus *Formes and Qualities* explicitly defends the sufficiency of a qualitative approach to analysis and synthesis because of its powerful demonstrative effect against Aristotelianism; regardless of the quantitative outcome of the experiment, the qualitative features are decisive arguments against Boyle's targets.

> For, even in such Experiments, it appears, that when the Form of a Natural Body is abolish'd, and its parts violently scatter'd; by the bare Reunion of some parts after the former manner, the very same Matter, the destroy'd was before made of, may, without Addition of other Bodies, be brought again to constitute a Body of the *like Nature* with the former, though not of *equal Bulk*.[111]

For Boyle's purposes it was enough to show the identical *qualities* of starting and initial products. There was no need to determine how much each separated component weighed or to attend to the *quantitative* identity of initial and final products. The Aristotelian theory of substantial forms, Boyle's main target, had itself made only qualitative claims. In addition, a strict mechanical chymistry of producing anything from anything merely by changes of corpuscular form and texture implied that gravimetric analyses of mixed bodies would be either futile or inconclusive.

Boyle's primary intellectual commitment to replacing the "vulgar doctrine of qualities" with the mechanical philosophy meant that Van Helmont's gravimetric techniques of analysis remained relatively unimportant or nongermane to his goals, even though Boyle did adopt a considerable array of other practical and theoretical items from Helmontian chymistry. This is by no means to claim that Boyle did not weigh things—we have the overwhelming testimony of his innumerable and precise specific gravity determinations, his interest in fine balances, and so on, and in terms of chymical processes, one need think only of his observation that metals gain

109. Ibid., 5:355.
110. Ibid., 5:372.
111. Ibid.

weight upon calcination.[112] But the fact remains that Boyle's primary interest does not lie in the use of gravimetric analysis either to discover "what is contained in a thing, and how much of it there is," or to put bodies back together in order to reveal their composition. The techniques developed in the seventeenth century that became so widely employed and so crucial in the eighteenth (and ever since), are not central to Boyle's endeavor. This observation may provide one avenue (of many) for approaching the problem of elucidating Boyle's real impact on subsequent chymistry. Although it cannot be our goal here to determine the precise nature of Boyle's influence on eighteenth-century chemists, his decision to opt for a qualitative approach to "redintegration" in order to refute Aristotelian physics diverts his chymical contributions from what would ultimately prove to be the more productive route for chemistry.

Lavoisier himself—like many eighteenth-century chemists—was notably uninterested in mechanical explanations and microstructural speculations. The development of chemical theory and practice in the early eighteenth century—at the hands of Homberg, Lemery, Geoffroy, and others—began with a silent or explicit dismissal—or at least limitation—of attention to the "ultimate" matter in favor of the kind of permanent chymical species that Boyle's mechanical chymistry played down. Homberg's chymistry, for example, although expressly corpuscularian and mechanical, is more "conservative" (or perhaps, more "chemical") than Boyle's, for he focuses on the mechanical interactions not of corpuscles of an indeterminate universal matter, but of corpuscles of chymically distinct species—"les principes plus materiels & plus sensibles"—Mercury, Sulphur, and so forth.[113] Only in this way were chymists able to have confidence in a "constant composition" of mixed bodies that enabled them to set about the kind of quantitative analyses that we have seen in Homberg and that formed the basis of Lavoisier's chemical practice.

At present it would be premature to insist too strongly on the relative importance of Homberg in particular in the transmission of seventeenth-century chymical traditions to Lavoisier.[114] There remains a vast amount of further investigation to be done in this area, and several such studies are currently under way. No doubt there are numerous routes for the transmis-

---

112. Boyle, "New Experiments to make Fire and Flame Ponderable" in *Essays of Effluviums,* in *Works,* 7:305–22, and "A Discovery of the Perviousness of Glass" in *Essays of Effluviums,* in *Works,* 7:323–33.

113. Homberg, "Essays de Chimie," *Mémoires de l'Académie Royale des Sciences* 4 (1702): 33; Principe, "Wilhelm Homberg," 538–39.

114. The young Lavoisier of course found inspiration in the works of many earlier chemists; see the list of chemists cited by the early Lavoisier in Palmer, "Early Scientific Work," 241–42.

sion of Helmontian ideas into eighteenth-century chemistry; the full impact of Van Helmont's contribution in this period has yet to be thoroughly assessed. The paths whereby the endeavors and practices of Helmontian chymists were absorbed into the analytical projects of their eighteenth-century descendants remain little-known alleys in the history of science. If we may develop this metaphor further, however, it is such routes and the sites along them that provide the setting for understanding the real life of a large and bustling city. Looking at only a select few topics in the history of eighteenth-century chemistry, such as phlogiston and the developing study of "airs," is like a commuter forming his urban impressions from brief glances out of a rear-view mirror while speeding along a metropolitan expressway. We must turn to the avenues and alleys, and better yet the sidewalks, if we wish to understand the development of "chymistry" into the modern science of chemistry.

Conclusion

Chemistry is not an armchair activity. Perhaps more than other disciplines within the "exact" sciences, it is closely linked to laboratory practice and to the explication of phenomena that are not universally deducible from thought-experiments or calculations alone. Moreover, chemistry has long been preoccupied with *production*. Whether turning lead into gold, poisons into medicines, Sulphurs into Salts, coal tar into dyestuffs, or air into fertilizer and explosives, the transformation of one substance into another (generally a more desirable or salable one) is one of the constant themes of alchemy/chymistry/chemistry from late antiquity down to the present day. This "artisanal" facet of chemistry again ties it closely to the activities of the laboratory or workshop. The close linkage between chemistry as an intellectual discipline to the practical work of hand and eye in the laboratory requires that we as historians pay particular attention to laboratory practice in the discipline. While the centrality of such practice has been well established in studies of later chemistry, this emphasis has not always been evident in those focusing on early modern chymistry.

The historical assessments of early chymists have, until quite recently, been oddly bifurcated: different schools of historical interpretation have tended to place early chymists at opposite extremes of the spectrum regarding their involvement with practical laboratory affairs. According to some views, early chymists (generally the ones denominated "alchemists") were partly or wholly removed from the realm of laboratory practice, being caught up either in airy speculations, reveries, or hallucinations without an appreciation of the real properties of the substances in their flasks. Yet according to views at the other extreme, early chymists were involved in a rather arbitrary empiricism without the guidance of reasoned or consistent theoretical frameworks. Alongside the latter category, but still separated from it, fell the "artisanal" workers—assayers, metallurgists, miners, distillers, and so forth—who produced goods, but without the motivation or mental wherewithal to derive useful scientific principles from their practices. Interestingly enough, as we showed in the first chapter, the roots of such divisions are not wholly modern creations, but lie partly in depictions of chymistry that stem from early modern figures—including Robert Boyle himself. These characterizations, although they may serve particular rhetorical functions, do not fairly represent the state of affairs in the early modern period. For although specific examples from both these extremes can be found within the wide spectrum of the history of chymistry and that of chemical technology more broadly construed, we cannot overlook those important workers who did in fact have both principles and practice and who forced the two to interact on a daily basis.

In the present book we have re-joined some of the dissevered parts of the broad chymical tradition and argued for early modern chymistry's vitality and independence as a discipline. Our contention that rational laboratory practice and methodology provide a locus of continuity between "alchemy" and "chemistry" rather than acting as a new, defining characteristic of the latter may surprise some readers. Other indications of the continuity between alchemy and chemistry have been presented elsewhere by us and by others. But the laboratory was perhaps the least obvious place—according to several still widely held conceptions of alchemy as a predominantly contemplative, spiritual, or simply empirical practice—to find such a sharing of approach. George Starkey's notebooks bear this out with striking clarity. While Starkey labored on a variety of chymical topics—metallic transmutation, iatrochemistry, and chymical production—he nonetheless maintained the same formalized methodology and insistence upon experimental trials in the fire, whether he was producing new perfumes, analyzing mercury, developing a new chemical pharmacy, or trying to prepare the Philosophers' Stone. Far from being either a deluded dreamer or a doting empiric, Starkey used clear-sighted methods and techniques to pursue his wide range of chymical goals, from the common to the extraordinary. Whether the majority of these goals were, in the light of present chemical knowledge, attainable or not, the theoretical structure of Helmontian chymistry provided a powerful and reasoned justification for pursuing them, as well as a practical methodology for carrying out the quest.

We cannot overstress the point that Starkey's methodology was consistently reasonable and rigorous across the many projects to which he devoted himself. Within Starkey's work there is no methodological or epistemological division between the quest for the transmutatory Philosophers' Stone and his other works. He did not exclude the possibility of divine enlightenment when carrying out the most mundane and repetitive of laboratory tasks, nor did he abandon *ratio* (reason) when seeking the advice of his "good genius" in preparing the *arcana maiora* of Helmontian chymistry. It is not as though Starkey worked as a rational "chemist" one day on one project, and then as a "mystic alchemist" the next. For him—

and for the majority of his contemporaries—endeavors in transmutation, chymical medicine, or the more efficient isolation of essential oils were all part of the same project of "uncovering Nature's secrets." All of these projects were party to the same methodological approach and scrutiny, and Starkey subjected all of them to the same practical tests and trials. Indeed, we have seen that even in the reading and writing of the most allusive chrysopoetic texts, Starkey worked by reasonable, consistent methods, demanded that his interpretations be demonstrable in the laboratory, and avoided arbitrary or merely picturesque metaphors in favor of those that were duly decipherable into laboratory practice and/or theory. Indeed, it is worth stressing yet again that the author of the straightforward laboratory notebooks studied here is also the author of the allusive and often seemingly extravagant corpus of Eirenaeus Philalethes. Thus we cannot simplistically divide up seventeenth-century activities into "alchemy" and "chemistry" based upon practices or beliefs, but must seek rather to understand chymistry as a whole and as it was actually practiced in the seventeenth century. This is by no means to say, of course, that seventeenth-century chymistry was wholly analogous to modern chemistry—no more than one would wish to argue that Kepler's astronomical goals, methods, and commitments remain those of today's Keck Observatory. But the outlines and content of any branch of natural philosophy in the seventeenth century, and especially chymistry, can only be rightly assessed in the light of how and why it was practiced.

It is also noteworthy that Starkey drew upon so wide a range of intellectual and practical traditions to develop his own investigational style. Formal Scholastic training, experience in early "chymical industry," and medical practice were all combined by Starkey into a unified style for the investigation of nature and its deployment toward specific practical goals. His notebooks therefore present a remarkable testimony of clear-sighted and tenacious laboratory practice in the mid–seventeenth century. By this very token one might ask how typical Starkey's activities were for his period. Did other chymists, chrysopoeians or Helmontians, deploy laboratory methodologies as developed as those of Starkey? This is still a difficult question; our intimate knowledge of Starkey's laboratory practice derives from the fortunate survival of his notebooks, and at present we know of very few comparable documents for other workers. Searches of archival deposits may well reveal the notebooks of other chymical practitioners, and these would prove valuable for supplementing this study. Yet even at this point we can say two things. First, our sense is that Starkey's innate character, his education, and the circumstances in which he worked led to something unusual—his successes, not least the creation and popularity of the Philalethes corpus, as well as his notebooks themselves bear witness to this fact. We therefore suggest that Starkey represents a practitioner at the "high end" of the spectrum. But while acknowledging Starkey's unusual prowess as a chymist, we do not believe that he represents a wholly anomalous case. Starkey's own successes should not overshadow the fact that the young American drew directly upon various existing chymical traditions that had themselves already reached a fairly high level. As we have argued throughout this study, one of these developed bodies of knowledge and method was Helmontianism.

Joan Baptista Van Helmont emerges from this study as a remarkably influential figure. Although Van Helmont has already been the subject of numerous scholarly treatments, there is much still to be learned regarding his impact and influence on subsequent generations, particularly in areas outside of medicine. While the Helmontian influence on Starkey is sufficiently explicit to need no uncovering, and some of Boyle's early debts to Van Helmont have long been recognized, this study has pointed out Van Helmont's longer-term impacts on Boyle's mature chymistry, and on the generation of chymists after the English natural philosopher. Undoubtedly, far more Helmontian influence remains to be uncovered. We have identified Van Helmont as a key point of intersection between the developed quantitative traditions of alchemy/assaying dating back to the High Middle Ages and the *spagyria*—the analysis and synthesis of mixed bodies—emphasized by the followers of Paracelsus. The combination of these traditions and their attendant battery of practical operations provided Van Helmont and his heirs with a core methodology for work in the chymical laboratory that stressed weight determinations, mass balance, and compositional analysis—issues familiar to chemists ever since.

In chapter 2 we spent considerable time examining Van Helmont's attitude to mathematics. Besides clarifying the Belgian chymist's complex position, this examination, when taken together with the rest of our study, carries a further historiographical message. In the 1950s and 1960s, the history of science as a discipline adopted the position that the great contributions of the Scientific Revolution were largely due to the quantification and mechanization of nature. Indeed, the chronological termini of the period we now call the Scientific Revolution were set precisely by reference to levels of mathematization. The classical "Master Narrative" for the period is based upon the most mathematical sciences—physics and astronomy—and accordingly traverses the century and a half from 1550 to 1700 by passing from Copernicus to Kepler and Galileo and finally to the consummate synthesis of Newton, sometimes stopping off for a brief look at Descartes or Boyle. According to this model, astronomy and mathematical physics were seen as the archetypes that all the other "progressive" sciences sought to emulate, and indeed, the "less mathematical" sciences like biology, chemistry, geology, and natural history have until recently been ancillary or even absent from standard narratives.

But it should be obvious that not all sciences use mathematics in the same way. Not only are the needs of various scientific disciplines different, but the branches of mathematics are themselves very different from one another, some better accommodated to one branch of the sciences than others. We saw that Van Helmont explicitly condemned the use of Scholastic mathematics in natural philosophy and medicine. A superficial reading of Van Helmont's critical attitude toward mathematics might well have been sufficient to relegate the Belgian to a marginal position in the "main story" of seventeenth-century science. But in spite of Van Helmont's rejection of Scholastic "geometrical methods" and proportionalities, he routinely used quantitative methods in his laboratory practice—as tests, as guides, and as important elements in his own apodictic demonstrations. While rejecting more speculative traditions in mathematics, Van Helmont nonetheless deployed the arithmetical parts of practical mathematics applicable to his own work. These were not advanced mathematical concepts to be sure; still less were they the mathematical abstractions of the natural world devised for planetary or free fall motions. But on the other hand, how sophisticated did mathematical methods need to be even for the striking advances of the eighteenth-century chemical revolution? Van Helmont used no less mathematics than most modern-day chemists. Synthetic organic chemists, the majority within the most populous community of modern scientists, do very well with very little mathematics at all. They weigh starting materials and final products, calculate yields and compositions—predominantly the same sorts of things that Van Helmont, Starkey, Homberg, and eventually Lavoisier did. The use of higher mathematics in chemistry is a development pertaining primarily to the more physical branches of the science, whose origins in its modern form date predominantly from the late nineteenth century. Practical chemists bent on productive processes—and this represents both the historical and the current majority of chemists—simply do not have the need for complex mathematical formulae or the levels of geometrical or algebraic abstraction that have come to characterize popular conceptions of the "exact sciences." Yet it would seem utterly incongruous to consider modern synthetic chemical work as "less scientific" than, say, quantum chemistry, simply because it makes use of simpler mathematics.

In this study, we have also tried to provide a view of chymistry in terms of an independent, long-term, and continuous development. First, by penetrating beneath the rhetoric of novelty deployed by Boyle in order to divide himself from earlier chymical traditions, and by recognizing the depth and persistence of his actual debts to that tradition, we have been able to diminish the retrospective divisions rather artificially positioned in the historiography of chemistry. Starkey and Boyle are not worlds apart, and neither are the traditions they have come to represent. Second, chymistry now appears less dependent upon extradisciplinary borrowings than has sometimes been thought. For example, the Helmontian appreciation of mass balance was an independent development arising mainly out of preexisting traditions in Paracelsian iatrochemistry, medieval assaying technology, and traditional transmutatory alchemy. These traditions either predated the renewed interest in the mathematical sciences or acted as their coeval twins. Thus we need not seek out an external source for Lavoisier's balance sheet method when it can be seen as the product of gradual, incremental improvements to an idea that had been internal to the practice of chemistry for many generations—being found in rudimentary form even in the Late Middle Ages, and slowly elaborated through the various chymical practitioners treated in this study.

Much remains to be said about early modern chymistry—not to say early modern natural philosophy in general—in terms of both its development and its practical aspects. We hope that these new studies will be carried out without the preconceptions about the nature of chymistry and its practitioners that this study has endeavored to dispel. Likewise, we would also hope that chemistry over the longue durée may be seen and studied on its own terms, not by the measure of other branches of natural philosophy, nor in terms of extradisciplinary borrowings and contributions. In this way, the developments recounted in this book, and further aspects of early modern chymistry yet to be explored, may again acquire the significance attached to them by their contemporaries and form an integral part of our accounts of the history of science in the early modern period.

WORKS CITED

Ahonen, Kathleen. "Johann Rudolph Glauber: A Study of Animism in Seventeenth-Century Chemistry." Ph.D. diss., University of Michigan, 1972.

Anthony, Francis. *Medicinae chymicae, et veri potabilis auri assertio.* Cambridge, 1610.

Aristotle. *De generatione et corruptione.*

———. *Physics.*

Aylmer, G. E. *The State's Servants: The Civil Service of the English Republic 1649–1660.* London: Routledge, 1973.

Bacon, Roger. *Sanioris medicinae magistri D. Rogeris Baconi.* Frankfurt, 1603.

Barnard, T. C. *Cromwellian Ireland: English Government and Reform in Ireland 1649–1660.* Oxford: Oxford University Press, 1975.

Barton, Tamsyn. *Ancient Astrology.* London: Routledge, 1994.

Bazan, Bernardo, et al. *Les questions disputées et les questions quodlibetiques dans les facultes de théologie, de droit et de medecine.* Turnhout: Brepols, 1985.

Beretta, Marco. *The Enlightenment of Matter.* Canton, Mass.: Science History, 1993.

Berthelot, Marcellin. *La chimie au moyen âge.* Paris, 1893; reprint, Osnabrück: Otto Zeller, 1967.

Bligh, E. W. *Sir Kenelm Digby and His Venetia.* London: S. Low, Marston, 1932.

Boerhaave, Hermann. *A New System of Chemistry.* London, 1727.

Boizard, Jean. *Traite des monoyes.* Paris, 1692.

Borel, Pierre. *A New Treatise, Proving the Multiplicity of Worlds.* Tr. D. Sashott. London, 1658.

Bougard, Michel. *La chimie de Nicolas Lemery.* Turnhout: Brepols, 1999.

Boyle, Robert. *The Correspondence of Robert Boyle.* Ed. Michael Hunter, Antonio Clericuzio, and Lawrence M. Principe. 6 vols. London: Pickering and Chatto, 2001.

———. *The Works of Robert Boyle.* Ed. Michael Hunter and Edward B. Davis. 14 vols. London: Pickering and Chatto, 1999–2000.

Braddick, M. J., and M. Greengrass. *The Letters of Sir Cheney Culpeper (1641–1657).* Camden Miscellany 33. Cambridge: Cambridge University Press, 1996.

Brinkmann, A. A. A. M. *De Alchemist en de Prentkunst.* Amsterdam: Rodopi, 1982.

Brock, William H. *The Norton History of Chemistry.* New York: Norton, 1993.

Broeckx, C. "Le premier ouvrage de J.-B. Van Helmont." *Annales de l'académie d'archéologie de Belgique* 10 (1853): 327–92; 11 (1854): 119–91.

Browne, Alice. "J. B. Helmont's Attack on Aristotle." *Annals of Science* 36 (1979): 575–91.

Butterfield, Herbert. *The Origins of Modern Science, 1300 –1800*. New York: Macmillan, 1951.

Campbell, Joseph. *The Flight of the Wild Gander: Explorations in the Mythological Dimension*. New York: Harper-Perennial, 1990; 1st ed., 1951.

———. *The Masks of God: Primitive Mythology*. Harmondsworth: Penguin, 1982; 1st ed., 1959.

Cardilucius, Johann Hiskias. *Magnalia medico-chymica*. Nuremberg, 1676.

Charas, Moyses. *Nouvelles experiences sur la vipere*. Paris, 1669.

Charleton, Walter. *A Ternary of Paradoxes*. London, 1650. (A loose translation of three essays from Van Helmont's *Ortus medicinae*.)

Clagett, Marshall. *The Science of Mechanics in the Middle Ages*. Madison: University of Wisconsin Press, 1959.

Clericuzio, Antonio. *Elements, Principles, and Corpuscles: A Study of Atomism and Chemistry in the Seventeenth Century*. Dordrecht: Kluwer, 2000.

———. "From van Helmont to Boyle: A Study of the Transmission of Helmontian Chemical and Medical Theories in Seventeenth-Century England." *British Journal for the History of Science* 26 (1993): 303–34.

———. "A Redefinition of Boyle's Chemistry and Corpuscular Philosophy." *Annals of Science* 47 (1990): 561–89.

———. "Robert Boyle and the English Helmontians." In *Alchemy Revisited*, ed. Z. R. W. M. van Martels, 192–99 (Leiden: Brill, 1990).

Clucas, Stephen. "The Correspondence of a XVII-Century 'Chymical Gentleman': Sir Cheney Culpeper and the Chemical Interests of the Hartlib Circle." *Ambix* 40 (1993): 147–70.

Clulee, Nicholas H. "*Astronomia Inferior:* Legacies of Johannes Trithemius and John Dee." In *Secrets of Nature: Astrology and Alchemy in Early Modern Europe*, ed. William R. Newman and Anthony Grafton, 173–233 (Cambridge: MIT Press, 2001).

———. *John Dee's Natural Philosophy: Between Science and Religion*. London: Routledge, 1988.

Comstock, J. L. *Elements of Chemistry*. Hartford: D. F. Robinson, 1831.

Coudert, Allison. *Alchemy: The Philosopher's Stone*. London: Wildwood House, 1980.

Crombie, A. C. *Robert Grosseteste and the Origins of Experimental Science (1100 – 1700)*. Oxford: Clarendon, 1953.

Crosland, Maurice. *Historical Studies in the Language of Chemistry*. New York: Dover, 1978.

Darmstaedter, Ernst. "Arznei und Alchemie: Paracelsus-Studien." *Studien zur Geschichte der Medizin* 20 (1931): 1–77.

———. "Berg-, Probir- und Kunstbüchlein." *Münchener Beiträge zur Geschichte und Literatur der Naturwissenschaften und Medizin* 2/3 (1926): 101–206.

Davidson, Jane P. *David Teniers the Younger*. London: Thames and Hudson, 1980.

Davis, Tenney L. "Boerhaave's Account of Paracelsus and Van Helmont." *Journal of Chemical Education* 5 (1928): 671–81.

de Waard, C. *Correspondence du P. Marin Mersenne*. 17 vols. Paris: Presses Universitaires de France, 1946.

Dear, Peter. *Mersenne and the Learning of the Schools.* Ithaca: Cornell University Press, 1988.

Debus, Allen G. *The Chemical Philosophy.* 2 vols. New York: Science History, 1977.

———. "Fire Analysis and the Elements in the Sixteenth and Seventeenth Centuries." *Annals of Science* 23 (1967): 127–47.

———. "Mathematics and Nature in the Chemical Texts of the Renaissance." *Ambix* 15 (1968): 1–28.

———. *Science and Education in the Seventeenth Century: The Webster-Ward Debate.* London: MacDonald, 1970.

———. "Solution Analyses Prior to Robert Boyle." *Chymia* 8 (1962): 41–61.

Dee, John. *Monas hieroglyphica.* In *Theatrum chemicum* (6 vols.), 2:178–215 (Strasbourg, 1659–61; reprint, Torino: Bottega d'Erasmo, 1981).

Des Chene, Dennis. *Physiologia: Natural Philosophy in Late Aristotelian and Cartesian Thought.* Ithaca: Cornell University Press, 1996.

Dickson, Donald R. *Thomas and Rebecca Vaughan's Aqua Vitae: Non Vitis.* Tempe: Arizona Center for Medieval and Renaissance Studies, 2001.

Digby, Kenelm. *Two Treatises.* Paris, 1644.

Dobbs, Betty Jo Teeter. *The Foundations of Newton's Alchemy, or The Hunting of the Greene Lyon.* Cambridge: Cambridge University Press, 1975.

———. "From the Secrecy of Alchemy to the Openness of Chemistry." In *Solomon's House Revisited,* ed. Tore Frängsmyr, 75–94 (Canton, Mass.: Science History, 1990).

———. "Studies in the Natural Philosophy of Sir Kenelm Digby." *Ambix* 18 (1971): 1–25; 20 (1973): 143–63; 21 (1974): 1–28.

Donne, John. *The Complete Poems,* ed. Alexander B. Grosart. Printed for private circulation, 1873.

Donovan, Arthur. *Antoine Lavoisier: Science, Administration, and Revolution.* Oxford: Blackwell, 1993.

———. "Lavoisier and the Origins of Modern Chemistry." *Osiris,* 2d ser., 4 (1988): 214–31.

Du Clos, Samuel Cottereau. *Sur les eaux minerales.* In *Histoire et mémoires de l'Académie Royale des Sciences* (11 vols.), vol. 4 (Paris, 1729–33).

Eamon, William. *Science and the Secrets of Nature: Books of Secrets in Medieval and Early Modern Culture.* Princeton: Princeton University Press, 1994.

Eastwood, B. S. "Medieval Empiricism: The Case of Robert Grosseteste's Optics." *Speculum* 43 (1968): 306–21.

Evans-Pritchard, E. E. *Witchcraft among the Azande.* Oxford: Clarendon, 1937.

Feingold, Mordechai. "English Ramism: A Reinterpretation." In *The Influence of Petrus Ramus,* ed. Mordechai Feingold et al., 127–76 (Basel: Schwabe, 2001).

———. *The Mathematicians' Apprenticeship: Science, Universities, and Society in England, 1560–1640.* Cambridge: Cambridge University Press, 1984.

Fiering, Norman. "Solomon Stoddard's Library at Harvard in 1664." *Harvard Library Bulletin* 20 (1972): 255–69.

Fontenelle, Bernard de. "Eloge de M. Homberg." *Histoire de l'Académie Royale des Sciences* 17 (1715): 82–93.

Fowler, Alistair, ed. *The New Oxford Book of Seventeenth-Century Verse.* Oxford: Oxford University Press, 1991.

Frank, Robert G., Jr. *Harvey and the Oxford Physiologists: A Study of Scientific Ideas and Social Interaction.* Berkeley and Los Angeles: University of California Press, 1980.

———. "The John Ward Diaries." *Journal of the History of Medicine* 29 (1974): 147–79.

Frye, Northrop. *Anatomy of Criticism.* Princeton: Princeton University Press, 1957.

Galen. *On the Natural Faculties.* Tr. Arthur John Brock. London: Heinemann, 1947.

Galison, Peter. *How Experiments End.* Chicago: University of Chicago Press, 1987.

Gaubius, Hieronymus David. *Oratio inauguralis, qua ostenditur, chemiam artibus academicas jure esse inserendam.* Leiden, 1732.

Gelman, Zahkar E. "Angelo Sala, An Iatrochemist of the Late Renaissance." *Ambix* 41 (1994): 121–34.

Geoffroy, Etienne-François. "Maniere de recomposer le souphre commun par la réunion de ses principes." *Mémoires de l'Académie Royale des Sciences* 6 (1704): 278–86.

Giglioni, Guido. *Immaginazione e malattia: Saggio su Jan Baptiste van Helmont.* Milan: Francoangeli, 2000.

———. "Per una storia del termine Gas da van Helmont a Lavoisier: costanza e variazione del significato." *Annali della Facoltà di Lettere e Filosofia dell'Università di Macerata* 25–26 (1992–93): 431–68.

Gillispie, Charles Coulston. *The Edge of Objectivity.* Princeton: Princeton University Press, 1961.

Glanvill, Joseph. *Plus Ultra.* London, 1668.

Glaser, Christophle. *Traite de chymie.* Paris, 1673.

Glauber, Johann Rudolph. 2 vols. *Continuatio operum chymicorum.* Frankfurt, 1659.

———. *Works.* Tr. Christopher Packe. London, 1689.

Golinski, Jan. "Chemistry in the Scientific Revolution: Problems of Language and Communication." In *Reappraisals of the Scientific Revolution,* ed. David Lindberg and Robert Westman, 367–96 (Cambridge: Cambridge University Press, 1990).

Goltz, Dietlinde. "Versuch einer Grenzziehung zwischen 'Chemie' und 'Alchemie.'" *Sudhoffs Archiv* 52 (1968): 30–47.

Gooding, David, Trevor Pinch, and Simon Schaffer, eds. *The Uses of Experiment: Studies in the Natural Sciences.* Cambridge: Cambridge University Press, 1989.

Gordon-Grube, Karen Joyce. "The Alchemical 'Golden Tree' and Associated Imagery in the Poems of Edward Taylor." 2 vols. Ph.D. diss., Free University of Berlin, 1990.

Grafton, Anthony. *Joseph Scaliger: A Study in the History of Classical Scholarship.* Vol. 1. Oxford: Clarendon, 1983.

———. *Joseph Scaliger: A Study in the History of Classical Scholarship.* Vol. 2. Oxford: Clarendon, 1993.

Guerlac, Henry. *Lavoisier—The Crucial Year: The Background and Origin of His First Experiments on Combustion in 1772.* New York: Gordon and Breach, 1961.

Haberling, Wilhelm. "Alexander von Suchten, ein Danziger Arzt und Dichter des 16. Jahrhunderts." *Zeitschrift des Westpreussischen Geschichtsverein* 69 (1929): 177–230.

Hall, A. Rupert, and M. B. Hall. *The Correspondence of Henry Oldenburg.* 9 vols. Madison: University of Wisconsin Press, 1965–73.

Hall, Marie Boas. "Acid and Alkali in Seventeenth-Century Chemistry." *Archives internationales d'histoire des sciences* 9 (1956): 13–28.

———. *Robert Boyle and Seventeenth-Century Chemistry.* Cambridge: Cambridge University Press, 1958.

Halleux, Robert. "L'alchimiste et l'essayeur." In *Die Alchemie in der europäischen Kultur- und Wissenschaftsgeschichte,* ed. Christoph Meinel (Wiesbaden: Otto Harrasowitz, 1986).

———. "Helmontiana II: Le prologue de L'Eisagoge, la conversion de Van Helmont au Paracelsisme, et les songes de Descartes." *Academiae Analecta, Klasse der Wettenschappen* 49 (1987): 20–36.

———. "Methodes d'essai et d'affinage des alliages aurifères dans l'Antiquité et au Moyen Age." *Cahiers Ernest Babelos* 2 (1985): 39–77.

———. *Les textes alchimiques.* Turnhout: Brepols, 1979.

———. "Theory and Experiment in the Early Writings of Johan Baptist Van Helmont." In *Theory and Experiment,* ed. Diderik Batens, 93–101 (Dordrecht: Rediel, 1988).

Hannaway, Owen. *The Chemists and the Word: The Didactic Origins of Chemistry.* Baltimore: Johns Hopkins University Press, 1975.

———. "Laboratory Design and the Aim of Science: Andreas Libavius versus Tycho Brahe." *Isis* 77 (1986): 585–610.

Hartley, E. N. *Ironworks on the Saugus.* Norman: University of Oklahoma Press, 1957.

Harwood, John T. *The Early Essays and Ethics of Robert Boyle.* Carbondale: Southern Illinois University Press, 1991.

Haynes, George H. "The Tale of Tantiusques." *American Antiquarian Society, Proceedings,* n.s., 14 (1901): 471–97.

Heinecke, Berthold. "The Mysticism and Science of Johann Baptista Van Helmont (1579–1644)." *Ambix* 42 (1995): 65–78.

———. *Wissenschaft und Mystik bei J. B. van Helmont (1579–1644).* Bern: Peter Lang, 1996.

Hill, C. R. "The Iconography of the Laboratory." *Ambix* 22 (1975): 102–10.

Hoff, Hebbel E. "Nicolaus of Cusa, van Helmont, and Boyle: The First Experiment of the Renaissance in Quantitative Biology and Medicine." *Journal of the History of Medicine and Allied Sciences* 19 (1964): 99–117.

Holmes, Frederic L. *Antoine Lavoisier: The Next Crucial Year, Or The Sources of His Quantitative Method in Chemistry.* Princeton: Princeton University Press, 1998.

———. "The Communal Context for Etienne-François Geoffroy's 'Table des Rapports.'" *Science in Context* 9 (1996): 289–311.

———. "Do We Understand Historically How Experimental Knowledge Is Acquired?" *History of Science* 30 (1992): 119–36.

Holmyard, E. J. *Alchemy.* New York: Dover, 1990; 1st ed., 1957.

Homberg, Wilhelm. "Essay de l'analyse du souphre commun." *Mémoires de l'Académie Royale des Sciences* 5 (1703): 31–40.

———. "Essays de Chimie." *Mémoires de l'Académie Royale des Sciences* 4 (1702): 33–52.

———. "Observations sur les dissolvans du mercure." *Mémoires de l'Académie Royale des Sciences* 2 (1700): 190–196 [1719 second ed.].

———. "Suite de l'article trois des Essais de Chimie." *Mémoires de l'Académie Royale des Sciences* 8 (1706): 260–72.

———. "Suite des Essays de Chimie. Article Troisiéme, du Souphre Principe." *Mémoires de l'Académie Royale des Sciences* 7 (1705): 88–96.

———. "Suite des observations sur les dissolvans du mercure." *Mémoires de l'Académie Royale des Sciences* 2 (1700): 196–212 [1719 second ed.].

Hubicki, Włodzimierz. "Alexander von Suchten." *Sudhoffs Archiv* 44 (1960): 54–63.

———. "Michael Sendivogius' Nitre Theory: Its Origins and Significance for the History of Chemistry." In *Actes du Xe Congrès International d'Histoire des Sciences*, 2:829–33 (Paris: Hermann, 1964).

Hugh of Saint Victor. *Didascalicon.* Ed. Jerome Taylor. New York: Columbia University Press, 1991.

Hunter, Michael. "Boyle versus the Galenists: A Suppressed Critique of Seventeenth-Century Medical Practice and Its Significance." *Medical History* 47 (1997): 322–61.

———. "How Boyle Became a Scientist." *History of Science* 33 (1995): 59–103.

———, ed. "Psychoanalyzing Robert Boyle." Special number of *British Journal for the History of Science.* Vol. 32 (1999): 257–324.

———. *Robert Boyle (1627–1691): Scrupulosity and Science.* Woodbridge: Boydell, 2000.

———. *Robert Boyle by Himself and His Friends.* London: Pickering, 1994.

Hunter, Michael, and Charles Littleton. "The Work-Diaries of Robert Boyle: A Newly Discovered Source and Its Internet Publication." *Notes and Records of the Royal Society* 55 (2001): 373–90.

Hunter, Michael, and Lawrence M. Principe. "The Lost Papers of Robert Boyle." *Annals of Science* 60 (2003).

Hutton, Sarah, ed. *The Conway Letters: The Correspondence of Anne, Viscountess Conway, Henry More, and Their Friends, 1642–1684.* Rev. ed. edited by Marjorie Hope Nicolson. Oxford: Clarendon, 1992.

Innes, Stephen. *Creating the Commonwealth: The Economic Culture of Puritan New England.* New York: Norton, 1995.

Johns, Adrian. *The Nature of the Book.* Chicago: University of Chicago Press, 1999.

Johnson, L. W., and M. L. Wolbarsht, "Mercury Poisoning: A Probable Cause of Isaac Newton's Physical and Mental Ills," *Notes and Records of the Royal Society* 34 (1970): 1–9.

Johnson, Samuel. *The Lives of the Most Eminent English Poets.* 2 vols. Charlestown: Etheridge, 1810.

Joly, Bernard. "Alchimie et Rationalité: La Question des Critères de Démarcation entre Chimie et Alchimie au XVIIe Siècle." *Sciences et techniques en perspective* 31 (1995): 93–107.

———. "L'alkahest, dissolvant universal, ou quand la théorie rend pensible une pratique impossible." *Revue d'histoire des sciences* 49 (1996): 308–30.

———. "Qu'est-ce qu'un laboratoire alchimique?" *Cahiers d'histoire et de philosophie des sciences* 40 (1992): 87–102.

Jung, Carl Gustav. "Die Erlösungsvorstellungen in der Alchemie." In *Eranos-*

Jahrbuch 1936: Gestaltung der Erlösungsidee in Ost und West, 13–III (Zurich: Rhein-Verlag, 1937).

Kargon, Robert H. *Atomism from Hariot to Newton.* Cambridge: Cambridge University Press, 1966.

Kendall, George. *An Appendix to the Unlearned Alchimist.* London, [1663].

Kittredge, George Lyman. "Dr. Robert Child the Remonstrant." *Colonial Society of Massachusetts, Transactions* (1919): 1–146.

Klein, Ursula. "Robert Boyle: Der Begründer der neuzeitlichen Chemie?" *Philosophia naturalis* 31 (1994): 63–106.

Kraus, Paul. *Jabir ibn Hayyan: Contribution à la histoire des idées scientifiques dans l'Islam.* Cairo: Institut d'Egypte, 1943.

Kuhn, Thomas. "Robert Boyle and Structural Chemistry in the Seventeenth Century." *Isis* 43 (1952): 12–36.

Laszlo, Pierre. *Qu'est-ce que l'alchimie?* Paris: Hachette, 1996.

Lavoisier, Antoine Laurent. *Elements of Chemistry.* Tr. Robert Kerr. Edinburgh: William Creech, 1790; reprint New York: Dover, 1965.

———. *Essays Physical and Chemical.* Tr. Thomas Henry (1776; London: Frank Cass, 1970).

———. *Oeuvres de Lavoisier: Correspondence.* Ed. René Fric. Paris: Albin Michel, 1955.

Lawn, Brian. *The Rise and Decline of the Scholastic "Quaestio disputata," with Special Emphasis on Its Use in the Teaching of Medicine and Science.* Leiden: Brill, 1993.

Leff, Gordon. *Paris and Oxford Universities in the Thirteenth and Fourteenth Centuries.* Huntington, N.Y.: Krieger, 1975.

Lemery, Nicolas. *Cours de chymie.* Paris, 1675.

[Libavius, Andreas]. *Commentariorum alchymiae ... pars prima.* In Andreas Libavius, *Alchymia* (Frankfurt: Joannes Saurius, 1606).

Linden, Stanton J. *Darke Hieroglyphicks: Alchemy in English Literature from Chaucer to the Restoration.* Lexington: University Press of Kentucky, 1996.

Lindey, Alexander. *Plagiarism and Originality.* New York: Harper, 1952.

Little, A. G., and F. Pelster. *Oxford Theology and Theologians.* Oxford: Oxford University Press, 1935.

Long, Pamela O. "The Openness of Knowledge: An Ideal and Its Context in 16th-Century Writings on Mining and Metallurgy." *Technology and Culture* 32 (1991): 318–55.

———. *Openness, Secrecy, Authorship: Technical Arts and the Culture of Knowledge from Antiquity to the Renaissance.* Baltimore: Johns Hopkins University Press, 2001.

Long, Pamela O., and Alex Roland. "Military Secrecy in Antiquity and Early Medieval Europe: A Critical Reassessment." *History and Technology* 11 (1994): 259–90.

[Longueville, Thomas]. *The Life of Sir Kenelm Digby.* London: Longmans, Green, 1896.

Lundgren, Anders. "The Changing Role of Numbers in 18th-Century Chemistry." In *The Quantifying Spirit in the 18th Century,* ed. Tore Frängsmyr, J. L. Heilbron, and Robin E. Rider, 245–66 (Berkeley and Los Angeles: University of California Press, 1990).

Macquer, Pierre Joseph, *A Dictionary of Chemistry*. 2 vols. London, 1771.

Maddison, R. E. W. "Studies in the Life of Robert Boyle, FRS. Part 6. The Stalbridge Period, 1645–1655, and the Invisible College." *Notes and Records of the Royal Society* 18 (1963): 104–24.

Magnus, Elisabeth M. "Originality and Plagiarism in *Areopagitica* and *Eikonoklastes*." *English Literary Renaissance* 21 (1991): 87–101.

Maier, Annelies. *An der Grenze von Scholastik und Naturwissenschaft*. Rome: Edizioni di Storia e Letteratura, 1952.

Mallon, Thomas. *Stolen Words: Forays into the Origins and Ravages of Plagiarism*. New York: Ticknor and Fields, 1989.

Mandelbrote, Scott. Review of *Gehennical Fire*, by William R. Newman. *British Journal for the History of Science* 30 (1997): 109–11.

Martin, Luther. "A History of the Psychological Interpretation of Alchemy." *Ambix* 22 (1975): 10–20.

Matthew, Richard. *The Unlearned Alchymist his Antidote*. London, 1660.

McEvoy, John E. "Continuity and Discontinuity in the Chemical Revolution." *Osiris*, 2d ser., 4 (1988): 195–213.

McKie, Douglas. *Antoine Lavoisier*. New York: Henry Schuman, 1952.

McVaugh, Michael. "The Development of Medieval Pharmaceutical Theory." In *Arnaldi de Villanova opera medica omnia*, vol. 2, *Aphorismi de gradibus*, ed. Michael McVaugh, 1–136 (Granada-Barcelona: Seminarium Historiae Medicae Granatensis, 1975).

Meinel, Christoph. "Early Seventeenth-Century Atomism: Theory, Epistemology, and the Insufficiency of Experiment." *Isis* 79 (1988): 68–103

———. "Theory or Practice? The Eighteenth Century Debate on the Scientific Status of Chemistry." *Ambix* 30 (1983): 121–32.

Meyerson, Emile. *Identity and Reality*. New York: Gordon and Breach, 1989.

Michael, Emily. "Sennert's Sea Change: Atoms and Causes." In *Late Medieval and Early Modern Corpuscular Matter Theories*, ed. Christoph Lüthy, John E. Murdoch, and William R. Newman, 331–62 (Leiden: Brill, 2001).

Milton, John. *Complete Prose Works*. Ed. Douglas Bush et al. New Haven: Yale University Press, 1962.

Molitor, Carl. "Alexander von Suchten, ein Arzt und Dichter aus der Zeit Herzogs Albrecht." *Altpreussische Monatschrift* 19 (1882): 480.

Momigliano, Arnaldo. "The First Political Commentary on Tacitus." In *Contributo alla storia degli studi classici*, 37–59 (Rome: Storia e Letteratura, 1955).

Moran, Bruce T. *Chemical Pharmacy Enters the University: Johannes Hartmann and the Didactic Care of Chymiatria in the Early 17th Century*. Madison, Wis.: American Institute for the History of Pharmacy, 1991.

Morison, Samuel Eliot. *Harvard College in the Seventeenth Century*. 2 vols. Cambridge: Harvard University Press, 1936.

Multhauf, Robert. *The Origins of Chemistry*. London: Oldbourne, 1966.

Nève de Mévergnies, P. *Jean Baptist Van Helmont, Philosophe par le Feu*. In *Bibliothèque de la Faculté de Philosophie et Lettres, Université de Liège*, fasc. 59 (Paris: Droz, 1935).

Newell, Margaret. "Robert Child and the Entrepreneurial Vision: Economy and Ideology in Early New England." *New England Quarterly* 68 (1995): 223–56.

Newman, William R. "The Alchemical Sources of Robert Boyle's Corpuscular Philosophy." *Annals of Science* 53 (1996): 567–85.

———. "Alchemical Symbolism and Concealment: The Chemical House of Libavius." In *The Architecture of Science*, ed. Peter Galison and Emily Thompson, 59–77 (Cambridge: MIT Press, 1999).

———. "Alchemy, Assaying, and Experiment." In *Instruments and Experimentation in the History of Chemistry*, ed. Frederic L. Holmes and Trevor H. Levere, 35–54 (Cambridge: MIT Press, 2000).

———. "Arabic Forgeries in the Seventeenth Century: The Case of the *Summa Perfectionis*." In *The "Arabick" Interest of the Natural Philosophers in Seventeenth-Century England*, ed. G. A. Russell, 278–96 (Leiden: Brill, 1994).

———. "The Authorship of the *Introitus Apertus ad Occlusum Regis Palatium*." In *Alchemy Revisited*, ed. Z. R. W. M. von Martels, 139–44 (Leiden: Brill, 1990).

———. "Boyle's Debt to Corpuscular Alchemy." In *Robert Boyle Reconsidered*, ed. Michael Hunter, 107–18 (Cambridge: Cambridge University Press, 1994).

———. "Corpuscular Alchemy and the Tradition of Aristotle's *Meteorology*, with Special Reference to Daniel Sennert." *International Studies in the Philosophy of Science* 15 (2001): 145–53.

———. "Decknamen or 'Pseudochemical Language'? Eirenaeus Philalethes and Carl Jung." *Revue d'histoire des sciences* 49 (1996): 159–88.

———. "Experimental Corpuscular Theory in Aristotelian Alchemy: From Geber to Sennert." In *Late Medieval and Early Modern Corpuscular Matter Theories*, ed. Christoph Lüthy, John E. Murdoch, and William R. Newman, 291–329 (Leiden: Brill, 2001).

———. *Gehennical Fire: The Lives of George Starkey, an American Alchemist in the Scientific Revolution*. Cambridge: Harvard University Press, 1994.

———. "L'influence de la *Summa perfectionis* du pseudo-Geber." In *Alchimie et philosophie à la renaissance*, ed. Jean-Claude Margolin and Sylvain Matton, 65–77 (Paris: Vrin, 1993).

———. "Newton's *Clavis* as Starkey's *Key*." *Isis* 78 (1987): 564–74.

———. "The Occult and the Manifest among the Alchemists." In *Tradition, Transmission, Transformation: Proceedings of Two Conferences on Pre-modern Science Held at the University of Oklahoma*, ed. F. Jamil Ragep and Sally P. Ragep, 173–98 (Leiden: Brill, 1996).

———. "The Philosophers' Egg: Theory and Practice in the Alchemy of Roger Bacon." *Micrologus* 3 (1995): 75–101.

———. "The Place of Alchemy in the Current Literature on Experiment." In *Experimental Essays: Versuche zum Experiment*, ed. Michael Heidelberger and Friedrich Steinle, 9–33 (Baden-Baden: Nomos, 1998).

———. "Prophecy and Alchemy: The Origin of Eirenaeus Philalethes." *Ambix* 37 (1990): 97–115.

———. "The *Summa Perfectionis* and Late Medieval Alchemy." 4 vols. Ph.D. diss., Harvard University, 1986.

———. *The Summa Perfectionis of Pseudo-Geber*. Leiden: Brill, 1991.

———. "Technology and Alchemical Debate in the Late Middle Ages." *Isis* 80 (1989): 423–45.

Newman, William R., and Lawrence M. Principe. "Alchemy vs. Chemistry: The Etymological Origins of a Historiographic Mistake." *Early Science and Medicine* 3 (1998): 32–65.

———. *The Laboratory Notebooks and Correspondence of George Starkey.* Forthcoming.

Newton, Isaac. *The Correspondence of Isaac Newton.* Ed. H. W. Turnbull. 7 vols. Cambridge: Cambridge University Press, 1960.

Nicholas of Cusa. *Nicolai de Cusa opera omnia.* Hamburg: Felix Meiner, 1983.

Noll, Richard. *The Aryan Christ.* New York: Random House, 1997.

———. *The Jung Cult.* Princeton: Princeton University Press, 1994.

O'Brien, J. J. "Samuel Hartlib's Influence on Robert Boyle's Scientific Development." *Annals of Science* 21 (1965): 1–14, 257–76.

Obrist, Barbara. *Les débuts de l'imagerie alchimique (XIVe–XVe siècles).* Paris: Le Sycomore, 1982.

Oreovicz, Cheryl Z. "Edward Taylor and the Alchemy of Grace." *Seventeenth-Century News* 34 (1976): 33–36.

———. "Eirenaeus Philoponos Philalethes: *The Marrow of Alchemy.*" Ph.D. diss., Pennsylvania State University, 1972.

———. "Investigating 'The *America* of Nature': Alchemy in Early American Poetry." In *Puritan Poets and Poetics,* ed. Peter White, 99–110 (University Park: Pennsylvania State University Press, 1985).

Osler, Margaret J., ed. *Rethinking the Scientific Revolution.* Cambridge: Cambridge University Press, 2000.

Oster, Malcolm R. "The 'Beaume of Diuinity': Animal Suffering in the Early Thought of Robert Boyle." *British Journal for the History of Science* 22 (1989): 151–79.

Pagel, Walter. *Joan Baptista van Helmont.* Cambridge: Cambridge University Press, 1982.

———. "The Religious and Philosophical Aspects of van Helmont's Science and Medicine." *Bulletin of the History of Medicine,* supp. 2 (1944).

Palmer, Louise. "The Early Scientific Work of Antoine Laurent Lavoisier: In the Field and in the Laboratory, 1763–1767." Ph.D. diss., Yale University, 1998.

Paracelsus [Theophrastus von Hohenheim]. *Sämtliche Werke.* Ed. Karl Sudhoff. Munich: Oldenbourg, 1930.

Partington, J. R. *A History of Chemistry.* 4 vols. London: MacMillan, 1961.

———. *A History of Greek Fire and Gunpowder.* Cambridge: Heiffer, 1960.

———. "Joan Baptista van Helmont." *Annals of Science* 1 (1936): 359–84.

Patterson, T. S. "Van Helmont's Ice and Water Experiments." *Annals of Science* 1 (1936): 462–67.

Paull, H. M. *Literary Ethics: A Study in the Growth of the Literary Conscience.* Port Washington, N.Y.: Kennikat, 1968. Reissue of 1928 ed.

Petersson, R. T. *Sir Kenelm Digby: The Ornament of England, 1603–1665.* Cambridge: Harvard University Press, 1956.

[Philadept]. *An Essay Concerning Adepts.* London, 1698.

Porto, Paulo Alves. "Michael Sendivogius on Nitre and the Preparation of the Philosophers' Stone." *Ambix* 48 (2001): 1–16.

———. "'Summus Atque Felicissimus Salium': The Medical Relevance of the *Liquor Alkahest.*" *Bulletin of the History of Medicine* 76 (2002): 1–29.

————. *Van Helmont e o Conceito de Gás: Quimica e Medicina no século XVII*. São Paulo: Educ/Edusp, 1995.

Powers, John C. "'Ars sine Arte': Nicholas Lemery and the End of Alchemy in Eighteenth-Century France." *Ambix* 45 (1998): 163–89.

Praz, Mario. *Studies in Seventeenth-Century Imagery*. Rome: Storia e Litteratura, 1975.

Principe, Lawrence M. "The Alchemies of Robert Boyle and Isaac Newton: Alternate Approaches and Divergent Deployments." In *Rethinking the Scientific Revolution*, ed. Margaret J. Osler, 201–20 (Cambridge: Cambridge University Press, 2000).

————. "Apparatus and Reproducibility in Alchemy." In *Instruments and Experimentation in the History of Chemistry*, ed. Frederic L. Holmes and Trevor H. Levere, 55–74 (Cambridge: MIT Press, 2000).

————. *The Aspiring Adept: Robert Boyle and His Alchemical Quest*. Princeton: Princeton University Press, 1998.

————. "Chacun à Son Goût: Experimental Continuity and Theoretical Diversity in Sixteenth- to Eighteenth-Century Chymistry." *Sudhoff's Archiv: Beihefte*, forthcoming.

————. "'Chemical Translation' and the Role of Impurites in Alchemy: Examples from Basil Valentine's *Triumph-Wagen*." *Ambix* 34 (1987): 21–30.

————. "Newly Discovered Boyle Documents in the Royal Society Archive: Alchemical Tracts and His Student Notebook." *Notes and Records of the Royal Society* 49 (1995): 57–70.

————. "Robert Boyle's Alchemical Secrecy: Codes, Ciphers, and Concealments." *Ambix* 39 (1992): 63–74

————. "Style and Thought of the Early Boyle: Discovery of the 1648 Manuscript of *Seraphic Love*." *Isis* 85 (1994): 247–60.

————. "Virtuous Romance and Romantic Virtuoso: The Shaping of Robert Boyle's Literary Style." *Journal of the History of Ideas* 56 (1995): 377–97.

————. "Wilhelm Homberg: Chymical Corpuscularianism and Chrysopoeia in the Early Eighteenth Century." In *Late Medieval and Early Modern Corpuscular Matter Theories*, ed. Christoph Lüthy, John Murdoch, and William Newman, 535–56 (Leiden: Brill, 2001).

Principe, Lawrence M., and William R. Newman. "Some Problems with the Historiography of Alchemy." In *Secrets of Nature: Astrology and Alchemy in Early Modern Europe*, ed. William R. Newman and Anthony Grafton, 385–434 (Cambridge: MIT Press, 2001).

Pseudo-Aristotle. *De perfecto magisterio*. In *Bibliotheca chemica curiosa*, ed. J. J. Manget (2 vols.), 1:638–59 (Geneva, 1702).

Pseudo-Avicenna. *De anima in arte alkimiae*. In *Artis chemicae principes, Avicenna atque Geber*, 1–471 (Basel: Petrus Perna, 1572).

Randall, John Herman, Jr. *The School of Padua and the Emergence of Modern Science*. Padua: Antenore, 1961.

Reidy, John, ed. *Thomas Norton's Ordinal of Alchemy*. London: Oxford University Press, 1975.

Reti, Ladislao. "Van Helmont, Boyle, and the Alkahest." In *Some Aspects of Seventeenth-Century Medicine and Science* (Berkeley and Los Angeles: University of California Press, 1969).

Ricketts, Mac Linscott. *Mircea Eliade: The Romanian Roots, 1907–1945.* Boulder, Colo.: East European Monographs, 1988.

Roberts, Brynley F. *Edward Lhuyd: The Making of a Scientist.* Cardiff: University of Wales Press, 1980.

Roberts, Gareth. *The Mirror of Alchemy.* London: British Library, 1994.

Rosenthal, Laura G. *Playwrights and Plagiarists in Early Modern England.* Ithaca: Cornell University Press, 1996.

Ruska, Julius. "Die Alchemie des Avicenna." *Isis* 21 (1934): 14–51.

———. *Das Buch der Alaune und Salze: Ein Grundwerk der spätlateinischen Alchemie.* Berlin: Chemie, 1935.

Ruysschaert, Jose. *Juste Lipse et Les annales de Tacite.* Turnhout: Brepols, 1949.

Sala, Angelus. *Opera medico-chymica.* Rouen, 1650.

———. *Processus de auro potabile.* Strasbourg, 1630.

Sargent, Rose-Mary. *The Diffident Naturalist: Robert Boyle and the Philosophy of Experiment.* Chicago: University of Chicago Press, 1995.

Schipperges, Heinrich. "Die arabische Medizin als Praxis und Theorie." *Sudhoffs Archiv* 43 (1959): 317–28.

Schmitt, Charles B. *The Aristotelian Tradition and Renaissance Universities.* London: Variorum Reprints, 1984.

———. *Aristotle and the Renaissance.* Cambridge: Harvard University Press, 1983.

Schmitt, Charles B. "Experience and Experiment: A Comparison of Zabarella's View with Galileo's in *De Motu.*" *Studies in the Renaissance* 16 (1969): 80–138. Reprint in Charles B. Schmitt, *Studies in Renaissance Philosophy and Science* (London: Variorum Reprints, 1981).

Schuler, Robert M. *Alchemical Poetry, 1575–1700.* New York: Garland, 1995.

Sendivogius, Michael. *Novum lumen chemicum.* In *Musaeum hermeticum,* 553–600 (Frankfurt, 1678; reprint, Graz: Akademische Druck, 1970).

Sennert, Daniel. *De chymicorum.* Wittenberg, 1619.

Shapin, Steven. "The House of Experiment in Seventeenth-Century England." *Isis* 79 (1988): 373–404.

———. *A Social History of Truth.* Chicago: University of Chicago Press, 1994.

Sherlock, T. P. "The Chemical Work of Paracelsus." *Ambix* 3 (1948): 33–63.

Siegel, Thomas Jay. "Governance and Curriculum at Harvard College in the 18th Century. Ph.D. diss., Harvard University, 1990.

Sisco, Anneliese, and Cyril S. Smith. *Lazarus Ercker's Treatise on Ores and Assaying.* Chicago: University of Chicago Press, 1951.

Smith, Pamela H. *The Business of Alchemy: Science and Culture in the Holy Roman Empire.* Princeton: Princeton University Press, 1994.

———. "Vital Spirits: Redemption, Artisanship, and the New Philosophy in Early Modern Europe." In *Rethinking the Scientific Revolution,* ed. Margaret J. Osler, 119–35 (Cambridge: Cambridge University Press, 2000).

Soukup, Rudolf Werner, and Helmut Mayer. *Alchemistisches Gold, Paracelsische Pharmaka: Chemiegeschichtliche und archaeometrische Untersuchungen am Inventar des Laboratoriums von Oberstockstall/Kirchberg am Wagram.* Vienna: Boehlau, 1997.

Spargo, P. E., and C. A. Pounds. "Newton's 'Derangement of the Intellect': New Light on an Old Problem." *Notes and Records of the Royal Society* 34 (1970): 11–32.

Starkey, George. *The Admirable Efficacy and almost Incredible Virtue of true Oyl, which is made of Sulphur-Vive, set on Fire.* London, 1660.

———. *A Brief Examination and Censure of Certain Medicines.* London, 1664.

———[Eirenaeus Philalethes, pseud.]. *Enarratio methodica trium Gebri medicinarum.* London, 1678.

———. *George Starkey's Pill Vindicated.* London, [1663].

———[Eirenaeus Philalethes, pseud.]. *Introitus apertus ad occlusum regis palatium.* In *Musaeum hermeticum,* 648–99 (Frankfurt, 1678; reprint, Graz: Akademische Druck, 1970).

———. *Liquor Alchahest.* London, 1675.

———[Eirenaeus Philoponus Philalethes, pseud.]. *The Marrow of Alchemy.* London, 1654–55.

———. *Natures Explication and Helmont's Vindication.* London, 1657.

———. *Pyrotechny Asserted and Illustrated.* London, 1658.

———[Eirenaeus Philalethes, pseud.]. *Ripley Reviv'd.* London, 1678.

———[Eirenaeus Philalethes, pseud.]. *Sir George Riplye's Epistle to King Edward Unfolded.* In *Chymical, Medicinal, and Chyrurgical Addresses,* ed. Samuel Hartlib, 19–47 (London, 1655).

Stock, John T. *Development of the Chemical Balance.* London: Her Majesty's Stationery Office, 1969.

Stroup, Alice. "Wilhelm Homberg and the Search for the Constituents of Plants at the 17th-Century Académie Royale des Sciences." *Ambix* 26 (1979): 184–202.

Suchten, Alexander von. *Of the Secrets of Antimony in Two Treatises.* "Translated out of the High-Dutch by Dr. C. a Person of Great Skill in Chymistry. . . ." London, 1670.

———. *Tractatus secundus de antimonio vulgari.* In *Mysteria gemina antimonii* (Leipzig, 1604).

Szabadváry, Ferenc. *History of Analytical Chemistry.* Oxford: Pergamon, 1966.

Szydlo, Zbigniew. "The Alchemy of Michael Sendivogius: His Central Nitre Theory." *Ambix* 40 (1993): 129–46.

———. "The Influence of the Central Nitre Theory of Michael Sendivogius on the Chemical Philosophy of the 17th Century." *Ambix* 43 (1996): 80–97.

———. *Water Which Does Not Wet Hands.* Warsaw: Polish Academy of Sciences, 1994.

Taylor, Edward. *Edward Taylor's Minor Poetry.* Ed. Thomas M. Davis and Virginia L. Davis. Boston: Twayne, 1981.

Taylor, F. Sherwood. *The Alchemists: Founders of Modern Chemistry.* New York: Schuman, 1949.

Taylor, Jerome, ed., *The Didascalicon of Hugh of Saint Victor.* New York: Columbia University Press, 1991.

Thomson, S. Harrison. "The Text of Michael Scot's *Ars Alchemie.*" *Osiris* 5 (1938): 523–59.

Thorndike, Lynn. *A History of Magic and Experimental Science.* 8 vols. New York: Columbia University Press, 1958.

Toellner, Richard. "Medicina Theoretica-Medicina Practica: Das Problem des Verhältnisses von Theorie und Praxis in der Medizin des 17. und 18. Jahrhunderts." *Studia Leibnitiana,* supp. 22 (1982): 69–73.

Trevisan, Bernard. *De secretissimo philosophorum opere chemico.* In BCC, 2:388–99.

Turnbull, George H. *Hartlib, Dury, and Comenius: Gleanings from Hartlib's Papers.* London: University Press of Liverpool, 1947.

Van Helmont, Joan Baptista. *Aufgang der Artzney-Kunst.* Ed. and tr. Christian Knorr von Rosenroth. 2 vols. Sulzbach, 1683; reprint, Munich: Kösel, 1971. (German translation and edition of *Ortus medicinae.*)

———. *Dageraad.* Rotterdam: Naeranus, 1660.

———. *Eisagoge in artem medicam a Paracelso restitutam.* 1607. In C. Broeckx, "Le premier ouvrage de J.-B. Van Helmont," *Annales de l'académie d'archéologie de Belgique* 10 (1853): 327–92; 11 (1854): 119–91.

———. *Opuscula medica inaudita.* Amsterdam, 1648. Reprint, Brussels: Culture et Civilisation, 1966.

———. *Oriatrike, or Physicke Refined.* Tr. John Chandler. London, 1662. (A translation of *Ortus medicinae.*)

———. *Ortus medicinae.* Amsterdam, 1648. Reprint, Brussels: Culture et Civilisation, 1966.

van Lennep, Jacques. "L'Alchimiste." *Revue Belge d'archéologie et de l'histoire d'art* 35 (1966): 149–88.

Vaughan, Thomas [Eugenius Philalethes, pseud.]. *Aula lucis.* London, 1652.

———[Eugenius Philalethes, pseud.]. *Euphrates.* London, 1655.

Voelkel, James R. *The Composition of the Astronomia Nova: The Context and Content of Kepler's New Astronomy.* Princeton: Princeton University Press, 2001.

Walden, Paul. *Mass, Zahl und Gewicht in der Chemie der Vergangenheit.* In *Sammlung chemischer und chemisch-technischer Vorträge,* n.s., vol. 8 (Stuttgart: Ferdinand Enke, 1931).

Walker, Jeffrey. "Anagrams and Acrostics: Puritan Poetic Wit." In *Puritan Poets and Poetics,* ed. Peter White, 247–57 (University Park: Pennsylvania State University Press, 1985).

Walton, Michael T. "Boyle and Newton on the Transmutation of Water and Air, from the Root of Helmont's Tree." *Ambix* 27 (1980): 11–18.

Wanamaker, Melissa C. *Discordia Concors: The Wit of Metaphysical Poetry.* Port Washington, N.Y.: Kennikat, 1975.

Watson, Patricia A. *The Angelical Conjunction.* Knoxville: University of Tennessee Press, 1991.

Webster, Charles. "Benjamin Worsley: Engineering for Universal Reform." In *Samuel Hartlib and Universal Reformation,* ed. Mark Greengrass et al., 213–35 (Cambridge: Cambridge University Press, 1994).

———. *The Great Instauration: Science, Medicine, and Reform, 1626–1660.* London: Duckworth; New York: Holmes and Meier, 1975.

———. "Water as the Ultimate Principle of Nature: The Background to Boyle's *Sceptical Chymist.*" *Ambix* 13 (1966): 96–107.

Westfall, Richard S. *Never at Rest: A Biography of Isaac Newton.* Cambridge: Cambridge University Press, 1980.

———. "The Role of Alchemy in Newton's Career." In *Reason, Experiment, and Mysticism in the Scientific Revolution,* ed. M. L. Righini Bonelli and W. R. Shea, 189–232 (New York: Science History, 1975).

Weyer, Jost. *Graf Wolfgang II. von Hohenlohe und die Alchemie.* Sigmaringen: Jan Thorbecke, 1992.

White, Harold Ogden. *Plagiarism and Imitation during the English Renaissance.* Cambridge: Harvard University Press, 1935.

Wilkinson, Ronald S. "Bibliographical Puzzles Concerning George Starkey." *Ambix* 20 (1973): 235–44.

———. "The Hartlib Papers and Seventeenth-Century Chemistry. Pt. 2." *Ambix* 17 (1970): 85–110.

Wilson, William J. "Robert Child's Chemical Book List of 1641." *Journal of Chemical Education* 20 (1943): 123–29.

Winthrop, John, Jr. *Winthrop Papers.* Boston: Massachusetts Historical Society, 1943–92.

Wojcik, Jan J. "Pursuing Knowledge: Robert Boyle and Isaac Newton." In *Rethinking the Scientific Revolution*, ed. Margaret J. Osler, 183–200 (Cambridge: Cambridge University Press, 2000).

Young, John T. *Faith, Medical Alchemy, and Natural Philosophy: Johann Moriaen, Reformed Intelligencer, and the Hartlib Circle.* Brookfield, Vt.: Ashgate, 1998.